Engineering Viscoelasticity

Danton Gutierrez-Lemini

Engineering Viscoelasticity

 Springer

Danton Gutierrez-Lemini
Special Products Division
Oil States Industries, Inc.
Arlington, TX
USA

Additional material to this book can be downloaded from http://extras.springer.com/

ISBN 978-1-4899-7849-3   ISBN 978-1-4614-8139-3   (eBook)
DOI 10.1007/978-1-4614-8139-3
Springer New York Heidelberg Dordrecht London

Danton Gutierrez-Lemini

Engineering Viscoelasticity

 Springer

Danton Gutierrez-Lemini
Special Products Division
Oil States Industries, Inc.
Arlington, TX
USA

Additional material to this book can be downloaded from http://extras.springer.com/

ISBN 978-1-4899-7849-3   ISBN 978-1-4614-8139-3   (eBook)
DOI 10.1007/978-1-4614-8139-3
Springer New York Heidelberg Dordrecht London

Printed on acid-free paper

Springer is part of Springer Science+Business Media (www.springer.com)

To Lolita, the love of my life; whose support, faith and perseverance have made this book possible

To my daughters, Gabriela and Paula, who made parenthood such a blessing!

Preface

Elastic solids and viscous fluids are two types of engineering materials whose response to loads, almost everyone, either seems to understand or takes for granted. Then there are materials whose response to loads combines the features of both elastic solids and viscous fluids. Not surprisingly, these materials are called viscoelastic and are a little trickier to understand than elastic solids or viscous fluids. The engineering discipline that developed to provide a rigorous mathematical framework to describe the behavior of such materials is called viscoelasticity. This book presents a comprehensive treatment of the theory and applications of viscoelasticity.

Polymers are viscoelastic materials. The term polymer has been around since Berzelius used the word "polymeric" in 1832; at a time when chemists were still unsure of the structure of even the simplest of molecules. Today, it is hard to imagine our world without polymers. Polymers and polymeric-based products are commonplace in virtually every industry. This is unquestionably true of the aerospace, rubber, oil, automotive, electronics, construction, piping, and appliances industries; and many more. Yet, despite the fact that viscoelasticity has been taught in universities for several decades, providing the necessary tools to design with polymers, today many polymeric-based products are still designed as if the materials involved were elastic. One reason for this practice is that viscoelasticity has been taught exclusively at the graduate level, yet most practicing professionals lack an advanced degree in engineering; and those with advanced degrees, never studied viscoelasticity, because the subject is usually an elective one.

If one thinks about it, the basic design courses, such as machine design, structural steel design, reinforce concrete design, and so on, are taught at the undergraduate level. The foundation of all these courses is mechanics of materials—the strength of materials of old—whose mastery requires a background in statics and some differential and integral calculus. If truth be told, the derivations of the design equations for viscoelastic materials are the same as for elastic solids; and the resulting expressions, virtually identical. The difference lies in that for viscoelastic materials the relationship between stress and strain is not an algebraic product of a constant modulus and the strain, as it is for elastic solids, but is given by a special type of product—called convolution—between a function which represents the modulus, and the time derivative of the strain. The point being made here is that it is just as demanding—perhaps only a tad more—to learn the art of designing with viscoelastic materials, as it is to learn the mechanics of elastic solids.

This book is intended to help Academia close the gap that exists between current practice and the proper way to designing with polymers, as well as to serve as a text on the theory of viscoelasticity. The book accomplishes these goals by presenting a self-contained, rigorous, and comprehensive treatment of all the topics that are relevant to the mechanical behavior of viscoelastic materials, and by providing all the background in mathematics and mechanics that are central to understanding the subject being presented.

As will be seen, Chaps. 1–7 could be used to teach a complete course on viscoelasticity at the undergraduate level. These chapters cover the theory in the one-dimensional form needed for design. The same chapters, complemented with Appendix A—which provides the mathematical background used in the derivations of the theory—contain all that a practicing professional would need to master the art of designing with polymers.

All equations in these chapters are developed from first principles, without presuming previous knowledge of the subject matter being presented. This approach is followed for two reasons: first, because it is necessary for readers who have no formal training in mechanics of materials; and second, because it provides a method to follow when the use of popular engineering shortcuts, like the use integral transform techniques, might not be clear.

The contents of Appendix B—which provides an introduction to tensors and an overview of solid mechanics together with Chaps. 8–11, are written with the graduate student in mind. A graduate-level course in viscoelasticity could be taught with this material, and perhaps selected sections of earlier chapters, as a continuation to an undergraduate course.

Arlington, TX, USA   Danton Gutierrez-Lemini

Acknowledgments

I would like to express my deep appreciation to Michael Luby, Engineering Editor at Springer, for believing in the manuscript; and because his easy and efficient business manner gave me the encouragement to pursue the writing of this book. I would also like to thank Merry Stuber, Michael's Assistant Editor, for her kindness, patience, and professional advice during the editorial process. Thanks are also due to Scott Sykes, for his invaluable help in formatting the original manuscript. Special thanks go to the reviewers of the original manuscript, for having endured its reading before I had a chance to review it myself, after letting it simmer for a while.

Contents

Fundamental Aspects of Viscoelastic Response

1

Abstract

This chapter describes the molecular structure of amorphous polymers, whose mechanical response to loads combines the features of elastic solids and viscous fluids. Materials that respond in such manner are called viscoelastic, and their mechanical properties have an intrinsic dependence on the time and temperature at which the response is measured. To put this into context, the chapter compares the nature of the response of elastic, viscous, and viscoelastic materials to several types of loading programs, examining the physical nature of their mechanical properties, their behavior regarding energy conservation, and the phenomenon of aging. The topics treated in this chapter provide the neophyte and casual reader with a good understanding of what viscoelastic materials are all about.

Keywords

Aging · Amorphous · Compliance · Creep · Cross-linked · Crystalline · Delta function · Dirac · Elastic · Energy · Equilibrium · Fluid · Glass · Glassy · Heaviside · Hooke · Long term · Modulus · Newton · Polymer · Relaxation · Strain · Stress · Temperature · Transition · Viscous · Viscoelastic

1.1 Introduction

As our intuition tells us, different materials respond differently to external agents. Our experience also indicates that there are materials which, depending on how the stimuli are applied, can respond either as solids or fluids, or can display behavior that combines the characteristics of both. Silly putty, for instance, will bounce just like an elastic solid if thrown against a hard surface, but will extend slowly and continually when held solely under the action of its own weight. Materials whose mechanical response to external agents combines the characteristics of both elastic

D. Gutierrez-Lemini, *Engineering Viscoelasticity*, DOI: 10.1007/978-1-4614-8139-3_1,   1
© Springer Science+Business Media New York 2014

solids and viscous fluids are called viscoelastic. Not surprisingly, viscoelastic materials are a little trickier to understand than elastic solids or viscous fluids.

The engineering discipline that developed to provide a mathematical framework to describe the behavior of viscoelastic materials is called viscoelasticity. This book presents a comprehensive treatment of the theory of viscoelasticity, explaining, in the present chapter, the nature of the response of viscoelastic materials to external loads, and in subsequent chapters, how to model that response mathematically, how to design with viscoelastic materials, and how to establish the material properties needed to describe their mathematical models.

This chapter begins by discussing an important class of materials, known as amorphous polymers, whose mechanical response to external stimuli combines the features of elastic solids and viscous fluids. The terms polymer and polymeric material have been in use since Berzelius coined the word "polymeric" in 1832, at a time when chemists were still unsure of the structure of even the simplest of molecules. Today, it is hard to imagine our world without polymers. Polymers and polymeric-based products are commonplace in virtually every industry. This is unquestionably true of the rubber, oil, aerospace, automotive, electronics, construction, piping, and appliances industries and many more.

The mechanical properties of polymers, such as tensile modulus, or tensile strength, have an intrinsic dependence on the time and temperature at which the response is measured. As will be seen, most viscoelastic properties exhibit a steep gradient in the neighborhood of a temperature termed the glass transition temperature. So much so that the graphs of property functions for this type of materials are typically displayed in double-logarithmic scales to allow encompassing the two or more orders of magnitude difference between their extreme values.

The nature of the response of elastic, viscous, and viscoelastic materials to several types of loading programs is examined next. It is then learned that viscoelastic solids and fluids will respond markedly differently only at temperatures above the glass transition temperature, or when their response is measured after a sufficiently long-time-following application of the load. It is learned that viscoelastic solids have non-zero long-term, equilibrium modulus and compliance, while viscoelastic fluids have to have zero long-term modulus, or, correspondingly, infinite compliance, to allow for viscous flow under sustained load. Two tests are identified—constant step strain and constant step stress—which can be used in an elementary manner to establish, respectively, the relaxation modulus and creep compliance of viscoelastic substances, at fixed temperature. The response of viscoelastic materials to constant rate loading is also compared to that of elastic solids and viscous fluids.

The behavior of elastic solids, viscous fluids, and viscoelastic materials regarding energy conservation and dissipation is also examined in some detail. The discussion will show that while elastic solids have the capacity to store, in the form of fully recoverable energy, all the work put into them, and viscous fluids dissipate all of the work of the external agents, viscoelastic materials store part of the energy that is put into them, and dissipate the rest as thermal energy, by raising their internal temperature.

Finally, it will be observed that viscoelastic materials may undergo aging, which is a continuous change in properties with elapsed time. Since materials are usually not put to use right after manufacture, aging materials make it necessary to keep two time scales: one to measure age and the other to measure the time at which load application starts.

1.2 The Nature of Amorphous Polymers

There are many materials, especially the so-called organic amorphous polymers,[1] whose behavior is of viscoelastic type. An amorphous polymer is made up of long-chain molecules. A typical polymeric chain may be comprised of several thousand molecules, strung together in a linear, chain-like fashion [1]. Amorphous polymers may be subdivided into uncross-linked and cross-linked, depending on the way in which their molecules are connected.

In uncross-linked amorphous polymers, such as unvulcanized natural rubber and hard and soft plastics, the individual long-chain molecules are randomly intertwined with each other but are not chemically bonded together, as indicated in part a of Fig. 1.1. The worm-like structure of uncross-linked polymers is, therefore, not permanent. As temperature increases, some chain disentanglement takes place and whole molecules, or segments of polymeric molecules tend to slide past each other. This allows the polymer to experience large deformations and, possibly, viscous flow [2].

There are several mechanisms by which polymer chains can be connected to one another to form a continuous network. Vulcanization utilizes sulfur as the bonding agent, which randomly attaches a chain to a number of neighboring

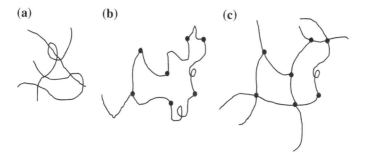

(a) **(b)** **(c)**

Fig. 1.1 Schematic representations of amorphous polymers. **a** Uncross-linked polymer. **b** End-linked polymer. **c** Cross-linked polymer

[1] Although polymers may be classified as crystalline and amorphous, the mechanical response of crystalline polymers cannot be described by the theory in this book; thus, only amorphous polymers are treated here.

chains, possibly at several points along its length, by means of strong covalent bonds. This process results in a relatively permanent, three-dimensional network structure which restrains molecules from freely slipping past each other, thus eliminating viscous flow. In all molecular networks, some loose ends of molecules attach to the network only at single points. In general, however, chemical bonding of polymeric molecules results in several chains, typically three or four, joining at the same locations, as illustrated in parts b and c of Fig. 1.1.

1.3 Mechanical Response of Viscoelastic Materials

According to the previous discussion, a material may be classified as viscoelastic if its response to external stimuli combines the characteristics of elastic and viscous behavior. Hence, the manner in which a viscoelastic substance would respond to an external agent can be guessed by examining the response of two identical specimens—one, made of an elastic solid and the other, of a viscous fluid—to the same stimulus, and by imagining that the behavior of the viscoelastic substance would lie somewhere in between.

For simplicity and clarity, one-dimensional specimens are used to examine the response of elastic solids and viscous fluids: a uniaxial bar is used for the elastic solid, and a hydraulic piston, or dashpot,[2] for the viscous fluid, as shown in Fig. 1.2. Also, in our experiment, we would apply a force to either of these specimens, as suggested in the figure, and use the change in their lengths as a measure of their response. However, for convenience, we normalize these quantities, and let σ stands for normal stress (force reckoned per unit of original cross-sectional area of the specimen) and use ε to denote normal strain (change in length per unit original length). In addition, we use E for the elastic modulus of the solid, and η for the coefficient of viscosity of the fluid, and assume their values to be constant.[3]

Fig. 1.2 Mechanical models of elastic solid (uniaxial *bar*), and viscous fluid (uniaxial *dashpot*), with externally applied force (*F*)

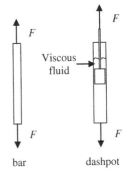

---

[2] A dashpot is a mechanical device which resists motion by viscous friction. The resisting force is directly proportional to the velocity; and, acting in the opposite direction, slows the motion and absorbs energy.

[3] The values of E and η of real materials typically depend on temperature, among other things.

On the previous definitions and assumptions, we introduce the one-dimensional versions of the stress–strain laws for the elastic solid and the viscous fluid. For the linearly elastic solid under one-dimensional conditions, we use Hooke's stress–strain law:

$$\sigma(t) = E \cdot \varepsilon(t) \tag{1.1}$$

Alternatively, Newton's law of viscosity is used to define the constitutive behavior of a linearly viscous fluid:

$$\sigma(t) = \eta \cdot \frac{d}{dt}\varepsilon(t) \tag{1.2}$$

1.3.1 Material Response to Step-Strain Loading

We consider the response to a constant strain history of a one-dimensional article which is made either of an elastic, viscous, or viscoelastic substance. A strain of magnitude ε_o is assumed suddenly applied at time $t = 0$ and held constant thereafter. Mathematically, we express such a strain history by means of the Heaviside or unit step function, $H(\cdot)$, which is identically zero for all negative values of its argument and equals one when its argument is positive (cf. Appendix A) and write:

$$\varepsilon(t) = \varepsilon_o H(t) \tag{a}$$

According to (1.1), the response of a straight bar of a linearly elastic solid to a step strain would be

$$\sigma(t) = E \cdot \varepsilon_o H(t) \tag{b}$$

In words, a uniaxial bar of an elastic solid would respond instantaneously to the suddenly applied strain, and the corresponding stress would remain in the material for as long as the strain is held.

The response of a linearly viscous fluid—in a dashpot, say—can be established by using (1.2) and taking the time derivative of (a). In doing so, note that the generalized derivative of the unit step function is the Dirac delta function (cf. Appendix A), so that:

$$\sigma(t) = \eta \cdot \varepsilon_o \delta(t) \tag{c}$$

The delta function is zero everywhere and becomes infinite where its argument vanishes. Thus, (c) indicates that to produce an instant strain of finite magnitude on a viscous fluid would require an infinitely large stress, and that the stress would disappear immediately after application of the strain.

Per our elementary definition of viscoelastic behavior, we would expect the response of a viscoelastic substance to lie somewhere between those of the elastic solid and the viscous fluid. On the grounds of its elastic component, it is logical to expect that a viscoelastic bar will respond with an instantaneous stress proportional to the applied strain; and, that to accommodate its viscous behavior, it also seems logical that the initial stress in the bar should decay, rather than remain constant—as it would in a purely elastic system—but that such decay should perhaps not be instantaneous, as in a linearly viscous fluid, but should depend on the elapsed time. The experiment just described is presented in Fig. 1.3.

Fig. 1.3 Qualitative material responses to step-strain loading. **a** Step strain. **b** Elastic solid. **c** Viscous fluid. **d** Viscoelastic material

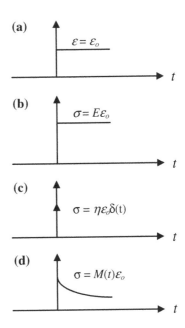

Since the applied stimulus in our experiment is constant, the response of the viscoelastic bar to a step-strain history could be mathematically described in the following form:

$$\sigma(t) = M(t)\varepsilon_o \tag{1.3}$$

According to the previous discussion, $M(t)$ must be a decreasing function of time, or at least non-increasing function of time, as suggested in Fig. 1.4. This material property function is called the relaxation modulus.

As will be discussed later on, not only is the relaxation modulus, M, a function of elapsed time, t, as indicated in (1.3), but it also depends very strongly on temperature, T. Thus, in general, $M = M(t, T)$. However, just as with E and η, temperature dependence is omitted from (1.3) because it has been assumed, for the time being only, that a uniform constant temperature is maintained in all experiments.

Fig. 1.4 Qualitative time
dependence of relaxation
modulus

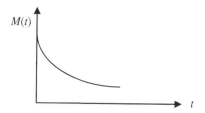

The constant strain history prescribed in (a) is known as a stress relaxation test
and is typically employed to establish the relaxation modulus, $M(t)$; since from
(1.3):

$$M(t) \equiv \frac{\sigma(t)}{\varepsilon_o} \qquad (1.4)$$

Specifically, to establish the relaxation modulus at a constant temperature:
- Select a displacement that would lead to an adequate strain for the test
  specimen.[4]
- Apply the selected target displacement as rapidly as possible and hold it
  constant.
- Record the force history necessary to maintain the prescribed displacement.
- Use (1.4) with the corresponding definitions of stress and strain for the speci-
  men to compute the relaxation modulus as a function of elapsed test time.

According to this terminology and the response shown in Fig. 1.3, elastic
materials do not relax. Thus, their characteristic relaxation time—the time it takes
the material to complete its relaxation process—is infinite. Viscous fluids, on the
other hand, relax completely and instantly: their relaxation times are zero. On this
basis, one can expect that viscoelastic materials will have finite, non-zero relaxation times. Also, by analogy with elastic solids and viscous fluids, viscoelastic
solids are expected to relax to non-zero stress, while viscoelastic fluids should
relax to zero stress, as indicated in the Fig. 1.5.

1.3.2   Material Response to Step-Stress Loading

Here, we consider the response of the elastic and viscous one-dimensional test
pieces to a constant stress of magnitude, σ_o, which is suddenly applied at time
$t = 0$ and held constant thereafter.

---

[4] By definition, the relaxation modulus of linear viscoelastic substances is assumed independent
of strain level. Thus, the magnitude of the prescribed displacement to use in a relaxation modulus
test should not be important. However, the relaxation modulus of materials capable of large
strains typically depends on strain level, and the selection of the test displacement should be
based on the end use of the material.

Fig. 1.5 Distinction between viscoelastic solids and fluids. **a** Step-strain history. **b** Viscoelastic solid. **c** Viscoelastic fluid

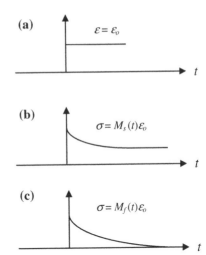

$$\sigma(t) = \sigma_o H(t) \tag{d}$$

According to Hooke's law, Eq. (1.1), an elastic solid would respond instantly, with the step strain:

$$\varepsilon(t) = \frac{\sigma_o}{E} H(t) \tag{e}$$

Per Eq. (1.2), a viscous fluid would continue to strain for as long as the applied stress is sustained:

$$\varepsilon(t) = \frac{\sigma_o}{\eta} t \tag{f}$$

Now the response of a viscoelastic specimen is expected to lie between that of the elastic and viscous test specimens. It should, therefore, exhibit an instantaneous strain proportional to the applied stress, like a solid would; but, similar to a fluid, its strain would grow with the passage of time.[5] The behavior just described is shown in Fig. 1.6.

It seems appropriate then, to call a viscoelastic substance a solid if, under constant load it creeps to a non-zero, finite strain, and to call it a fluid if its strain response does not seem to approach a finite limit. The distinction between viscoelastic solids and fluids is shown schematically in Fig. 1.7.

---

[5] At this point we avoid the temptation to describe viscoelastic response to sustained stress using the reciprocal of the relaxation modulus, as, say: $\varepsilon(t) = [1/M(t)]\sigma_o$. Suffice it to say that doing so would simply define viscoelastic behavior as little more than elastic.

Fig. 1.6 Qualitative material responses to step-stress loading. **a** Step-stress history. **b** Elastic solid. **c** Viscous fluid. **d** Viscoelastic material

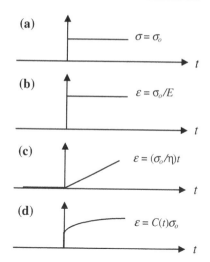

Fig. 1.7 Distinction between viscoelastic solids and fluids. **a** Step-stress history. **b** Viscoelastic solid. **c** Viscoelastic fluid

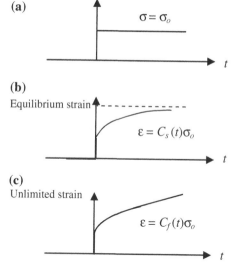

Since the applied load in this experiment is constant, the time variation of the response of a uniaxial bar of viscoelastic material can be mathematically described in the form:

$$\varepsilon(t) = C(t) \cdot \sigma_o \tag{1.5}$$

According to the foregoing arguments the function $C(t)$, called the creep compliance, must be an increasing, or, at the very least a non-decreasing function of time; as shown in Fig. 1.8.

Fig. 1.8 Conceptual time dependence of creep compliance

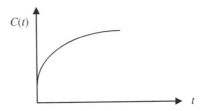

The constant stress history prescribed in (d) is known as a creep test, and is typically employed to establish the creep compliance, $C(t)$; since from (1.5):

$$C(t) = \frac{\varepsilon(t)}{\sigma_o} \tag{1.6}$$

Specifically, to establish the creep compliance at a constant temperature:
- Select a force that would lead to an adequate stress level for the test specimen.[6]
- Apply the selected target force as rapidly as possible and then hold it constant.
- Record the displacement history produced by the prescribed load.
- Use (1.6) with the corresponding definitions of stress and strain for the specimen to compute the creep compliance as a function of elapsed test time.

As will be discussed at length in a later chapter, the creep compliance and the relaxation modulus are not reciprocals of each other, except under special conditions, but they are not completely independent of each other, either.

1.3.3   Material Response to Cyclic Strain Loading

We now examine the stress response of a linear elastic solid and a linear viscous fluid to the sinusoidal, cyclic strain history:

$$\varepsilon(t) = \varepsilon_o sin\omega t \tag{g}$$

As before, we use expressions (1.1) and (1.2) to establish, respectively, the responses of the elastic solid and viscous fluid, as:

$$\sigma(t) = E \cdot \varepsilon_o \sin(\omega t) \tag{h}$$

$$\sigma(t) = \eta \cdot \varepsilon_o \omega \cdot \cos(\omega t) \tag{i}$$

---

[6] By definition, the creep compliance of linear viscoelastic materials is assumed independent of stress level. Thus, the magnitude of the stress to use in a creep test should not be important. However, the compliance of nonlinear materials depends on stress level, so that the selection of the test stress level should be made carefully.

Fig. 1.9 Qualitative material responses to cyclic strain. **a** Cyclic strain history. **b** Elastic solid. **c** Viscous fluid. **d** Viscoelastic material

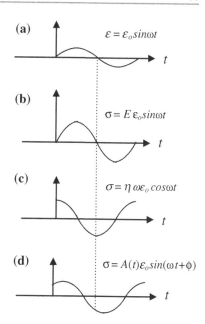

These expressions indicate that, while the stress in the elastic solid is in phase with the strain, the stress in the viscous material is exactly 90° out of phase with it. It is then reasonable to expect that the response of a viscoelastic material to cyclic harmonic loading will lie anywhere between 0° and 90° out of phase with its loading. This observation is indicated in Fig. 1.9.

1.3.4 Material Response to Constant Strain Rate Loading

We next examine the stress response of the linear elastic solid and linear viscous fluid to the constant strain rate history:

$$\varepsilon(t) = R \cdot t \tag{j}$$

Using this relation and (1.1), the response of the solid is seen to be:

$$\sigma(t) = E \cdot R \cdot t \tag{k}$$

The response of the linear viscous fluid follows from (j) and (1.2), as:

$$\sigma(t) = R \tag{l}$$

These expressions indicate that while the stress in the elastic solid is directly proportional to the strain, the stress in the viscous material is equal to the rate of straining. One may then argue that the stress response of a linear viscoelastic

Fig. 1.10 Qualitative material responses to constant strain rate loading: comparison of responses at one loading rate. **a** Strain history. **b** Elastic solid. **c** Viscous fluid. **d** Viscoelastic material

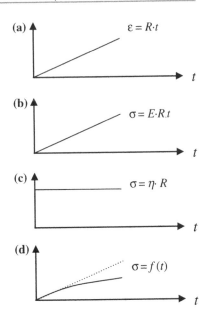

(a) $\varepsilon = R \cdot t$

(b) $\sigma = E \cdot R \cdot t$

(c) $\sigma = \eta \cdot R$

(d) $\sigma = f(t)$

material, having to lie between a constant and a linear function of time, would have to be described by a function that increases less than linearly with time, as indicated in Fig. 1.10.

Clearly, without knowing more about the material at hand, a graphic response in a form such as that shown in Fig. 1.10d would not suffice to definitely identify a material as viscoelastic. That type of response is also characteristic of nonlinear elastic behavior. The point to emphasize here, regarding elastic and viscoelastic response to constant strain rate loading is that, an elastic material will react with the same stress to a given strain, irrespective of how long it takes to reach that strain. A viscoelastic material, on the other hand, will respond with a stress that depends on how long it takes to apply the strain. This aspect of material behavior is depicted in Fig. 1.11.

1.4   Energy Storage and Dissipation

As with other aspects of viscoelastic behavior, we should require that the response of any viscoelastic material regarding energy conservation should lie between those of an elastic solid and a viscous fluid. As will be shown shortly, the total amount of work performed on an elastic solid by external agents is stored in it in the form of internal energy, which is fully recoverable upon removal of the external agents.[7] By

---

[7] This means that if the external agents that put work into an elastic solid were completely removed, the internal energy stored in the solid could be used to perform an amount of work—on another system, say—equal to the work that was put into the solid in the first place.

Fig. 1.11 Qualitative material responses to constant strain rate loading: comparison of responses at two loading rates. **a** Strain history. **b** Elastic solid. **c** Viscous fluid. **d** Viscoelastic material

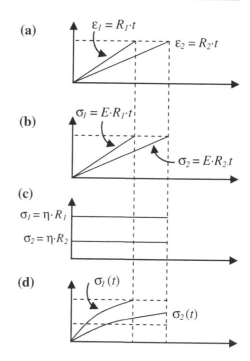

contrast to an elastic solid, a viscous fluid has no capacity to store energy, and all work performed on viscous fluids is lost or, more properly, dissipated.

By our definition of viscoelastic behavior, a viscoelastic material, be it a solid or a fluid, should be expected to be able to store, and have available for recovery, at least part of the energy put into it, while it should dissipate the rest. It therefore seems logical to postulate that, irrespective of the constitution of the material at hand, "the total work performed on a body by external agents, W_{ext}, should be equal to the work of the internal forces, W_{int}, minus the work, W_{diss}, dissipated in the process." This statement embodies the first law of thermodynamics, on the conservation of energy, that: "Energy can neither be created nor destroyed, but only transformed."

According to the previous discussion, linearly elastic solids must exhibit no dissipation, while linearly viscous fluids must dissipate all energy put into them. That is:

$$W_{diss} = \begin{cases} 0; & for \quad linearly \quad elastic \quad solids \\ W_{ext}; & for \quad linearly \quad viscous \quad fluids \end{cases} \tag{1.7}$$

The energy balance equation may be cast in the form:

$$W_{ext} = W_{int} - W_{diss} \tag{1.8}$$

Although we present energy conservation and dissipation in a broader sense in Appendix B, in what follows we examine the balance of energy using one-dimensional models: the bar as a linearly elastic solid, and the dashpot as a linearly viscous fluid.

We consider an experiment in which the test piece is fixed at one end, while its other end is first pulled a certain amount and then moved back to its original position.[8] Under these circumstances, the work of the external forces is exactly zero, irrespective of the material constitution; since, by definition:

$$W_{ext} \equiv \int_{P_1}^{P_2} F_{ext} \cdot du_{ext} = \int_{P_1}^{P_1} F_{ext} \cdot du_{ext} \equiv 0 \qquad (a)$$

where, F and u denote force and displacement, respectively. In this case (1.8) takes the simpler form:

$$W_{diss} = W_{int} \qquad (b)$$

To calculate the amount of energy dissipated in the process, we note that, similar to W_{ext}, the work of the internal forces is given by:

$$W_{int} \equiv \int_{P_1}^{P_2} F_{int} \cdot du_{int} = \int_{\varepsilon_1}^{\varepsilon_2} (A\sigma) \cdot (Ld\varepsilon) = V \int_{\varepsilon_1}^{\varepsilon_2} \sigma \cdot d\varepsilon \qquad (c)$$

This expression uses that, in a one-dimensional system: $\sigma = F/A$, $d\varepsilon = du/L$, and $V = A \cdot L$, for the stress, strain, and system's volume, respectively.

To apply (c) to the bar of a linearly elastic solid, we use the stress–strain Eq. (1.1), which leads to:

$$W_{int} \equiv V \int_{\varepsilon_1}^{\varepsilon_2} \sigma \cdot d\varepsilon = V \cdot E \int_{\varepsilon_1}^{\varepsilon_1} \varepsilon \cdot d\varepsilon = V \cdot \frac{1}{2} E \cdot \left[\varepsilon^2\right]_{\varepsilon_1}^{\varepsilon_1} \equiv 0 \qquad (d)$$

Taking this into (b) yields the first of (1.7): $W_{diss} = 0$; that a linearly elastic solid absorbs all external work as internal energy and does not dissipate any energy.

Noting that the constitutive Eq. (1.2) for a linearly viscous fluid involves strain rate, and not strain, we first transform (c) to accommodate this fact. In doing so, it is assumed, without loss of generality, that the total duration of the experiment is t^*. Thus:

$$W_{int} \equiv V \int_{\varepsilon_1}^{\varepsilon_2} \sigma \cdot d\varepsilon = V \cdot \eta \int_{t=0}^{t=t^*} \frac{d\varepsilon}{dt} \cdot \frac{d\varepsilon}{dt} dt \equiv V \cdot \eta \cdot \int_{t=0}^{t=t^*} \left(\frac{d\varepsilon}{dt}\right)^2 dt \qquad (e)$$

---

[8] In general, the loading and unloading paths do not have to be the same, as long as the test piece is returned to its original position.

Fig. 1.12 Example 1.1

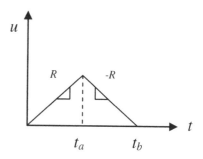

The quantity under the integral sign in (e) is never negative, vanishing only for the trivial case when the strain rate is identically zero in the interval of integration. Therefore:

$$W_{\text{int}} \equiv V \cdot \eta \cdot \int_{t=0}^{t=t^*} \left(\frac{d\varepsilon}{dt}\right)^2 dt \geq 0 \tag{f}$$

This result, together with (b) leads to:

$$W_{\text{diss}} = W_{\text{int}} = V \cdot \eta \cdot \int_{t=0}^{t=t^*} \left(\frac{d\varepsilon}{dt}\right)^2 dt \geq 0 \tag{g}$$

Thus, as asserted in the second part of (1.7): a linearly viscous fluid dissipates all energy put into it.

Example 1.1 Determine the total energy that would be dissipated by applying the displacement history shown in Fig. 1.12 to a dashpot with a linearly viscous fluid of viscosity η.

Solution:

To calculate the total energy dissipated in the process we evaluate the evolution of the strain rate, $d\varepsilon/dt = (1/L)du/dt$ during the process:

$$\frac{d\varepsilon}{dt} = \begin{cases} R; & 0 \leq t < t_a \\ -R; & t_a \leq t \leq t_b \end{cases}$$

and take the result into (g), to get:

$$W_{\text{diss}} = V \cdot \eta \cdot \int_{t=0}^{t=t_a} \left(\frac{d\varepsilon}{dt}\right)^2 dt + \int_{t=t_a}^{t=t_b} \left(\frac{d\varepsilon}{dt}\right)^2 dt = V\eta R^2 t_b$$

Fig. 1.13 Typical relaxation modulus as a function of temperature; showing the regions of viscoelastic behavior for typical viscoelastic solids and fluids

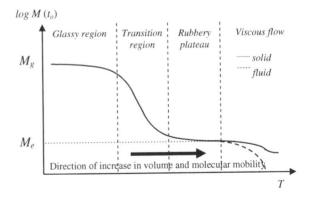

1.5   Glass Transition and Regions of Viscoelastic Behavior

Experiments with viscoelastic materials indicate that:

- The reported values of the relaxation modulus and creep compliance depend on the time scale of the observations (10 s modulus, 10 min modulus, etc.).
- Most properties of viscoelastic materials depend strongly on temperature.
- Some properties change very drastically when the temperature is near a critical value, called the "glass transition temperature," usually denoted as T_g.

The influence of temperature and time of observation on viscoelastic properties—the time and temperature dependence of functions such as the modulus, M, or the creep compliance, C, of previous discussions—is emphasized with the explicit notations $M(t, T)$ and $C(t, T)$. The relaxation modulus, M, of a typical amorphous polymer is used in Fig. 1.13 to exemplify the temperature dependence of viscoelastic properties in general. A logarithmic scale is employed to accommodate the strong dependence of modulus with temperature. In such graphical representations, it is always necessary to indicate exactly to what test time the reported property corresponds ($t = t_o$, in this case).

The transition region falls in a narrow temperature range; and the temperature in the middle of this region is called the glass transition temperature, T_g. At temperatures well below the glass transition, in the so-called glassy region, an amorphous polymer is an organic glass: a hard and brittle plastic with a high modulus, M_g, called the glassy modulus. At such low temperatures, the polymer chains are essentially "frozen" in fixed positions. At temperatures around T_g, in the glass-transition region, whole molecules, or segments of polymeric molecules are somewhat free to "jump" from one site to another, and the polymer responds with a modulus that changes very sharply with temperature. At temperatures above the glass transition but below the melting point, in what is called the rubbery plateau, molecular mobility increases and segments of polymeric chains reorient relative to each other. In this region, cross-linked polymers would relax to a more-or-less constant "equilibrium" modulus, M_e, behaving like rubbery elastic solids

Fig. 1.14 Fixed-time creep compliance as a function of temperature, showing the regions of viscoelastic behavior for typical solid and fluid polymers

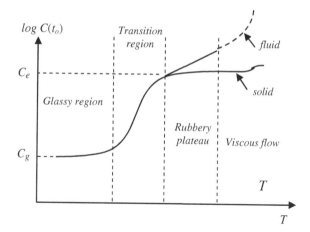

of low modulus. In fact, M_e may be several orders of magnitude smaller than the glassy value, M_g. In this region, uncross-linked polymers eventually disentangle, with entire molecular segments sliding past one another, giving rise to viscous flow [3].

Similar observations apply to the creep compliance function, C (t, T). The temperature dependence of the creep compliance function of a polymer looks very much like the mirror image of the relaxation modulus about a line parallel to the temperature axis, as indicated in Fig. 1.14.

The creep compliance function of an uncross-linked polymer typically lacks the rubbery plateau, exhibiting continued creep, followed by failure. This is indicated by a dotted line in the figure.

Also, to the elastic glassy and equilibrium moduli, M_g and M_e, correspond glassy and equilibrium compliances, C_g, and C_e, respectively. Since modulus and compliance of any linearly elastic material are reciprocals of each other, it follows that both: $C_g = 1/M_g$, and $C_e = 1/M_e$. In the transition region, however: $C(t, T) \neq 1/M(t, T)$; although the two functions are inverses of each other, in a sense to be explained in Chap. 2. In fact, it will be proven there, that in general, $C(t, T) \cdot M(t, T) \leq 1$ [4].

As pointed out in Sect. 1.3, the relaxation modulus is a decreasing—or, at least, non-increasing—function of time. In a double-logarithmic plot, $log(M)$ *versus* $log(t)$, the shape of its graph resembles that of $log(M)$ *versus* T. Entirely similar observations hold for the creep compliance, as indicated in Fig. 1.15.

The fact that the graphs of the material functions M and C versus temperature have the same general shape as those versus elapsed time suggests a relationship between time and temperature for viscoelastic substances. The detailed nature of this relationship is contained in the time–temperature superposition principle that: *in amorphous polymers, time and temperature are interchangeable*. This principle is explained in Chap 6. In loose terms, as far as material property functions are concerned, the principle implies that short test times correspond to low

Fig. 1.15 Fixed-temperature relaxation modulus and creep compliance as functions of test time, showing regions of viscoelastic behavior for a typical cross-linked polymer

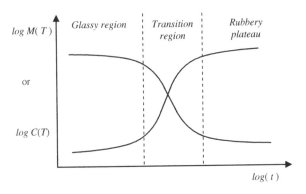

temperatures and long test times to high temperatures, and vice versa. This means that reducing the test temperature is equivalent to shortening the test time; and conversely, increasing test temperature is equivalent to extending the test time.

Two additional qualitative observations regarding polymers were reported by R. Gough, in 1805, and confirmed experimentally by J. P. Joule, in 1857, at Lord Kelvin's insistence, [5] that:

- Rubber heats up on stretching.
- A loaded rubber band contracts on heating.

The first observation shows that rubber dissipates energy in the form of heat, as our discussion of energy dissipation suggested viscoelastic materials would. The second observation, known as the Gough-Joule effect, indicates that, unlike metals, the modulus of rubber experiences a relative increase—stiffening up—with temperature.

Example 1.2 Draw a sketch of the response of a linear viscoelastic solid to the step-strain program shown in Fig. 1.16. Consider: (a) response in the glassy region, and (b) response in the transition region.

Solution:

(a) *Response in the glassy region*

In the glassy region, the behavior of any viscoelastic substance is elastic, with glassy modulus, M_g, and glassy compliance, $C_g = 1/M_g$. For this reason, its response must be directly proportional to the applied step strain. This is indicated in Fig. 1.17.

(b) *Response in the transition region*

Here, we use that: the response of any viscoelastic substance to a suddenly applied stimulus—instantaneous loading or instantaneous unloading—is always elastic with glassy modulus M_g, and glassy compliance, $C_g = 1/M_g$. Additionally, its response to a constant strain in the transition region follows the shape of the relaxation modulus $M(t)$, which is a decreasing function of the time elapsed since the strain was applied. On unloading, the same is true, only

Fig. 1.16 Example 1.2: step-strain load–unload event

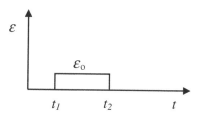

Fig. 1.17 Example 1.2: response of a linear viscoelastic solid to a step-strain load–unload event in the glassy region

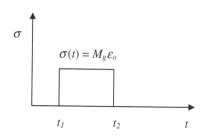

Fig. 1.18 Example 1.2: response of a linear viscoelastic solid to a step-strain load–unload event in the transition region

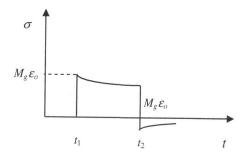

the sign of the response is reversed. Thus, the stress first drops by $M_g\varepsilon_o$—becoming negative—to then undergo a reverse relaxation, recovering monotonically toward zero. This behavior is indicated in Fig. 1.18.

Example 1.3 Draw a sketch of the response of a linear viscoelastic solid to the step-stress program shown in Fig. 1.19. Consider: (a) response in the equilibrium zone, and (b) response in the transition region.

Solution:

(a) *Response in the rubbery equilibrium region*
   In the rubbery region, a viscoelastic material will respond like an elastic solid with modulus, M_e, and compliance, $C_e = 1/M_e$. Therefore, its response must be directly proportional to the applied loading. This is indicated in Fig. 1.20.

Fig. 1.19 Example 1.3: step-stress load–unload event

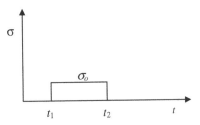

Fig. 1.20 Example 1.3: response of a linear viscoelastic solid to a step-stress load–unload event in the rubbery region

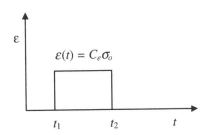

(b) *Response in the transition region*

Here, we use that: the response of any viscoelastic substance to a suddenly applied stimulus—instantaneous loading or instantaneous unloading—is always elastic with glassy modulus M_g, and glassy compliance, $C_g = 1/M_g$. Additionally, its response to a constant stress in the transition region follows the shape of the creep compliance, $C(t)$, which is an increasing function of the time elapsed since the stress was applied. On unloading, the same is true, only the sign of the response is reversed: the strain first drops by first drops by $C_g\sigma_o$, and then "recovers" monotonically toward zero. This behavior is indicated in Fig. 1.21.

1.6 Aging of Viscoelastic Materials

Aging is a phenomenon observed in many viscoelastic substances. It may be defined as any change in constitutive or failure properties with time. There are two types of aging: physical, reversible aging, which is due to thermodynamic

Fig. 1.21 Example 1.3: response of a linear viscoelastic solid to a step-stress load–unload event in the transition region

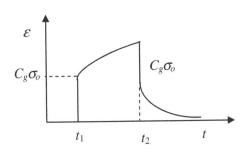

processes; and chemical aging, which is caused by irreversible chemical reactions in the material. Although their origins differ, the macroscopic manifestations of physical and chemical aging are very similar. From a mechanical point of view, material aging frequently manifests itself as an increase in modulus and a reduction of creep strain [6].

As indicated earlier in this chapter, the response of a non-aging viscoelastic substance to an external stimulus changes with the passage of time because its material properties, such as relaxation modulus and creep compliance—depend on the time at which the observations are made during an experiment. However, such response is independent of the time at which the experiment is started, and is the same every single time the same experiment is repeated, provided the experiment is the same each time.[9] Put another way, the material properties of a non-aging material are fully described by functions which depend on a single time scale, which measures only the time the material is under load, irrespective of when it was manufactured.

By contrast, the properties of a material susceptible to aging change with the passage of time even in the absence of external agents. That is, the response of an aging material, measured at a specified time during an experiment, depends on the time elapsed since the material was manufactured and the time it spends under load; and hence, on the time at which the experiment is started. For this reason, two time scales are required to unambiguously describe the material property functions—and response—of an aging material. One time scale is needed to keep track of the time since the manufacturing of the material, and the other time scale is needed to keep track of the time the material is under load.

References

1. A.N. Gent (ed.), *Engineering with Rubber, How to Design Rubber Components* (Jos. C. Huber KG, Diessen/Ammersee, 2002), pp. 38–40
2. L.E. Malvern, *Introduction to the Mechanics of a Continuous Medium* (Prentice-Hall Inc., 1969), pp. 306–327
3. J.J. Aklonis, *Introduction to Polymer Viscoelasticity* (Wiley, 1983), pp. 44–52
4. A.C. Pipkin, *Lectures on Viscoelasticity Theory* (Springer, 1986), pp. 14–15
5. H. Morawetz, *Polymers: The Origins and Growth of a Science* (Dover, 1985), pp. 35–36
6. A.D. Drosdov, *Finite elasticity and Viscoelasticity, A course in the Nonlinear Mechanics of Solids* (World Scientific, Singapore, 1996), pp. 229–234

[9] Assuming it were possible to produce exactly the same non-ageing material every time—without lot-to-lot variability—this result would hold true even if the specimens came from material lots of different age.

Constitutive Equations in Hereditary Integral Form

2

Abstract

Materials respond to external load by deforming and straining, and by developing stresses. The internal stresses corresponding to a given set of strains depend on the constitution of the material itself. For this reason, the rules that permit calculation of internal stresses from known strains, or vice versa, are called constitutive laws or constitutive equations. There are two equivalent ways to describe the mathematical relationships between stresses and strains for viscoelastic materials. One form uses integrals to define the constitutive relations, while the other relates stresses and strains by means of differential equations. Starting from Boltzmann's superposition principle, this chapter develops the integral form of the one-dimensional constitutive equations for linearly viscoelastic materials. This is followed by a discussion of the principle of fading memory, which helps to define the acceptable analytical forms of the material property functions. It is then shown that the closed-cycle condition (i.e., that the steady-state response of a non-aging viscoelastic material to a periodic excitation be periodic) requires that the material property functions depend only on the difference of their arguments. The chapter also examines the relationships between the relaxation modulus and creep compliance functions in the physical time domain as well as in Laplace-transformed space. Various alternative forms of the integral constitutive equations often encountered in practice are discussed as well.

Keywords

Boltzmann · Constitutive · Convolution · Creep · Cycle · Equilibrium · Fading · Glassy · Hereditary · Isothermal · Laplace · Long-term · Matrix · Memory · Operator · Principle · Relaxation · Symbolic

2.1 Introduction

Materials respond to external stimuli by deforming and straining, that is by changing their shape or size, and by developing stresses. The internal stresses corresponding to a given set of strains depend on the constitution of the material itself. For this reason, the rules that permit calculation of internal stresses from known strains, or vice versa, are called constitutive laws, or, constitutive equations—when such relationships are known in analytical form. The terms stress–strain or strain–stress relations or equations, are widely used to emphasize that the first variable is expressed in terms of the second.

There are two equivalent ways to describe the mathematical relationships between stress and strain for linear viscoelastic materials. One way uses integrals to define these relations, while the other relates stresses and strains through linear ordinary differential equations. In this chapter, we develop the integral form of constitutive equations, leaving for Chap. 3 the discussion of their differential counterparts. All the developments are presented in great mathematical detail but to motivate the proofs, some physical insight is also provided. The level of mathematical detail used to present the subject matter and the exercises in this chapter is intended to give the reader the confidence necessary to engage in independent research, irrespective of the field of interest.

For clarity of presentation, only non-aging materials under isothermal conditions are treated in this and subsequent chapters, until Chap. 6, where the dependence of material properties on temperature is examined. All material functions referred to here are thus presumed independent of age and available at the constant temperature implied in the discussions. The dependence of material property functions on temperature will be omitted but assumed understood.[1]

This chapter starts from Boltzmann's superposition principle and develops the integral form of the one-dimensional constitutive equations for a linearly viscoelastic substance. This is followed by a discussion of the principle of fading memory, which helps to define the acceptable forms of relaxation and compliance functions. It is then shown that the closed-cycle condition (that the steady-state response of a non-aging viscoelastic material to a periodic excitation be periodic) requires that the material property functions depend only on the difference of their arguments, and all transients die out. The chapter also examines various relationships between the relaxation modulus and creep compliance functions, both in the time domain and in Laplace-transformed space. Alternative forms of constitutive equations often encountered in practice are also discussed. We conclude the chapter with a discussion of how to evaluate the work done by external agents acting on a linear viscoelastic material. This topic of great practical use, since, as shown in Chap. 1, viscoelastic materials dissipate as heat, some of the energy that is put into them, and hence polymeric materials are often used in industry to dissipate energy.

---

[1] On this assumption, for instance, $M(t)$ and $C(t)$ will be used for $M(t,T)$ and $C(t,T)$, respectively.

Fig. 2.1 Stress response to a step strain applied at the time the test clock is started

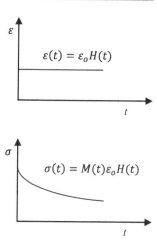

2.2 Boltzmann's Superposition Principle

By definition [c.f. Chap. 1], the tensile relaxation modulus, $M(t,T)$, at any time t, and fixed temperature T describes how the stress varies with time under a step-strain load. To fix ideas, imagine a one-dimensional bar of a linearly viscoelastic material after it is subjected to a strain of magnitude ε_o, suddenly applied at the start of an experiment and held constant thereafter. As seen in (Fig. 2.1), in accordance with Eq. (1.3), the stress response, $\sigma(t)$, of the bar to the applied step strain would be given by:

$$\sigma(t) = \begin{cases} 0, & \text{for } t < 0 \\ M(t)\varepsilon_o, & \text{for } t \geq 0 \end{cases} \tag{a}$$

By the definition of the Heaviside step function H, that: $H(t) = 0$, for negative values of its argument, while $H(t) = 1$, whenever its argument is zero or positive, one can rewrite (a) in the form [c.f. Appendix A]:

$$\sigma(t) = M(t) \cdot H(t)\varepsilon_o \tag{b}$$

Now assume that exactly the same experiment as that described by (a) or (b) were to be carried out using the same material but applying the loading t_1 units of time after "starting the clock." Also assume that all loading[2] and environmental conditions would be the same in both cases. If the material did not age, all its relevant property functions would be exactly the same in both experiments.

---

[2] The terms "load" and "loading" are used in their broader sense to include tractions, or stresses, as well as displacements, or strains. The exact meaning should be clear from the context in which the term is used.

Fig. 2.2 Stress response to a
step strain applied t_1 units of
time after the test clock is
started

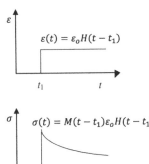

Consequently, exactly the same response would be observed in the second
experiment as in the first, but with a time delay t_1, as indicated in Fig. 2.2.

Similarly to (a) and (b), the stress response could now be expressed, respectively, as follows:

$$\sigma(t) = \begin{cases} 0, & \text{for } *\tau < \tau_\emptyset \\ M(t - t_1)\varepsilon_o, & \text{for } *\tau \geq \tau_\emptyset * \end{cases} \tag{c}$$

$$\sigma(t) = M(t - t_1)H(t - t_1)\varepsilon_o \tag{d}$$

It is an easy matter to extend these results to arbitrary load cases. As suggested
in Fig. 2.3, any piecewise continuous function of time may be approximated by a
series of step functions; with each subsequent step adding an incremental amount
to the previous step. Using (c), then, the response to the kth incremental step strain,
$\Delta\varepsilon_k$, which is taken to occur at time t_{k+1}, would be:

$$\Delta\sigma_k(t) = M(t - t_k)\Delta\varepsilon_k, \quad t \geq t_k \tag{e}$$

Fig. 2.3 Approximation of a
continuous function as a finite
series of incremental step
functions

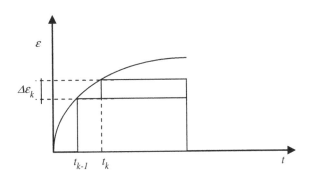

According to Boltzmann's principle, the response to each incremental load is independent of those due to the other incremental loads, and the response to the complete load history, as idealized through the series of incremental step-loads, equals the sum of the individual responses:

$$\sigma(t) \approx \sum_{k=1}^{N} \Delta\sigma_k(t) = \sum_{k=1}^{N} M(t - t_k)\Delta\varepsilon_k, \quad t \geq t_k \tag{f}$$

Dividing and multiplying the right-hand side of (f) by the time interval, $\Delta t_k = t_k - t_{k-1}$, between successive steps, and using the properties of the Heaviside step function, yields:

$$\sigma(t) \approx \sum_{k=1}^{N} \Delta\sigma_k(t) = \sum_{k=1}^{N} M(t - t_k)\frac{\Delta\varepsilon_k}{\Delta t_k}\Delta t_k; \quad t \geq t_k \tag{g}$$

Passing to the limit as N increases without bound and the size of successive intervals is made vanishingly small:

$$\sigma(t) = \lim_{\lim N \to \infty} \sum_{k=1}^{N} \Delta\sigma_k(t) = \lim_{\substack{N \to \infty \\ t_k \to \tau}} \sum_{k=1}^{N} M(t - t_k)\frac{\Delta\varepsilon_k}{\Delta t_k}\Delta t_k; \quad t \geq t_k \tag{h}$$

Since this process turns the discrete set t_k into a continuous spectrum, we use the letter τ to denote it and arrive at[3] (see, for instance, [1]):

$$\sigma(t) = \int_{0^+}^{t} d\sigma(t) = \int_{0^+}^{t} M(t - \tau)\frac{d}{\partial\tau}\varepsilon(\tau)d\tau \tag{i}$$

To allow the strain to have a step discontinuity at time $t = 0^+$, we add (a) and (i) and write:

$$\sigma(t) = M(t)\varepsilon(0^+) + \int_{0^+}^{t} M(t - \tau)\frac{d}{\partial\tau}\varepsilon(\tau)d\tau \tag{2.1a}$$

The term $M(t)\,\varepsilon(0^+)$ may be taken inside the integral, using that $\varepsilon(0^-) \equiv 0$, because:

$$\int_{0^-}^{0^+} M(t - \tau)\frac{d}{\partial\tau}\varepsilon(\tau)d\tau = M(t)\int_{0^-}^{0^+}\frac{d}{d\tau}\varepsilon(\tau)d\tau \equiv M(t)\varepsilon(0^+)$$

---

[3] The notation x^+ is used to signify a value of x that is just larger than x. Similarly, x^- means a value of x just less than x.

Hence, (2.1a) may be alternatively expressed as:

$$\sigma(t) = \int\limits_{0^-}^{t} M(t - \tau)\frac{d}{\partial\tau}\varepsilon(\tau)d\tau \qquad (2.1b)$$

Had we chosen the applied action to be a stress instead of strain history, entirely similar arguments would have led to the strain–stress forms:

$$\varepsilon(t) = C(t)\sigma(0^+) + \int\limits_{0^+}^{t} C(t - \tau)\frac{d}{d\tau}\sigma(\tau)d\tau \qquad (2.2a)$$

$$\varepsilon(t) = \int\limits_{0^-}^{t} C(t - \tau)\frac{d}{d\tau}\sigma(\tau)d\tau \qquad (2.2b)$$

Equations in (2.1a, b) and (2.2a, b) show that the response of a viscoelastic substance at any point in time depends not only on the value of the action at that instant, but also on the integrated effect, or complete history of all past actions. In other words, the response at the present instant inherits the effects of all past actions. For this reason, viscoelastic materials are also frequently called hereditary materials; and viscoelasticity, hereditary elasticity.

Example 2.1 The (one-dimensional) viscoelastic response to a constant strain-rate loading, $\varepsilon(t) = R{\cdot}t$, may be expressed in the elastic form: $\sigma(t) = E_{\mathit{eff}}(t){\cdot}\varepsilon(t)$. Derive an expression for $E_{\mathit{eff}}(t)$, the constant-rate effective modulus, for a viscoelastic substance.

Solution:

Assume the relaxation modulus of the viscoelastic material to be $M(t)$ and compute its stress response with (2.1a), using that $d\varepsilon(s)/ds = d(Rs)/ds = R$, and introducing the change of variables $t - \tau = u$, to arrive at:

$$\sigma(t) = M(t)\varepsilon(0^+) + R\int\limits_{0^+}^{t} M(t - \tau)d\tau = R\int\limits_{0}^{t} M(u)du$$

Multiplying and dividing this expression by t, recalling that $\varepsilon(t) = R \cdot t$, and re-ordering:

$$\sigma(t) = Rt\frac{1}{t}\int_0^t M(u)du \equiv \left[\frac{1}{t}\int_0^t M(u)du\right]\varepsilon(t) \equiv E_{eff}(t)\varepsilon(t)$$

With the obvious definition of the constant-rate effective modulus, E_{eff}:

$$E_{eff}(t) \equiv \frac{1}{t}\int_0^t M(u)du \tag{2.3}$$

This expression can be used to evaluate the stress response of a viscoelastic material to constant strain-rate loading, by means of the elastic-like expression: $\sigma(t) = E_{eff}(t) \cdot \varepsilon(t)$.

Had the roles of strain and stress been reversed, we would have employed (2.2a) to derive the following definition of the constant-rate effective compliance:

$$D_{eff}(t) \equiv \frac{1}{t}\int_0^t C(u)du \tag{2.4}$$

As before, this can be used to determine the strain at any specified time, of a viscoelastic material subjected to constant-rate stress, using the elastic-like form: $\varepsilon(t) = D_{eff}(t) \cdot \sigma(t)$.

Example 2.2 Obtain the instantaneous response of a viscoelastic material with relaxation modulus, $M(t)$, to a general strain history $\varepsilon(t)$.

Solution:

We evaluate the stress response using expression (2.1a) at $t = 0$, to get:

$$\sigma(t) = M(0)\varepsilon(0^+) \equiv M_g\varepsilon(0^+) \tag{2.5}$$

In similar fashion, (2.1b) would yield the instantaneous strain response to an arbitrary stress history $\sigma(t)$, as:

$$\varepsilon(t) = C(0)\sigma(0^+) \equiv C_g\sigma(0^+) \tag{2.6}$$

This example indicates that the instantaneous, impact, or glassy response of a non-aging viscoelastic material is elastic, with operating properties equal to its glassy modulus, or its glassy compliance, depending on whether strain or stress, respectively, is the controlled variable.

Example 2.3 Obtain the equilibrium response of a viscoelastic substance with relaxation modulus, $M(t)$, to a general strain history $\varepsilon(t)$.

Solution:

We evaluate the stress response using expression (2.1a) as $t \rightarrow \infty$:

$$\sigma(\infty) = \lim_{t \to \infty} \int_{0^-}^{t} M(t - \tau) \frac{d\varepsilon}{d\tau} d\tau = \lim_{t \to \infty} \left[\int_{0^-}^{0^+} M(t - \tau) \frac{d\varepsilon}{d\tau} d\tau + \int_{0^+}^{t} M(t - \tau) \frac{d\varepsilon}{d\tau} d\tau \right]$$

Noting that $\varepsilon(t) \equiv 0$, $t < 0$:

$$\sigma(\infty) = M(\infty)\varepsilon(0^+) + M(\infty) \lim_{t \to \infty} \int_{0^+}^{t} \frac{d\varepsilon}{d\tau} d\tau$$

Or, after canceling like terms, since the integral evaluates to: $\varepsilon(\infty) - \varepsilon(0)$, and $M(\infty)$ is the equilibrium modulus M_e:

$$\sigma(\infty) = M_e \varepsilon(\infty) \tag{2.7}$$

By the same procedure, starting with (2.2a), it is found that the long-term strain response to an arbitrary stress history, $\sigma(t)$, is given as:

$$\varepsilon(\infty) = C_e \sigma(\infty) \tag{2.8}$$

This example indicates that the long-term response of a non-aging viscoelastic material is elastic, with operating properties equal to either its long-term or equilibrium modulus, or its long-term or equilibrium compliance, depending on whether strain or stress is the controlled variable.

2.3   Principle of Fading Memory

Loosely speaking, we say that a material has fading memory if the influence of an action on its response becomes less important as time goes by. Accordingly, the mathematical implications of the fading memory hypothesis—often called principle—can be established by loading and unloading a viscoelastic system, and monitoring its response after the load is removed. Before establishing the consequences of the principle of fading memory on a rigorous basis, we develop them by examining the response of a viscoelastic material to the relaxation and creep experiments; with which we are already familiar. The results of these experiments are the relaxation modulus and the creep compliance. As discussed in Chap. 1, the general shapes of these functions are as shown in Fig. 2.4.

Fig. 2.4 Functional forms of the stress *relaxation modulus* and *creep compliance* of a viscoelastic material used to explain the fading memory hypothesis

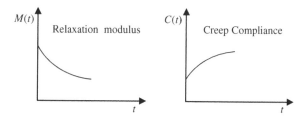

The functional forms shown in the figure indicate that the fading memory hypothesis should require that the relaxation modulus be a monotonically decreasing function of time, with monotonically decreasing slope. In similar fashion, the creep compliance should be a monotonically increasing function of time, with monotonically decreasing slope. We now proceed with the rigorous proofs of these statements. To do that, we will take the applied action to be a step strain of magnitude ε_o, applied to a one-dimensional viscoelastic system starting at time $t = 0$ and ending at time $t = t^*$: $\varepsilon(t) = \varepsilon_o[H(t) - H(t - t^*)]$.

Expression (2.1a) will be used to establish the corresponding response. Before we proceed, we put (2.1a) in a form more suitable to our purposes, integrating it by parts and writing the resulting derivative of the modulus in terms of the time difference, $t - \tau$; thus:

$$\sigma(t) = M(0)\varepsilon(t) + \int\limits_0^t \frac{\partial M(t - \tau)}{\partial(t - \tau)} \varepsilon(\tau)d\tau \qquad (2.9)$$

Inserting the step-strain load into this expression leads to the response after the load is removed ($t > t^*$):

$$\sigma(t) = M(0)\varepsilon_o[H(t) - H(t - t^*)] + \int\limits_0^{t^*} \frac{\partial M(t - \tau)}{\partial(t - \tau)} [H(\tau) - H(\tau - t^*)]\varepsilon_o d\tau; \quad t > t^*$$

$$\text{(j)}$$

By the definition of the Heaviside unit step function, the term inside the first set of brackets is zero. The other term in the expression may be evaluated using the mean-value theorem of integral calculus[4] [c.f. Appendix A]:

$$\sigma(t) = t^* \cdot \{\frac{\partial M(t - \lambda t^*)}{\partial(t - \lambda t^*)}\}[H(\lambda t^*) - H(\lambda t^* - t^*)]\varepsilon_o; \quad t > t^*; \quad 0 < \lambda < 1 \quad \text{(k)}$$

---

[4] The mean value theorem of integral calculus states that $\int_a^b f(x)dx = (b - a)f[a + \lambda(b - a)]$; $0 < \lambda < 1$.

Since $\lambda t^* < t^*$, the second Heaviside step function inside the brackets vanishes, so that:

$$\sigma(t) = t^* \cdot \{-\frac{\partial M(t - \lambda t^*)}{\partial(t - \lambda t^*)}\}\varepsilon_o; \quad t > t^*; \quad 0 < \lambda < 1 \tag{1}$$

For the influence of an action removed at $t = t^*$ to eventually disappear, so that $\sigma \to 0$, it is necessary that:

$$\lim_{t \to \infty}\{\frac{\partial M(t - \lambda t^*)}{\partial(t - \lambda t^*)}\} = 0; \quad \forall t^* < \infty; \quad 0 < \lambda < 1 \tag{m}$$

Or, equivalently:

$$\lim_{t \to \infty}\{\frac{\partial}{\partial t}M(t)\} = 0 \tag{2.10}$$

Otherwise, the material would retain permanent memory of the effect of the applied load, and the process would induce irreversible changes.

As may be seen from (2.9), the derivative, $\partial M(s)/\partial s$, of the relaxation function with respect to its argument acts as a weighting factor on the applied action, ε. For the effect of the action to be less and less pronounced with the passage of time, it is necessary that the weighting factor be a monotonically decreasing function of its argument. That is,

$$\left|\frac{\partial}{\partial t}M(t)\right|_{t=t_2} \leq \left|\frac{\partial}{\partial t}M(t)\right|_{t=t_1}; \quad t_2 > t_1 \tag{2.11}$$

Also, as experimental evidence shows [c.f. Chap. 1]:

$$|M(t)|_{t_2} \leq |M(t)|_{t_1}; \quad t_2 > t_1 \tag{2.12}$$

In similar fashion, repeating the previous arguments with a step stress applied at $t = 0$ and removed at $t = t^*$, leads to the following requirements for the creep compliance function:

$$\lim_{t \to \infty}\{\frac{\partial}{\partial t}C(t)\} = 0 \tag{2.13}$$

$$\left|\frac{\partial}{\partial t}C(t)\right|_{t=t_2} \leq \left|\frac{\partial}{\partial t}C(t)\right|_{t=t_1}; \quad t_2 > t_1 \tag{2.14}$$

$$|C(t)|_{t=t_2} \geq |C(t)|_{t=t_1}; \quad t_2 > t_1 \tag{2.15}$$

Geometrically, then, the fading memory hypothesis simply requires that the relaxation modulus and creep compliance be monotonically decreasing and increasing functions of their arguments, respectively, and also that the absolute values of their slopes decrease monotonically. In addition, as indicated in Chap. 1, experimental observations indicate that:

- The relaxation modulus decreases with observation time and is bounded by the glassy modulus for fast processes and by the equilibrium modulus for very slow processes.
- The creep compliance increases with observation time and is bounded by the glassy and equilibrium compliances for very fast and slow processes, respectively.

The fading memory principle embodied in (2.10)–(2.15), together with the experimental observations, requires that the general forms of the relaxation and creep compliance functions be as shown in Fig. 2.4.

Example 2.4 As an application of the fading memory principle, we evaluate the stress responses of a viscoelastic material to two arbitrary loading programs, $\varepsilon_1(t)$ and $\varepsilon_2(t)$, which reach the same constant value, ε^*, at time t^* and remain at that level from that point on, as indicated in Fig. 2.5.

Solution:

Use (2.1a) to evaluate the response as $t \to \infty$, splitting the integration interval from 0^+ to t^*, and t^* to ∞; and note that the derivatives of the strain histories $\varepsilon_1(t)$ and $\varepsilon_1(t)$ vanish after $t = t^*$ to write:

$$\sigma_1(\infty) = M(\infty)\varepsilon_1(0^+) + \lim_{t \to \infty} \int_{0^+}^{t^*} M(t - \tau)\frac{d\varepsilon_1}{d\tau}d\tau = M(\infty)\varepsilon^*$$

$$\sigma_2(\infty) = M(\infty)\varepsilon_2(0^+) + \lim_{t \to \infty} \int_{0^+}^{t^*} M(t - \tau)\frac{d\varepsilon_2}{d\tau}d\tau = M(\infty)\varepsilon^*$$

Fig. 2.5 Example 2.4: Two arbitrary loading histories which become identical and constant after a finite time

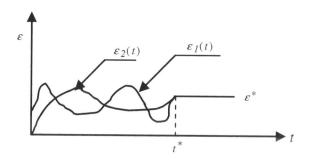

Since the integrals evaluate to $M(\infty)[\varepsilon_1(t^*) - \varepsilon_1(0^+)]$ and $M(\infty)[\varepsilon_2(t^*) - \varepsilon_2(0^+)]$, and also, $\varepsilon_1(t^*) = \varepsilon_2(t^*) = \varepsilon^*$, it follows that: $\sigma_1(\infty) = \sigma_2(\infty) = M(\infty) \cdot \varepsilon^*$. Or, using the alternate notations for the long-term or equilibrium modulus $M(\infty) \equiv M_\infty \equiv M_e$:

$$\sigma(\infty) = M_\infty \varepsilon^* \equiv M_e \varepsilon^* \qquad (2.16)$$

Proceeding in an entirely similar fashion, but using (2.2a), one would find that the long-term, equilibrium strain response to an arbitrary stress history, $\sigma(t)$ would be given by:

$$\varepsilon(\infty) = C_\infty \sigma^* \equiv C_e \sigma^* \qquad (2.17)$$

These expressions clearly show that a viscoelastic material would "remember" only that the loading got to ε^*—or σ^*, for that matter—but not how it got there. That is, after sufficiently long, a viscoelastic material will have effectively forgotten the details of the loading history; in agreement with the principle of fading memory.

2.4   Closed-Cycle Condition

This section examines the mathematical consequences of the physical expectation stated in Chap. 1, that the response of a linear viscoelastic material to harmonic loading ought to be harmonic, of the same frequency as the excitation, but out of phase with it. This so-called closed-cycle condition, that: "the steady-state response to harmonic loading also be harmonic," is satisfied by materials that do not age. That is, by materials whose property functions depend only on one timescale: the time measured from when the load was first applied, irrespective of the time elapsed since their manufacturing.

As will be shown in what follows, the closed-cycle condition requires that the kernels of the constitutive integrals, M and C, depend only on the difference of their arguments, and also, that all transients die out. In other words, the closed-cycle condition requires that $M(t, \tau) = M(t - \tau)$, and $C(t, \tau) = C(t - \tau)$, as has been assumed without proof in our derivations, so far. A physical proof of this implication of the closed-cycle condition can be constructed rewriting (2.1a), say, using $M(t, \tau)$, in place of $M(t - \tau)$, in order to remove the assumption made so far in our derivations that the kernel of the constitutive equation depends only on the difference of its arguments:

$$\sigma(t) = M(0)\varepsilon(t) - \int_0^t \frac{\partial}{\partial \tau}\{M(t, \tau)\}\varepsilon(\tau)d\tau \qquad (a)$$

The first term in this expression is simply the instantaneous value of the stress response. The second term, the hereditary component, is calculated as follows. In the time interval between τ and $\tau + d\tau$ of the past, the strain was $\varepsilon(\tau)$. Since the material is assumed to be linear, its memory of this past action should be proportional to the product $\varepsilon(\tau)$ and the duration of the action; that is: $\varepsilon(\tau) \, d\tau$; producing the stress: $\frac{\partial}{\partial \tau} M(t, \tau) \cdot \varepsilon(\tau) d\tau$. If the material does not age, its properties must be independent of the time when the experiment starts. For this to be the case, the kernel $M(t, \tau)$ can only be a function of the difference $t - \tau$. Clearly, the same is true of the creep compliance. In particular, and for this reason, such kernels are called difference kernels.

Proceeding now with the mathematical proof, we evaluate the stress response to a periodic strain of period $p : \varepsilon(t + p) = \varepsilon(t)$, using (a):

$$\sigma(t + p) = M(0)\varepsilon(t + p) - \int_0^{t+p} \frac{\partial}{\partial \tau} M(t + p, \tau)\varepsilon(\tau) d\tau \tag{b}$$

Next, introduce the change of variable $\tau = \tau' + p$ and use the stated periodicity of the applied strain, $p : \varepsilon(t + p) = \varepsilon(t)$, to write:

$$\sigma(t + p) = M(0)\varepsilon(t) - \int_{-p}^{t} \frac{\partial}{\partial \tau'} M(t + p, \tau' + p)\varepsilon(\tau') d\tau'$$

Splitting the interval of integration from $-p$ to 0, and from 0 to t; and afterward replacing the new variable of integration, τ' with the original symbol τ, for simplicity, get:

$$\sigma(t + p) = M(0)\varepsilon(t) - \int_{-p}^{0} \frac{\partial}{\partial \tau} M(t + p, \tau + p)\varepsilon(\tau) d\tau - \int_0^{t} \frac{\partial}{\partial \tau} M(t + p, \tau + p)\varepsilon(\tau) d\tau$$

$$\tag{c}$$

Now, use that: $\sigma(t) = M(0)\varepsilon(t) - \int_0^{t} \frac{\partial}{\partial \tau} M(t, \tau)\varepsilon(\tau) d\tau$, to cast (c) in the form:

$$\sigma(t + p) = \sigma(t) + \int_0^{t} \left[\frac{\partial}{\partial \tau} M(t, \tau) - \frac{\partial}{\partial \tau} M(t + p, \tau + p) \right] \varepsilon(\tau) d\tau$$

$$- \int_{-p}^{0} \frac{\partial}{\partial \tau} M(t + p, \tau + p)\varepsilon(\tau) d\tau \tag{d}$$

It then follows that, for the response to be periodic, that is, for $\sigma(t+p) = \sigma(t)$:

$$\int\limits_0^t \left[\frac{\partial}{\partial \tau} M(t,\tau) - \frac{\partial}{\partial \tau} M(t+p,\tau+p) \right] \varepsilon(\tau) d\tau = 0; \quad \forall \varepsilon(t) \tag{e}$$

Together with:

$$\int\limits_{-p}^0 \frac{\partial}{\partial \tau} M(t+p,\tau+p) \varepsilon(\tau) d\tau = 0 \tag{f}$$

Condition (e) implies that:

$$M(t,\tau) - M(t+p,\tau+p) = 0 \tag{g}$$

Differentiating this expression with respect to p, and setting $p = 0$, afterward, leads to[5]:

$$-\frac{\partial}{\partial(t+p)} M(t+p,\tau+p) \bigg|_{p=0} - \frac{\partial}{\partial(\tau+p)} M(t+p,\tau+p) \bigg|_{p=0} = 0 \tag{h}$$

The general solution of this equation is an arbitrary function of $t - \tau$, as is easily verified by direct substitution. Consequently:

$$M(t,\tau) = M(t - \tau) \tag{2.18}$$

According to the fading memory principle, condition (e) is met for arbitrary excitations only in the limit as $t \to \infty$, if the kernel $|\partial M/\partial t|$ of the integral is bounded, as indicated by relation (2.10). This means, additionally, that the lower limit in the integral in (b) must be taken as $-\infty$; and that the approximation:

$$\int\limits_0^t \frac{\partial}{\partial \tau} M(t+p,\tau+p) \varepsilon(\tau) d\tau \approx \int\limits_{-\infty}^t \frac{\partial}{\partial \tau} M(t+p,\tau+p) \varepsilon(\tau) d\tau \tag{2.19}$$

holds only for sufficiently long times. Otherwise, the response to a periodic excitation, even of a non-aging material, will be non-periodic.

---

[5] Here, use is made of the total derivative: $\frac{d}{dp} f(x,y) = \frac{\partial}{\partial x} f(x,y) \frac{dx}{dp} + \frac{\partial}{\partial y} f(x,y) \frac{dy}{dp}$.

Fig. 2.6 Side-by-side
comparison of relaxation
modulus and creep
compliance

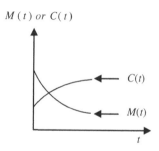

Summarizing: for the response of a viscoelastic material to a cyclic excitation to also be periodic, its material property functions must depend on the difference between current time and loading time.[6] That is, the closed-cycle condition (that the response to periodic excitation be periodic) can only be satisfied by non-aging materials.

2.5 Relationship Between Modulus and Compliance

Expressions (2.1a, b) and (2.2a, b) relate stresses to strains, through the corresponding relaxation modulus and creep compliance. This suggests that the two expressions may be combined in some form to obtain the relationship between the two property functions. Before we go into the mathematical details of this, we use what we have learned already about these two material functions and compare their forms side-by-side in Fig. 2.6.

As suggested by the figure, it is reasonable to expect that the values of C and M at $t = 0$, as well as at sufficiently long times, might be reciprocals of each other. Although one could argue that $C(t) \cdot M(t) \approx 1$ elsewhere, the figure shows that, in general, the creep compliance and the relaxation modulus are not reciprocals of each other. As will be shown subsequently, the values of the relaxation modulus and its creep compliance, for the extreme cases of glassy and equilibrium response are indeed reciprocals of each other, as they would be for elastic solids. However, unlike for elastic materials, the relaxation modulus is not the reciprocal of the creep compliance.

A relationship between relaxation modulus and creep compliance may be derived using (2.1b) to evaluate the stress response to a step-strain history $\varepsilon(t) = \varepsilon_o H(t)$, together with the fact that $dH(t)/dt = \delta(t)$ [c.f. Appendix A]; thus:

$$\sigma(t) = \int_{0^-}^{t} M(t - \tau)\varepsilon_o\delta(\tau)d\tau \equiv M(t)\varepsilon_o \qquad \text{(a)}$$

---

[6] Material property functions which depend on the difference between current and loading time are known as "difference" kernels.

Putting this result into (2.2b), taking ε_o outside the integral, and using that $\varepsilon(t) = \varepsilon_o H(t)$:

$$\varepsilon(t) = \left[\int_{0^-}^{t} C(t - \tau) \frac{d}{d\tau} M(\tau) d\tau \right] \varepsilon_o = \varepsilon_o H(t) \tag{b}$$

Canceling out ε_o produces the first form of the relationship between a relaxation modulus and its corresponding creep compliance:

$$\int_{0^-}^{t} C(t - \tau) \frac{d}{d\tau} M(\tau) d\tau = H(t) \tag{2.20}$$

Proceeding in the reverse order, applying a step stress $\sigma(t) = \sigma_o H(t)$, and then calculating the corresponding strain response, the result would be:

$$\int_{0^-}^{t} M(t - \tau) \frac{d}{d\tau} C(\tau) d\tau = H(t) \tag{2.21}$$

As stated earlier, (2.20) and (2.21) show that, in general, the relaxation modulus and creep compliance are not reciprocals of each other. Additional, practical information can be gained by examining the behavior of these expressions as time approaches 0 and ∞; as well as by invoking the consequences of the fading memory principle.

Before proceeding, we note that the integrals in (2.20) and (2.21) correspond to a special class of integrals known as Stieltjes convolutions. Convolution integrals are presented in the next section in the context of viscoelasticity and are fully discussed in Appendix A. As shown in the Appendix, by the commutative property of convolution integrals, Eqs. (2.20) and (2.21) are mathematically equivalent and either one could have been derived from the other.

2.5.1 Elastic Relationships

The relationships between a relaxation modulus and its creep compliance, corresponding to short and long term are obtained by taking the limit of either (2.20) or (2.21), as $t \to 0^+$ and $t \to \infty$, respectively. To do that, (2.20) is rewritten by splitting its integration interval into two intervals going from 0^- to 0^+, and 0^+ to t:

$$\int_{0^-}^{0^+} C(t - \tau) \frac{d}{d\tau} M(\tau) d\tau + \int_{0^+}^{t} C(t - \tau) \frac{d}{d\tau} M(\tau) d\tau = H(t) \tag{c}$$

The relationship between the short-term property functions is developed by letting $t \to 0$; noting that the first integral evaluates to $M(0^+)C(0^+)$ and that the second integral vanishes. Proceeding thus, and using the notation $M(0^+) = M_g$, and $C(0^+) = C_g$, to denote glassy quantities, leads to:

$$M_g = {}^1\!/C_g \qquad (2.22)$$

In similar fashion, taking the limit of either (2.20) or (2.21) as $t \to \infty$, and using the notation $M(\infty) = M_e$, and $C(\infty) = C_e$, to denote equilibrium properties:

$$M_e = {}^1\!/C_e \qquad (2.23)$$

It is left as an exercise for the reader to derive (2.23).

The last two expressions show that, as pointed out at the beginning of the section, in the extreme cases of short-term (or glassy) and long-term (or equilibrium) response, a relaxation function and its compliance counterpart are indeed reciprocals of each other, just as for elastic materials.

The monotonic nature of the modulus and compliance functions, stated in (2.12) and (2.15), can be used to establish a relationship between them which also shows that modulus and compliance are not, in general, simple inverses of each other [2].

Indeed, using (2.21), say, with the facts that $M(t)$ is a monotonically decreasing function of its argument, so that $M(t - \tau) \ge M(t)$, for all $\tau \ge 0$; and $M(t) = C(t) \equiv 0$, for $t < 0$, there results:

$$H(t) = \int_{0^-}^{t} M(t - \tau)\frac{\partial}{\partial \tau}C(\tau)d\tau \ge M(t) \int_{0^-}^{t} \frac{\partial}{\partial \tau}C(\tau) = M(t)C(t) \qquad (d)$$

That is:

$$M(t)C(t) \le 1 \qquad (2.24)$$

Which, as stated before, shows that in general: $M(t) \ne 1/C(t)$.

2.5.2 Convolution Integral Relationships

The mathematical relationships listed in (2.20) and (2.21) are known as Stieltjes convolution integrals [c.f. Appendix A]. Formally, the Stieltjes integral of two functions, ϕ and ψ, is defined as [3]:

$$\varphi(t - \tau) * d\psi(\tau) \equiv \int_{-\infty}^{t} \varphi(t - \tau) \frac{d}{d\tau}\psi(\tau)d\tau \equiv \varphi * d\psi \tag{e}$$

In which $\varphi(t)$ is assumed continuous in $[0,\infty)$; $\psi(t)$, vanishes at $-\infty$; and the form on the far right is used when the argument, t, is understood.

In line with the mathematical structure of relaxation and compliance functions, the further assumption is made that φ and ψ vanish for all negative arguments, which allows splitting the interval of integration from $-\infty$ to 0^-, and from 0^- to t, to write, more simply:

$$\varphi(t - \tau) * d\psi(\tau) \equiv \int_{0^-}^{t} \varphi(t - \tau) \frac{d}{d\tau}\psi(\tau)d\tau \tag{f}$$

Alternatively, integrating by parts:

$$\varphi(t - \tau) * d\psi(\tau) \equiv \varphi(t)\psi(0^+) + \int_{0^+}^{t} \varphi(t - \tau) \frac{d}{d\tau}\psi(\tau)d\tau \tag{g}$$

As shown in Appendix A, under the stated restrictions on the functions involved, the convolution integral is commutative, associative and distributive. Thus, for any three well-behaved functions, f, g, and h:

$$f * g = g * f$$

$$f * (g * h) = (f * g) * h = f * h * h \tag{h}$$

$$f * (g + h) = f * g + f * h$$

Based on their definition, the convolution integral allows writing viscoelastic constitutive equations in elastic-like fashion. Corresponding to (2.1a, b) or (2.2a, b), for instance, we write:

$$\sigma(t) = M(t - \tau) * d\varepsilon(\tau) \equiv M * d\varepsilon \tag{2.25}$$

$$\varepsilon(t) = C(t - \tau) * d\sigma(\tau) \equiv C * d\sigma \tag{2.26}$$

Additionally, corresponding to (2.20) and (2.21), above:

$$C(t - \tau) * dM(\tau) \equiv C * dM = H(t) \tag{2.27}$$

$$M(t - \tau) * dC(\tau) \equiv M * dC = H(t) \tag{2.28}$$

These expressions clearly show that the relaxation modulus and creep compliance are, in general, not mere inverses, but convolution inverses of each other. In addition, the viscoelastic relations in (2.25) and (2.26) look exactly like elastic constitutive equations, if the operation of multiplication is replaced by that of convolution. Using this fact, it is straightforward to write down the viscoelastic constitutive counterparts of any given elastic constitutive equations. This is done by simply replacing the elastic property of interest (modulus or compliance) with the corresponding viscoelastic property, and ordinary multiplication with the convolution operation between the material property function and the applied action (strain or stress).

Example 2.5 Write the viscoelastic version of the three-dimensional constitutive equations of a linear isotropic elastic solid which has its stress–strain equations split into a spherical and a deviatoric part as follows[7]: $\sigma_S = 3K\varepsilon_S$; $\sigma_{Dij} = 2G\varepsilon_{Dij}$; $i, j = 1, 3$

Solution:

Although three-dimensional constitutive equations will be discussed at length in Chap. 8, this exercise is meant to get the reader comfortable with writing the viscoelastic counterparts of elastic constitutive equations. So, whatever the meaning of the symbols involved, replace the elastic products with convolutions to write the results directly: $\sigma_S(t) = 3K(t - \tau) * d\varepsilon_S(\tau)$; $\sigma_{Dij}(t) = 2G(t - \tau) * d\varepsilon_{Dij}(\tau)$; $i, j = 1, 3$.

2.5.3 Laplace-Transformed Relationships

Since linear viscoelastic constitutive equations correspond to convolution integrals, one may apply the Laplace transform to convert them into algebraic equations. As explained in Appendix A, any piecewise continuous function, $f(t)$, of exponential order—that is, bounded by a finite exponential function—has a Laplace transform, $\bar{f}(s)$, defined as:

$$L\{f(t)\} \equiv \bar{f}(s) = \int_0^\infty e^{-st} f(t) dt \qquad (i)$$

---

[7] The 3×3 stress and strain matrices—indeed any square matrix of any order—may be split into a spherical and a deviatoric part. The spherical part is a diagonal matrix with each of its three non-zero entries equal to the average of the diagonal elements of the original matrix. Therefore, any one of its non-zero entries may be used to represent it. The deviatoric part of the matrix is, by definition, the matrix that is left over from such decomposition. This decomposition is discussed fully in Appendix B.

Some properties of the Laplace transform are presented in Appendix A. We list the following two and use them to transform convolution integrals in the time domain, t, into algebraic expressions in the transform variable, s.

$$\text{Transform of first derivative:} \quad L\{\frac{d}{dt}f\} = s\bar{f}(s) - f(0) \tag{j}$$

$$\text{Transform of the convolution:} \quad L\{f * g\} = \bar{f}(s)\bar{g}(s) \tag{k}$$

Indeed, applying these expressions to the convolution forms (2.25) and (2.26), respectively, results in the following algebraic form of the constitutive equations:

$$\bar{\sigma}(s) = s\bar{M}(s)\bar{\varepsilon}(s) \tag{2.29}$$

$$\bar{\varepsilon}(s) = s\bar{C}(s)\bar{\sigma}(s) \tag{2.30}$$

The same results would have been obtained if the Laplace transform had been applied to the original stress–strain and strain–stress equations, (2.1a, b) and (2.2a, b). For example, if the Laplace transform is applied to both sides of (2.1a), the relationship in (2.29) would be obtained, after collecting terms as follows:

$$\bar{\sigma}(s) = \bar{M}(s)\varepsilon(0^+) + \bar{M}(s) \cdot [s\varepsilon(s) - \varepsilon(0^+)] = s\bar{M}(s) \cdot \bar{\varepsilon}(s) \tag{l}$$

The advantage of taking the Laplace transform of viscoelastic constitutive equations is that the transformed expressions involve only products of the transform of the material property function of interest (modulus or compliance) and the Laplace transform of the input function—strain or stress, just like elastic constitutive equations do. In other words, the Laplace transform converts a viscoelastic constitutive equation into an elastic-like expression between transformed variables. Conversely, if each material property in an elastic constitutive relation is replaced by its Carson[8] transform and each input variable in it is replaced by its Laplace transform, the resulting expression must stand for the Laplace transform of the corresponding viscoelastic constitutive equation. Thus, as in the case of the convolution notation, this equivalence between elastic constitutive relations and the Laplace transform of viscoelastic equations allows one to write down the transformed viscoelastic constitutive equations directly from the elastic ones. This equivalence forms the basis of a so-called elastic–viscoelastic correspondence principle, which is presented in Chap. 9.

Example 2.6 Use the elastic–viscoelastic correspondence to write down the viscoelastic version of the three-dimensional constitutive equations of the linear isotropic elastic solid of Example 2.5.

---

[8] The s-multiplied Laplace transform of a function is simply called the Carson transform of the function.

Solution:

Using the elastic–viscoelastic correspondence, write the Laplace transform of the elastic expressions as: $\bar{\sigma}_S = 3s\bar{K}\bar{\varepsilon}_S$; $\bar{\sigma}D_{ij} = 2s\bar{G}\bar{\varepsilon}D_{ij}$; $i, j = 1, 3$. The viscoelastic constitutive equations are obtained taking the inverse Laplace transform of the forms given. Thus, $\sigma_S = 3K * d\varepsilon_S$; $\sigma D_{ij} = 2G * d\varepsilon D_{ij}$; $i, j = 1, 3$.

The relationship between the relaxation modulus, M, and the creep compliance, C, in the transformed plane, can be obtained either by applying the Laplace transform to (2.27) or (2.28), or by combining the algebraic expressions (2.29) and (2.30). In either case, there results:

$$\overline{M}(s)\overline{C}(s) = \frac{1}{s^2} \tag{2.31}$$

Example 2.7 The relaxation modulus of a viscoelastic solid is given by $M(t) = M_e + M_1 e^{-\alpha t}$. Use expression (2.31) and Laplace transform inversion to obtain its creep compliance, assuming the latter is a function of the form: $C(t) = C_e - C_1 e^{-\beta t}$.

Solution:

According to (2.31), the creep compliance function would be given by the inverse Laplace transform of the function $1/s^2\overline{M}(s)$. Hence, we first evaluate this function, then invert it, and equate it to the Laplace transform $\overline{C}(s) = C_e/s - C_1/(s + \beta)$, of the desired creep compliance. Proceeding thus, using the table of transforms included in Appendix A, and simplifying, there results: $s^2\overline{M}(s) = \dfrac{s + \alpha}{s[M_e\alpha + (M_e + M_1)s]}$. Expanding this rational function into its partial fractions, as explained in Appendix A; using the notation $M_e + M_1 = M_g$, simplifying and equating the result to the Laplace transform of $C(t)$, there results: $\dfrac{C_e}{s} - \dfrac{C_1}{(s + \beta)} = \dfrac{1/M_e}{s} + \dfrac{(M_e - M_g)/M_e M_g}{(s + \alpha M_e/M_g)}$. Equating coefficients of the corresponding powers of s yields: $C_e = 1/M_e$, $C_1 = (M_g - M_e)/(M_e M_g)$, $\beta = M_e \alpha/M_g$.

More general methods of approximate and exact inversion of material property functions given as sums of exponential functions are presented in Chap. 7.

2.6   Alternate Integral Forms

Depending on preference, and the application at hand, the integral constitutive equations for viscoelastic substances may be written in several different ways. The mathematical operations that are used to transform one constitutive form into another—most typically, integration by parts—require that the material property functions involved, and their time derivatives, be bounded. On occasion, the transformations also assume that the material property functions vanish identically for all negative time.

For ease of reference, we list Boltzmann's equation (2.1b), where it was noted that $\varepsilon(t) \equiv 0$, for $t < 0$, allowed us to write [4]:

$$\sigma(t) = \int_{0^-}^{t} M(t - \tau)\frac{d\varepsilon}{d\tau}d\tau \tag{2.32a}$$

On the physical expectation that $M(t)$ be bounded for all values of time, and requiring that $\varepsilon(t) \rightarrow 0$, as $t \rightarrow -\infty$, which is satisfied, since both $\varepsilon(t) \rightarrow 0$, and $d\varepsilon/dt \equiv 0$, for $t < 0$, one may extend the lower limit of integration to $-\infty$, in (2.32a), without altering its value. Thus,

$$\sigma(t) = \int_{-\infty}^{t} M(t - \tau)\frac{d\varepsilon}{d\tau}d\tau \tag{2.33a}$$

Another useful form is obtained integrating (2.32a) by parts and simplifying:

$$\sigma(t) = M(0)\varepsilon(t) - \int_{0}^{t} \frac{\partial}{\partial\tau}M(t - \tau)\varepsilon(\tau)d\tau \tag{2.34a}$$

Using the notation $M(0) \equiv M_g$, and introducing the normalized function $m(t) \equiv M(t)/M_g$:

$$\sigma(t) = M_g\left\{ \varepsilon(t) - \int_{0}^{t} \frac{\partial}{\partial\tau}m(t - \tau)\varepsilon(\tau)d\tau \right\} \tag{2.35a}$$

Using the notation $\sigma_g(t) \equiv M_g \cdot \varepsilon(t)$, and taking M_g inside the integral, produces the form:

$$\sigma(t) = \sigma_g(t) - \int_{0}^{t} \frac{\partial}{\partial\tau}m(t - \tau)\sigma_g(\tau)d\tau \tag{2.36}$$

An important application of this is in the derivation of constitutive equations for materials that are termed hyper-viscoelastic. Equations for hyper-viscoelastic materials are derived from those of hyper-elastic materials. A material is termed hyper-elastic, if there exists a potential function of the strains, say, W, such that each individual stress component in such a material may be computed as the derivative of W with respect to the corresponding strain [5]. Since both the glassy or short-term and the equilibrium or long-term responses of a viscoelastic solid are elastic, either can be used to define the potential function. We proceed by using the glassy response; thus

$$\sigma_g(t) = \frac{\partial}{\partial \varepsilon(t)} W_g(\varepsilon(t)) \qquad (2.37)$$

The stress–strain law in (2.36) would then take the equivalent form:

$$\sigma(t) = \frac{\partial}{\partial \varepsilon(t)} W_g(\varepsilon(t)) - \int_0^t \frac{\partial}{\partial \tau} m(t - \tau) \frac{\partial}{\partial \varepsilon(\tau)} W_g(\varepsilon(\tau)) d\tau \qquad (2.38)$$

Another form, which allows a generalization to non-linear viscoelasticity, is derived by introducing the strain relative to the configuration at time t: $\varepsilon_{rel}(t, \tau) = \varepsilon(t) - \varepsilon(\tau)$. Using that $\varepsilon_{rel}(t, 0) = \varepsilon(t)$ in (2.35a) yields:

$$\sigma(t) = M_g \left\{ \varepsilon_{rel}(t) + \int_0^t \frac{\partial}{\partial \tau} m(t - \tau) \varepsilon_{rel}(\tau) d\tau \right\} \qquad (2.39)$$

Constitutive Eqs. (2.34a, b), (2.35a, b), (2.36) and (2.39) are also frequently written in terms of integral operators, using convolution integral notation, but the exact form of the kernel (i.e., the derivative of the relaxation function) is not disclosed. With the obvious definitions, those equations would read:

$$\sigma(t) = M_g \{\varepsilon(t) - \Gamma(t - s) * \varepsilon(s)\} \qquad (2.40a)$$

$$\sigma(t) = M_g \{1 - \Gamma(t - s)*\}\varepsilon(t) \qquad (2.40b)$$

$$\sigma(t) = \sigma_g(t) - \Gamma(t - s) * \sigma_g(s) \qquad (2.41a)$$

$$\sigma(t) = \{1 - \Gamma(t - s)*\}\sigma_g(t) \qquad (2.41b)$$

$$\sigma(t) = M_g \varepsilon_{rel}(t, 0) + M_g \Gamma(t - s) * \varepsilon_{rel}(t, s) \qquad (2.42)$$

Example 2.8 Use (2.37) and (2.38) to develop the stress–strain law for a hyper-viscoelastic material having normalized relaxation function, $m = a + (1 - a)e^{-t/n}$, if it is known that its glassy response can be established from the potential function of the strains $W_g = \frac{1}{2}E\varepsilon^2(t)$.

Solution:

The hyper-viscoelastic form is derived by putting the given functions into (2.37) and (2.38) directly. Evaluating (2.37) first: $\sigma_g(t) = \frac{\partial}{\partial \varepsilon} W_g(\varepsilon(t)) = E \cdot \varepsilon(t)$.

Taking this result and m into (2.38) and re-arranging: $\sigma(t) = E \cdot \varepsilon(t) - \int_0^t [\frac{\partial}{\partial \tau} \{E[a +$

$(1 - a)]e^{-(t-\tau)/\eta}\} \cdot \varepsilon(\tau)]d\tau$.

Comparing this with (2.34a) shows that hyper-viscoelastic material in question is linearly viscoelastic with relaxation modulus: $M(t) = E[a + (1 - a)e^{-t/\eta}]$.

We end this section by presenting some of the constitutive equations in strain–stress form which are the exact counterparts of the foregoing expressions. These strain–stress forms are derived by reversing the roles of stress and strain in the arguments that led to the previous forms. For instance, the strain–stress equations analogous to (2.32a)–(2.35a) are:

$$\varepsilon(t) = \int_{0^-}^t C(t - \tau)\frac{d\sigma}{d\tau}d\tau \qquad (2.32b)$$

$$\varepsilon(t) = \int_{-\infty}^t C(t - \tau)\frac{d\sigma}{d\tau}d\tau \qquad (2.33b)$$

$$\varepsilon(t) = C(0)\varepsilon(t) - \int_0^t \frac{\partial}{\partial \tau}C(t - \tau)\sigma(\tau)d\tau \qquad (2.34b)$$

$$\varepsilon(t) = C_g\left\{\sigma(t) - \int_0^t \frac{\partial}{\partial \tau}c(t - \tau)\tau(\tau)d\tau\right\} \qquad (2.35b)$$

2.7   Work and Energy

Under the action of external agents, be these loads or displacements, a deformable body will change its configuration and, if not properly restrained, undergo large-scale motion. At any rate, as the points of application of the external agents move, work—defined as the product of force and displacement—is performed on the body. To develop an expression for the work performed on a body by the external agents, we use a uniaxial specimen of constant cross-sectional area, A, initial length, l, and volume, V, that is loaded at its ends by either a displacement u or a force F. With this, the rate of work of the external forces—that is, force times displacement rate—can then be expressed as:

$$\frac{dW}{dt} \equiv F\frac{du}{dt} \qquad (2.43)$$

Multiplying and dividing by the specimen's volume $V = A{\cdot}l$, and using that the strain is given $\varepsilon = u/l$, we cast the previous expression in the form:

$$\frac{dW}{dt} \equiv F\frac{du}{dt} = \frac{F}{A} A \cdot l \cdot \frac{d(u/l)}{dt} \equiv V \cdot \sigma \cdot \frac{d\varepsilon}{dt} \tag{2.44}$$

The total work performed during a time interval $(0, t)$ is given by the integral:

$$W|_0^t = \int_0^t \frac{dW}{ds}\,ds \equiv \int_0^t F(s)\frac{du}{ds}\,ds \tag{2.45}$$

Combining (2.44) and (2.45) and dividing the result by the specimen's volume, V, produces the work per unit volume, W_V, that is input into the system:

$$W_V|_0^t = (1/V)W|_0^t = \int_0^t \sigma(s)\frac{d\varepsilon(s)}{ds}\,ds \tag{2.46}$$

In practical applications, we insert an appropriate form or another of the constitutive equation, such as (2.1b) and write (2.46) as:

$$W_V|_0^t = \int_{s=0}^t \int_{\tau=0}^s M(s - \tau)\frac{d\varepsilon(\tau)}{d\tau}\,d\tau\,\frac{d\varepsilon(s)}{ds}\,ds \tag{2.47}$$

Suitable functions (bounded and piecewise continuous) allow interchanging the order of integration. Before doing this, we note that the relaxation modulus, $M(t)$, is defined only for positive values of time. One can continue it to negative values of its argument in an arbitrary manner. In particular, it is sometimes convenient to assume $M(t)$ either as an even or an odd function of time, that is,

$$M(t) = M(-t) \tag{2.48}$$

Fig. 2.7 Region of integration used for change of variables in work expression

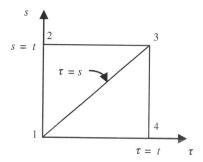

$$M(t) = -M(-t) \tag{2.49}$$

Here, we take M to be an even function of time. With this, the integral in (2.47) in the τ-s plane is taken over the area of a right triangle with base of length t, whose hypotenuse is the ray $\tau = s$, as indicated in Fig. 2.7.

Using (2.48) and noting that the square $\overline{1234}$ in the figure is made up of two triangles of equal area, and that on account of (2.48), the value of M is the same on points that are symmetrically located about the diagonal of the square, there results:

$$W_V|_0^t = \frac{1}{2} \int\limits_{\tau=0}^{t} \left[\int\limits_{s=0}^{t} M(s-\tau) \frac{d\varepsilon(s)}{ds} ds \right] \frac{d\varepsilon(\tau)}{d\tau} d\tau \tag{2.50}$$

The expressions derived here apply equally to any other one-dimensional pair or work-conjugate quantities, such as shearing force and deflection, torque and twist angle, or bending moment and rotation.

2.8   Problems

P.2.1 Determine the constant-rate effective modulus, $E_{eff}(t)$, of a one-dimensional solid made of a viscoelastic material whose relaxation modulus is: $M(t) = E_e + E_1 e^{-t/\tau_1}$.

$$\text{Answer}: \ E_{eff}(t) = E_e + \frac{E_1}{(1/\tau_1)} \left[1 - e^{-t/\tau_1} \right]$$

Hint: Use $M(t)$ with the defining expression derived in Example 2.1 and carry out the indicated integration.

P.2.2 Use convolution notation to derive the relationship between relaxation modulus and creep compliance.

$$\text{Answer}: \ M(t-\tau) * dC(\tau) = H(t)$$

Hint: Combine (2.25) and (2.26) to get $\sigma(t) = M(t-\tau) * d\varepsilon(\tau) \equiv M(t-\tau) * d\{C(\tau-s) * d\sigma(s)\}$; then, use that: $\sigma(t) = \sigma_o H(t)$ and thus $d\sigma(t) = \sigma_o \delta(t)$ to evaluate the convolution integral inside the braces, and obtain: $\sigma_o = M(t-\tau) * dC(\tau)\sigma_o$, from which the desired result follows.

P.2.3 As presented in Chap. 7, a popular analytical form used to represent relaxation functions consists of a finite sum of decaying exponentials, which in the literature is usually referred to as a Dirichlet–Prony series or, more simply, Prony series:

$$M(t) = M_e + \sum_{i=1}^{N} M_i e^{-t/\tau_i}; \quad M_e \geq 0; \quad and: M_i, \quad \tau_i > 0, \forall i$$

In this expression, M_e represents the equilibrium modulus, which is zero for a viscoelastic liquid [c.f. Chap. 1]. Also, although the τ_i's represent relaxation times of the material and are thus material properties, in practice, they, as well as M_e and the coefficients M_i, are all established by fitting the Prony series to experimental data. Prove that such forms satisfy the requirements of fading memory.

Hint:

(a) Evaluate the derivative of the series as $t \to \infty$ to show it satisfies (2.10).

$$\lim_{t \to \infty} \left\{ \frac{\partial}{\partial t} M(t) \right\} \equiv \lim_{t \to \infty} \left\{ -\sum_{1}^{N} \frac{M_i}{\tau_i} e^{-t/\tau_i} \right\} = 0$$

(b) Compare the values of the function at $t_2 > t_1$, to prove that the Prony series is a monotonically decreasing function, in accordance with (2.12).

$$M(t_2) = \sum_{1}^{N} M_i e^{-t_2/\tau_i} \leq \sum_{1}^{N} M_i e^{-t_1/\tau_i} = M(t_1); \quad \forall t_2 > t_1$$

(c) Evaluate the derivative of the series at $t_2 > t_1$ and prove that the absolute value of its derivative is also monotonically decreasing, satisfying (2.11).

$$\left| \frac{\partial}{\partial t} M(t) \right|_{t=t_2} \equiv \left| -\sum_{1}^{N} \frac{M_i}{\tau_i} e^{-t_2/\tau_i} \right| \leq \left| -\sum_{1}^{N} \frac{M_i}{\tau_i} e^{-t_1/\tau_i} \right| \equiv \left| \frac{\partial}{\partial t} M(t) \right|_{t=t_1}; \quad \forall t_2 > t_1$$

P.2.4 As discussed in Chap. 7, the power-law form: $M(t) = M_e + M_t(1 + \frac{t}{\alpha})^{-p}$ is also used to represent the relaxation function of viscoelastic solids. Show that this form satisfies the requirements of fading memory. In this expression, M_e, M_t, α and p, are all positive.

Hint:
(a) Proceed as in P.2.3 and evaluate the derivative of the given power-law form as $t \to \infty$ to show it satisfies (2.10).

$$\lim_{t \to \infty} \left\{ \frac{\partial}{\partial t} M(t) \right\} \equiv \lim_{t \to \infty} \left\{ -\frac{pM_t}{\alpha} \frac{1}{(1 + t/\alpha)^{(1+p)}} \right\} = 0$$

(b) Compare the values of the given function at $t_2 > t_1$, to prove that this power-law form is a monotonically decreasing function, in accordance with (2.12).

$$M(t_2) \equiv M_e + M_t(1 + \frac{t_2}{\alpha})^{-p} \leq M_e + M_t(1 + \frac{t_1}{\alpha})^{-p}; \quad t_2 > t_1$$

(c) Evaluate the derivative of the function at $t_2 > t_1$ and prove that the absolute value of its derivative is also monotonically decreasing, satisfying (2.11).

$$\left| \frac{\partial}{\partial t} M(t) \right|_{t_2} \equiv \left| -\frac{pM_t}{\alpha}(1 + \frac{t_2}{\alpha})^{-(1+p)} \right| \leq \left| -\frac{pM_t}{\alpha}(1 + \frac{t_1}{\alpha})^{-(1+p)} \right| \equiv \left| \frac{\partial}{\partial t} M(t) \right|_{t_1};$$

$$t_2 > t_1$$

P.2.5 Compute the steady-state response of the one-dimensional solid of P.2.1 if it is subjected to the cyclic strain history $\varepsilon(t) = \varepsilon_o \cos(\omega t)$.

$$\text{Answer}: \ \sigma(t) = \left[E_e + \frac{E_1(\omega\tau)^2}{1 + (\omega\tau)^2} \right] \varepsilon_o \cos(\omega t) - \frac{E_1(\omega\tau)}{1 + (\omega\tau)^2} \varepsilon_o \cos(\omega t)$$

Hint: Take the strain history into (2.1b); use integration-by-parts twice; simplify, and discard the transient term: $E_1\varepsilon_o \frac{(\omega\tau)}{1+(\omega\tau)^2} e^{-t/\tau}$ to obtain the desired result.

P.2.6 Repeat problem P.2.5 if the cyclic strain history is $\varepsilon(t) = \varepsilon_o \sin(\omega t)$.

$$\text{Answer}: \ \sigma(t) = \left[E_e + \frac{E_1(\omega\tau)^2}{1 + (\omega\tau)^2} \right] \varepsilon_o \sin(\omega t) + \frac{E_1(\omega\tau)}{1 + (\omega\tau)^2} \varepsilon_o \cos(\omega t)$$

Fig. 2.8 Problem 2.7

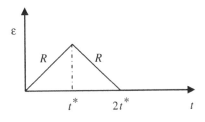

Hint: Take the strain history into (2.1b); use integration-by-parts twice; simplify, and discard the transient term $E_1 \varepsilon_o \frac{(\omega\tau)}{1+(\omega\tau)^2} e^{-t/\tau}$ to obtain the desired result.

P.2.7 A uniaxial bar of a viscoelastic solid with relaxation modulus $M(t)$ is subjected to a constant-rate load–unload strain history, as shown in Fig. 2.8. Prove that a non-zero stress will exist in the bar at the time when the strain reaches zero at the end of the load-unload cycle.

Hint:

Using that $\varepsilon(t) = \begin{cases} Rt, & t \le t^* \\ Rt^* + R(t - t^*), & t \ge t^* \end{cases}$, evaluate (2.1b) at $t = 2t^*$. Split the integration interval into two parts: from 0 to t^* and t^* to $2t^*$ and introduce a change of variables to arrive at the result: $\sigma(2t^*) = -R \int_0^{t^*} E(s)ds + R \int_{t^*}^{2t^*} E(s)ds$. Proceed as in Example 2.1 and multiply and divide this expression by t^* to cast the result into the form: $\sigma(2t^*) = Rt^* \left[-\frac{1}{t^*} \int_0^{t^*} E(s)ds + \frac{1}{t^*} \int_{t^*}^{2t^*} E(s)ds \right]$. The quantities inside the brackets are the average values of the relaxation function in the respective intervals of integration. Because the relaxation modulus is a monotonically decreasing function of time, the first integral inside the brackets is numerically larger than the second. This proves that, while $\varepsilon(2t^*)$ is zero, $\sigma(2t^*)$ is negative.

P.2.8 In Chap. 1, it was pointed out that in an elastic solid the stress corresponding to a given strain will always be the same, irrespective of the time it takes to apply the strain, and that contrary to this, the stress in a viscoelastic material will depend on the rate of straining, and hence, on the time it takes the strain to reach a specified value. Considering two constant strain-rate histories, $\varepsilon_1(t) = R_1 \cdot t$ and $\varepsilon_2(t) = R_2 \cdot t$, derive an expression for the duration, t_2, at which a viscoelastic material subjected to a strain history $\varepsilon_2(t) = R_2 \cdot t$ would develop the same stress response as it would after t_1 units of time under the strain history $\varepsilon_1(t) = R_1 \cdot t_1$.

$$\text{Answer}: t_2 = \frac{E_{eff}(t_1)R_1}{E_{eff}(t_2)R_2} t_1$$

Hint: Proceeding as in Example 2.1, evaluate (2.1b) for each constant strain-rate load and arrive at $\sigma(t_1) = E_{eff}(t_1)R_1 t_1$ and $\sigma(t_2) = E_{eff}(t_2)R_2 t_2$; where the average or effective modulus, $E_{eff}(t)$, is given by (2.3). The result follows from these relations.

P.2.9 The work per unit volume, $W_V(t)$, performed by external agents acting for t units of time on a uniaxial bar of a viscoelastic material is given by: $W_V(t) = \int_0^t \sigma(s)d\varepsilon(s)$. Evaluate the work per unit volume, done in a complete cycle, on a bar of a viscoelastic materials with relaxation modulus $M(t)$, if the applied excitation is $\varepsilon(t) = \varepsilon_o sin(\omega t)$.

$$\text{Answer}: \ W_V = \pi\sigma_o\varepsilon_o sin\delta$$

Hint: Using that the response to a periodic excitation will be periodic and of the same frequency as the excitation, but out of phase with it, let δ be the phase angle, and take the response to be $\sigma(t) = \sigma_o sin(\omega t + \delta)$. Insert the stress and strain into the expression for the work per unit volume, and write: $W_V = \int_t^{t+p} \sigma_0 \sin$ $(\omega s + \delta)\varepsilon_o\omega\cos(\omega s)ds$; where $p = 2\pi/\omega$ is the period. Now use trigonometric identities to expand the circular function $sin(\omega s + \delta) = sin(\omega s)\cos(\delta) + cos(\omega s)\sin(\delta)$, perform the integration, using the periodicity of the circular functions, and simplify to arrive at the result. As will be explained in Chap. 4, the phase angle, δ, is a characteristic of the material's relaxation modulus.

P.2.10 Repeat Problem P2.9 using the periodic strain history $\varepsilon(t) = \varepsilon_o cos(\omega t)$.

Answer: The result is the same as for a sine function history: $W_V = \pi\sigma_o\varepsilon_o sin\delta$

Hint: Proceed as in Problem 2.9.

References

1. A.D. Drosdov, Finite elasticity and viscoelasticity; a course in the nonlinear mechanics of solids, World Scientific, pp. 267–271, 279–283 (1996)
2. A.C. Pipkin, Lectures on viscoelasticity theory (Springer, New York, 1986), pp. 14–16
3. H.-P. Hsu, Fourier analysis (Simon and Schuster, New York, 1970) pp. 88–92
4. R.M. Christensen, Theory of viscoelasticity, 2nd edn. (Dover, New York, 2003), pp. 3–9
5. L.E. Malvern, Introduction to the mechanics of a continuous medium (Prentice-Hall, Englewood, 1963) pp. 278–290, 282–285

Constitutive Equations in Differential Operator Form

3

Abstract

The mechanical response of a viscoelastic material to external loads combines the characteristics of elastic and viscous behavior. On the other hand, as we know from experience, springs and dashpots are mechanical devices which exhibit purely elastic and purely viscous response, respectively. It is then natural to imagine that the equations that relate stresses to strains in a viscoelastic material could be represented with an appropriate combination of equations which relate stresses to strains in springs and dashpots. To develop this idea, Sect. 3.2 examines the response of the linear elastic spring and linear viscous dashpot to externally applied loads. The response equations for these simple mechanical elements are formalized in Sect. 3.3, with the introduction of so-called rheological operators. As it turns out, because combinations of springs and dashpots require the addition and multiplication of constant and first derivative operators, it turns out that the constitutive equation of general arrangements of springs and dashpots, such as are needed to reproduce observed viscoelastic behavior, must be represented by linear ordinary differential equations whose order depends on the number, type, and specific arrangement of the springs and dashpots. The physical significance of the coefficients in the resulting differential equations is examined also, and the proper form of the initial conditions established. As will be seen, the mere presence or absence of some of the coefficients of a differential equation reveals whether the particular arrangement of springs and dashpots it represents will model fluid or solid behavior, and whether it will exhibit instantaneous, elastic response. A general approach to establishing rheological models is presented in Sect. 3.4, and applied in Sects. 3.5 through 3.7 to develop the differential equations, and examine the behavior of simple and general rheological models.

Keywords

Compliance · Creep · Damper · Dashpot · Deviatoric · Fluid · Kelvin · Laplace ·

Maxwell · Model · Modulus · Pressure · Recovery · Relaxation · Spherical · Spring · Strain · Stress · Relaxation · Retardation · Solid · Spectrum

3.1   Introduction

As indicated in Chap. 1, the mechanical response of a viscoelastic material to external loads combines the characteristics of elastic and viscous behavior. On the other hand, as we know from experience, springs and dashpots are mechanical devices exhibiting purely elastic and purely viscous response, respectively. It is then natural to imagine that the equations that relate stresses to strains in a viscoelastic material could be represented with an appropriate combination of equations which relate stresses to strains in springs and dashpots. To develop this idea, Sect. 3.2 examines the response of both the linear elastic spring and linear viscous dashpot to externally applied loads. The response equations for these simple mechanical elements are formalized in Sect. 3.3, with the introduction of so-called rheological operators. Because the mathematical combinations of springs and dashpots require the addition and multiplication of constant and first-derivative operators, it turns out that the constitutive equation of general arrangements of springs and dashpots, which are needed to capture real viscoelastic behavior, must be represented by linear ordinary differential equations whose order depends on the number, type, and specific arrangement of the springs and dashpots. The physical significance of the coefficients in the resulting differential equations is also examined, and the proper form of the initial conditions is established. As will be seen, the mere presence or absence of some of the coefficients of a differential equation can reveal whether the particular arrangement of springs and dashpots it represents will model fluid or solid behavior and whether it will exhibit instantaneous, elastic response. A general approach to establishing rheological models is presented in Sect. 4 and then applied to develop the differential equations and examine the behavior of simple, generalized, and degenerate rheological models, in Sects. 3.5 through 3.7, successively.

3.2   Fundamental Rheological Models

There are two basic rheological models in linear viscoelasticity: the linear elastic spring and the linear viscous damper. The constitutive equations of elastic springs and viscous dashpots are expressed in terms of force and displacement, and force and displacement rate, respectively.[1] Here, an appropriate scale is assumed and forces are replaced with stresses and displacements with strains. Also, in

---

[1] We use axial springs and dashpots in the derivations, but emphasize that the corresponding relations for shear stress and shear strain have the same mathematical form.

Fig. 3.1 General loading
with step discontinuity

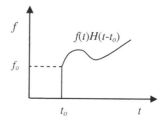

examining the response of the basic rheological elements to general loading (either stress or strain), it is assumed that the load may have a finite jump discontinuity at the time of its application, but is continuous thereafter, as depicted in Fig. 3.1.

Note that, by definition:

$$f(t)H(t - t_o) \equiv \begin{cases} 0, & t \leq t_o^- \\ f(t), & t \geq t_o^+ \end{cases}$$

3.2.1 Linear Elastic Spring

The physical (mechanical) model and free body diagram of the linear elastic spring are shown in Fig. 3.2. Its constitutive equation is presented next, followed by its response to general strain and stress loading.

3.2.1.1 Constitutive Equation

The mechanical behavior of a linear elastic spring is governed by Hooke's law: "stress is directly proportional to strain." Using σ to represent the stress, ε the strain, and E the modulus as the "constant of proportionality," we write the constitutive equation of the spring in "stress–strain" and strain–stress forms, as

$$\sigma = E \cdot \varepsilon \tag{3.1a}$$

$$\varepsilon = 1/E \cdot \sigma \tag{3.1b}$$

Fig. 3.2 Mechanical model
and free body diagram of
elastic spring

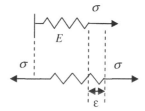

3.2.1.2 Response to Strain Loading

In this case, the controlled variable is strain, the loading takes the form $\varepsilon(t)=f(t)H(t - t_o)$, and the response of the elastic spring to such load is from (3.1a):

$$\sigma(t) = Ef(t)H(t - t_o) \tag{a}$$

This shows that the response of an elastic spring is instantaneous and remains non-zero for as long as the applied strain is non-zero: a known characteristic of the behavior of solid materials.

The relaxation modulus of an elastic spring, defined as the stress response to a step strain, $\varepsilon(t) = \varepsilon_o \cdot H(t - t_o)$, reckoned per unit applied strain, is obtained from (a), using $f(t) \equiv \varepsilon_o$

$$M(t - t_o) = \sigma(t)/\varepsilon_o = E \cdot H(t - t_o) \tag{3.2a}$$

or, simply

$$M(t) = E \cdot H(t). \tag{3.2b}$$

3.2.1.3 Response to Stress Loading

The controlled variable for this case is stress. The loading takes the form $\sigma(t)=f(t)H(t - t_o)$, and the response of the spring is obtained using (3.1b), as

$$\varepsilon(t) = (1/E) \cdot \sigma(t) \cdot H(t - t_o) \tag{b}$$

In other words, the stain response of an elastic spring is instantaneous, remaining non-zero for as long as the applied stress remains non-zero. This is a characteristic of solid behavior.

The creep compliance of an elastic spring, defined as the strain response to a step stress, $\sigma(t) = \sigma_o \cdot H(t - t_o)$, measured per unit of applied stress, is obtained from (b), using $f(t) \equiv \sigma_o$

$$C(t - t_o) \equiv \varepsilon(t)/\sigma_o \equiv (1/E)H(t - t_o) \tag{3.3a}$$

More simply,

$$C(t) = (1/E)H(t) \tag{3.3b}$$

In particular, the response of the linear elastic spring to stress and strain recovery tests is easily evaluated using (a) and (b), setting $f(t) = f_o[H(t - t_o) - H(t - t_1)]$, with f_o replaced by ε_o or σ_o, respectively. The two experiments and corresponding responses are depicted in Fig. 3.3.

Fig. 3.3 Recovery response of elastic spring. **a** Stress recovery. **b** Strain recovery

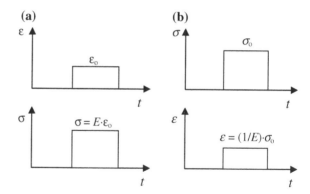

3.2.2 Linear Viscous Dashpot

The physical model and free body diagram of the linear viscous damper—or dashpot—are presented in Fig. 3.4. Its constitutive equation is discussed next and used afterward to examine the mechanical response of the dashpot to general stress and strain loading.

3.2.2.1 Constitutive Equation

The mechanical behavior of a linear dashpot follows Newton's law: "stress is directly proportional to strain rate." As before, σ and ε will denote stress and strain, respectively. Therefore, using η to represent the dashpot's constant viscosity, the stress–strain form of the constitutive equation for the linear dashpot becomes

$$\sigma(t) = \eta \cdot \frac{d}{dt}\varepsilon(t) \tag{3.4a}$$

Dispensing with integration, for the time being, we rewrite the constitutive equation for the linear dashpot in strain–stress form, by reversing the order of terms in the equation:

$$\frac{d}{dt}\varepsilon(t) = \frac{1}{\eta} \cdot \sigma(t) \tag{3.4b}$$

Fig. 3.4 Mechanical model and free body diagram of linear dashpot

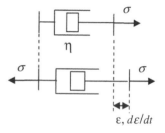

3.2.2.2 Response to Strain Loading

The response of the linear dashpot to the general strain history $\varepsilon(t) = f(t)H(t - t_o)$, shown in Fig. 3.1, is obtained by taking it into (3.4a) and performing the indicated differentiation using the properties of the unit step and delta functions[2] (see Appendix A):

$$\sigma = \eta \frac{d}{dt} \cdot \varepsilon(t) \equiv \eta \frac{d}{dt} \cdot [f(t)H(t - t_o)]$$

$$= \eta \left\{ \left[\frac{d}{dt} f(t) \right] H(t - t_o) + f(t)\delta(t - t_o) \right\} \tag{a}$$

$$= \eta \left[H(t - t_o) \frac{d}{dt} f(t) + f(t_o)\delta(t - t_o) \right]$$

Note that, because the response contains an impulse function at $t = t_o$, it would take an infinite stress to impose an instantaneous strain on a linear viscous dashpot. Once the strain is imposed, however, the stress would decay instantly to the value of the derivative of the strain history.

The relaxation modulus of the linear viscous dashpot, being defined as the stress response to a step-strain history $\varepsilon(t) = \varepsilon_o H(t - t_o)$, and measured per unit applied strain, is obtained from (a), noting that $f = f(t_o) = \varepsilon_o$ and $df/dt = 0$; hence,

$$M(t - t_o) = \eta \cdot \delta(t - t_o) \tag{3.5a}$$

or

$$M(t) = \eta \cdot \delta(t) \tag{3.5b}$$

In other words, the relaxation modulus of a linear viscous dashpot is an impulse function.

3.2.2.3 Response to Stress Loading:

The response of the linear viscous dashpot to a general stress history is obtained by taking the controlled variable, $\sigma(t) = f(t)H(t - t_o)$, into (3.4b), and integrating the resulting expression between 0 and t; using the shifting property of the Heaviside function[3] (see Appendix A):

$$\varepsilon(t) = \varepsilon(0) + \frac{1}{\eta} \cdot \int_0^t f(s)H(s - t_o)ds \equiv H(t - t_o) \frac{1}{\eta} \cdot \int_{t_o}^t f(s)ds$$

Using the initial condition, $\varepsilon(0) = 0$:

---

[2] In particular, that $dH(t)/dt = \delta(t)$; $f(t)\delta(t - t_o) = f(t_o)\delta(t - t_o)$.

[3] Specifically, that $\int_0^t f(s)H(s - t_o)ds = H(t - t_o)\int_{t_o}^t f(s)ds$, for $t_o > 0$.

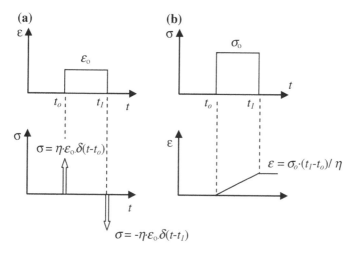

Fig. 3.5 Response of viscous damper. **a** Stress recovery. **b** Strain recovery

$$\varepsilon(t) = \frac{1}{\eta} \cdot H(t - t_o) \cdot \int_{t_o}^{t} f(s)ds \tag{b}$$

The creep compliance follows from this by setting $f(t) \equiv f(t) \equiv \varepsilon_o$ and performing the required integration:

$$C(t - t_o) = \frac{1}{\eta} \int_{t_o}^{t} \varepsilon_o \, ds \equiv \frac{(t - t_o)}{\eta} H(t - t_o) \tag{3.6a}$$

or

$$C(t) = \frac{t}{\eta}, \text{for } t > 0 \tag{3.6b}$$

The response of the linear viscous dashpot to the standard stress and strain recovery tests is evaluated using (a) and (b), setting $f(t) = f_o[H(t - t_o) - H(t - t_1)]$, with f_o replaced by ε_o or σ_o, respectively. The experiments and corresponding responses are depicted in Fig. 3.5.

3.3 Rheological Operators

Rheological operators arise naturally from the form of the constitutive equations of the spring and dashpot models. Indeed, these equations listed in stress–strain form as (3.1a, b) can be expressed symbolically as resulting from a mathematical operation on the strain history. For the linear elastic spring, this interpretation actually defines the modulus as the constant operator in question, $[E]$, and multiplication of real numbers (i.e., modulus times strain) as the operation. Similarly, the stress response of the linear viscous damper may be thought of as resulting from applying the first-order derivative operator, $[\eta d/dt]$ to the strain history, $\varepsilon(t)$. This idea is developed further in what follows.

3.3.1   Fundamental Rheological Operators

Following the previous reasoning, we rewrite the stress–strain constitutive equations of the spring and dashpot using the two fundamental rheological operators, $[E]$ and $[\eta d/dt]$, as

$$\sigma(t) = [E]\varepsilon(t) \tag{3.7a}$$

$$\sigma(t) = \left[\eta \frac{d}{dt}\right]\varepsilon(t) \tag{3.8a}$$

The usefulness of working with rheological operators stems from the fact that rheological operators may be considered as algebraic entities which follow the rules of addition and multiplication of real numbers and enjoy the properties of derivatives of functions of real variables. In more formal language, the basic rheological operators are as follows:

- Linear, because they are homogeneous and additive:
  $E[a\varepsilon_1(t) + b\varepsilon_2(t)] = aE\varepsilon_1(t) + bE\varepsilon_2(t)$;
  $\eta\partial_t[a\varepsilon_1(t) + b\varepsilon_2(t)] = a\,\eta\partial_t\varepsilon_1(t) + b\eta\partial_t\varepsilon_2(t)$; etc.
- Commutative, because they can be added or multiplied together in any order:

  $E\varepsilon(t) + \eta\partial_t\varepsilon(t) = [E + \eta\partial_t]\varepsilon(t) = [\eta\partial_t + E]\varepsilon(t) = \eta\partial_t\varepsilon(t) + E\varepsilon(t)$;
  $[E]\cdot[\eta\partial_t]\varepsilon(t) = [E][\eta\partial_t\varepsilon(t)] = E\eta\partial_t\varepsilon(t) = \eta E\partial_t\,\varepsilon(t) = \eta\partial_t[E\varepsilon(t)]$; etc.

- Associative, because they can be grouped together in any order:
  $E_1\varepsilon(t) + E_2\varepsilon(t) + a\,\eta\partial_t E\varepsilon(t) = [E_1 + aE\eta\partial_t]\varepsilon(t) + E_2\varepsilon(t)$
  $$= [E_1 + E_2]\varepsilon(t) + aE\eta\,\partial_t\varepsilon(t); \text{ etc.}$$
- Distributive, because their products distribute their sums and vice versa:
  $[E_1 + \eta_1\partial_t] \cdot [E_2 + \eta_2\partial_t]\varepsilon(t) = [E_1E_2 + E_1\eta_2\partial_t + \eta_1\,E_2\partial_t + \eta_1\eta_2\partial_t\partial_t]\varepsilon(t)$; etc.

Put another way, rheological operators can be added or multiplied together in any order; subtracted from one another; and, being careful about it, even symbolically divided by one another. The trick to performing these algebraic manipulations with rheological operators is to bear in mind that each operator or operator expression acts on a function and its meaning is deciphered proceeding from right to left. For instance, in the product of the operator $\eta \cdot \partial_t$ and the constant operator E, the result must equal $\eta \cdot E \cdot \partial_t \equiv E \cdot \eta \cdot \partial_t$, irrespective of which operator appears initially on the left. This is so, because the product should be interpreted as $\eta \cdot \partial_t(E(f)) \equiv E \cdot (\eta \cdot \partial_t(f))$, and not as the differential operator ∂_t applied first to the constant operator E and then to the function, f, which would, incorrectly, yield the zero operator $(\eta \cdot \partial_t E) = 0$, applied on f.

It is important to note that when the functions involved (stress or strain) depend both on position and on time, as is typical in two and three dimensions, it is more proper to use the partial derivative operator, ∂_t, in place of the total derivative operator $\frac{d}{dt}$. For ease of notation, we use the partial derivate symbol almost exclusively, even in the one-dimensional case. In keeping with this notation, $\frac{d^k}{dt^k}$ and ∂_t^k will be used to denote the kth-order derivative operator with respect to the independent variable, t.

With the above in mind, we divide (3.7a) and (3.8a) by the corresponding operator and arrive at the constitutive equations of the linear spring and linear viscous damper in strain–stress form:

$$\varepsilon(t) = \frac{1}{E}[\sigma(t)] \tag{3.7b}$$

$$\varepsilon(t) = \frac{1}{\eta \partial_t}[\sigma(t)] \tag{3.8b}$$

3.3.2   General Rheological Operators

Since the constitutive relations for the spring and dashpot involve derivatives of both the stress and the strain, it is logical to expect that the constitutive equations corresponding to rheological models made up of more elaborate combinations of linear springs and linear dashpots should include higher-order derivatives of the stress and the strain. That this should be so is due to the fact that, whatever the complexity of a given spring-and-dashpot arrangement, its rheological equation has to be constructed through the addition, subtraction, multiplication, and division of fundamental rheological operators, E_i and $\eta_j \partial_t$, whose parameters, E_i and η_j, depend on the specific properties of the basic elements they represent. The result is that the constitutive equation of a general rheological model is always an expression of the form:

$$p_o \sigma + p_1 \frac{d^1}{dt^1}\sigma + \cdots + p_m \frac{d^m}{dt^m}\sigma = q_o \varepsilon + q_1 \frac{d^1}{dt^1}\varepsilon + \cdots + q_n \frac{d^n}{dt^n}\varepsilon \tag{3.9a}$$

and, by the very process that leads to this expression, it may be argued that
- $m \leq n$ because the orders of the differential operators, P and Q are established by products of the fundamental rheological operators, one of which—the spring's—is balanced regarding differentiation; the other one—the dashpot's—contains a derivative of strain which is of order one higher than that of stress.

- If $q_o = 0$, the constitutive equation will contain only derivatives of strain, and at least the lowest of them must be non-zero, for a stress to develop in the system. Since the stress would then depend on strain rate, the rheological equation would represent fluid behavior.
- If $q_o \neq 0$, both the stress and the strain would approach finite values σ_∞ and ε_∞, as time increases without limit ($t \to \infty$), but all derivatives in (3.9a) would vanish. In the limit, a non-zero stress, $\sigma_\infty = (q_o/p_o)\varepsilon_\infty$, would remain, for as long as the strain does, and vice versa. Such rheological models represent elastic, solid behavior with long term, or equilibrium modulus, $M(\infty) = p_o/q_o \equiv M_\infty \equiv M_e$, and also equilibrium compliance $C(\infty) = p_o/q_o \equiv C_\infty \equiv C_e$, which is the reciprocal of the equilibrium modulus.

Expression (3.9a) may be rewritten in a number of forms as follows:

- Using the summation symbol and introducing the zero-derivative operator, $\frac{d^0}{dt^0} \equiv 1$:

$$\sum_{k=0}^{m} \left[p_k \frac{d^k}{dt^k} \right] \sigma(t) = \left[\sum_{j=0}^{n} q_j \frac{d^j}{dt^j} \right] \varepsilon(t) \tag{3.9b}$$

- Or, as we shall usually do, in symbolic form, using linear differential operator notation, introducing the stress and strain operators, P and Q, respectively, as

$$P[\sigma] = Q[\varepsilon], \text{or: } P\sigma = Q\varepsilon \tag{3.9c}$$

With the following obvious definitions of the operators P and Q,

$$P \equiv p_o \frac{d^0}{dt^0} + p_1 \frac{d^1}{dt^1} + p_2 \frac{d^2}{dt^2} + \cdots + p_m \frac{d^m}{dt^m} \tag{3.10a}$$

$$Q \equiv q_o \frac{d^0}{dt^0} + q_1 \frac{d^1}{dt^1} + q_2 \frac{d^2}{dt^2} + \cdots + q_n \frac{d^n}{dt^n} \tag{3.10b}$$

- Finally, by the properties of linear differential operators, stated earlier, we write the symbolic stress–strain and strain–stress equations of a general rheological model, as

$$\sigma = [Q/P]\varepsilon \tag{3.11a}$$

$$\varepsilon = [P/Q]\sigma \tag{3.11b}$$

3.3.3   Rheological Equations in Laplace Transformed Space

Under certain conditions of continuity and boundedness of the functions involved, it is possible and often advantageous to transform an ordinary linear differential equation into an algebraic one. As discussed in Appendix A, this can be done by means of an integral transformation such as the Laplace or the Fourier transform.[4] The Laplace transform is used in what follows to convert the constitutive equation (3.9a) into an algebraic equation between transformed stress and transformed strain. As usual, the letter s will be used to denote the transformed variable, and an over-bar will denote a transformed quantity.

Accordingly, applying the Laplace transform to the general constitutive equation (3.9a), using at-rest conditions for both the stress and the strain, yields (see Appendix A):

$$p_o\bar{\sigma} + p_1 s\bar{\sigma} + \cdots + p_m s^m \sigma = q_o\bar{\varepsilon} + q_1 s\bar{\varepsilon} + \cdots + q_n s^n\bar{\varepsilon} \tag{3.12a}$$

or after factoring out the transformed stress and strain and including the transform variable as argument of the functions involved, for emphasis

$$\bar{P}(s)\bar{\sigma}(s) = \bar{Q}(s)\varepsilon(s) \tag{3.12b}$$

where

$$\bar{P}(s) \equiv p_o + p_1 s + p_2 s^2 + \cdots + p_m s^m \tag{3.13a}$$

$$\bar{Q}(s) \equiv q_o + q_1 s + q_2 s^2 + \cdots + q_n s^n \tag{3.13b}$$

Since $\bar{P}(s)$ and $\bar{Q}(s)$ are algebraic quantities, the symbolic expressions (3.11a, b) take a true algebraic meaning in Laplace-transformed space, becoming, respectively:

$$\bar{\sigma}(s) = \frac{\bar{Q}(s)}{\bar{P}(s)}\bar{\varepsilon}(s) \tag{3.14a}$$

$$\bar{\varepsilon}(s) = \frac{\bar{P}(s)}{\bar{Q}(s)}\bar{\sigma}(s) \tag{3.14b}$$

3.3.4   Initial Conditions for Rheological Models

As expression (3.9a) indicates, the constitutive equation of a general rheological model is an ordinary linear differential equation of order m, if stress is the dependent

---

[4] The Fourier transform is used in later chapters as an efficient means of solving steady-state oscillation problems.

variable, and strain, the controlled variable, and of order n, if the roles of stress and strain are reversed. Its general solution requires a set of independently prescribed initial conditions, which must be equal in number to the order (m or n) of the equation [1]. Before we present the general procedure to establishing the correct initial conditions for general rheological differential equations, we examine a simple case.

Example 3.1 Establish the initial conditions for the rheological model $p_o\sigma(t) + p_1\partial_t\sigma(t) = q_1\partial_t\varepsilon(t)$ if it is subjected to a step strain history $\varepsilon(t) = \varepsilon_oH(t)$.
Solution:

In particular, note that we need to establish the value $\sigma(0^+)$ of the stress at $t = 0^+$. To do this, we integrate the differential equation of the model in the interval $(0^-, 0^+)$ and write $p_o\int_{0^-}^{0^+}\sigma(s)ds + p_1[\sigma(0^+) - \sigma(0^-)] = q_1[\varepsilon(0^+) -\varepsilon(0^-)]$.

Using that the first integral in this expression is zero, for any bounded stress, and that so are $\sigma(0^-)$ and $\varepsilon(0^-)$, leads to the initial condition: $\sigma(0^+) = (q_1/p_1)\varepsilon(0^+) \equiv (q_1/p_1)\varepsilon_o$.

We note that, each term in the general rheological equation is of the form $\frac{d^k}{dt^k}f$, where f stands for either stress or strain. Also, although $m \leq n$, we assume the differential equation is of order n. In other words, we assume $p_n \neq 0$, but set $p_n = 0$, in case $m < n$.

We also allow the dependent variable (the one which is the object of the differential equation) to have jump discontinuities whenever the controlled variable has them. This is convenient for analytical purposes, as it permits the derivation of succinct mathematical expressions for creep and stress relaxation experiments.

With the previous notes in mind, we proceed as follows:

1. Assume that the controlled variable (either σ or ε) and all its time derivatives, up to and including its nth derivative, are given at $t = t_o^+$ and are zero at $t = t_o^-$.

$$f(t_o^+), \partial_t^1 f(t_o^+), \partial_t^2 f(t_o^+), \ldots, \partial_t^n f(t_o^+), \neq 0; \quad f = \sigma \text{ or } \varepsilon \qquad (a)$$

2. Assume that the dependent variable (either σ or ε) and all its time derivatives, up to and including the nth derivative, may have a non-zero value immediately after the controlled variable is applied. That is,

3. Integrate the differential equation n times, between t_o^- and t_o^+. In doing this, note that after n-iterated integrations, all derivatives of order $k \leq n$ become continuous. This means that, after n integrations, the respective values of each integrated function are the same at t_o^- and t_o^+, and so their integrals between t_o^- and t_o^+ vanish. Then, after n integrations, obtain that

$$p_n\sigma(t_o^+) = q_n\varepsilon(t_o^+) \qquad (b)$$

There are two cases to consider:

Case 1 $m = n$, and therefore, $p_n \neq 0$.
(a) If stress is the dependent variable, the initial condition on stress is

$$\sigma(t_o^+) = \frac{q_n}{p_n} \varepsilon(t_o^+) \tag{c}$$

This implies that to a strep strain of magnitude $\varepsilon(t_o^+)$, the material will respond with a step stress of magnitude $\frac{q_n}{p_n} \varepsilon(t_o^+)$. In other words, the material exhibits instantaneous elasticity with glassy (impact) modulus $M(0) = \frac{q_n}{p_n} \equiv M_o \equiv M_g$.
(b) If strain is the dependent variable, the initial condition on it is

$$\varepsilon(t_o^+) = \frac{p_n}{q_n} \sigma(t_o^+) \tag{d}$$

This implies that to a step stress of magnitude $\sigma(t_o^+)$, the material will respond with a step strain of magnitude $\frac{p_n}{q_n} \sigma(t_o^+)$. The material exhibits instantaneous elasticity, with compliance: $C(0) = \frac{p_n}{q_n} \equiv C_o \equiv C_g$.

Clearly, the instantaneous modulus and compliance are reciprocals of each other.

Case 2 $m < n$, that is: $p_n = 0$.
(a) if stress is the dependent variable, the result after m integrations would be

$$p_m \sigma(t_o^+) = q_m \varepsilon(t_o^+) + q_{m+1} \frac{d}{dt} \varepsilon(t_o^+) \tag{e}$$

For a discontinuous strain, this implies that the stress would be infinite at the point of application of the strain, because the derivative of a step function is the Dirac delta function (see Appendix A):

$$\sigma(t_o^+) \rightarrow \infty \tag{f}$$

In other words, if the higher-order derivative of the stress is less than that of the strain, and the stress is the dependent variable, the material will not be able to respond to an instantaneously applied strain.
(b) If strain is the dependent variable, we would integrate the differential equation m times to get the initial condition on strain. The resulting functions of stress would be continuous and would integrate to zero in the interval between t_o^- and t_o^+. The proper initial condition on strain would then have to be that

$$\varepsilon(t_o^+) = 0 \tag{g}$$

In this case, the material has zero instantaneous elastic response and does not respond instantaneously to an instantaneously applied stress.

4. Having established the initial value of the dependent function, $\sigma(t_o^+)$ or $\varepsilon(t_o^+)$, we now integrate the differential equation $n - 1$ times, between the same two limits t_o^- and t_o^+, and arrive at the following equation, in which the only unknown is the first derivative of the dependent variable (stress or strain):

$$_{n-1}\sigma(t_o^+) + p_n \frac{d}{dt}\sigma(t_o^+) = q_{n-1}\varepsilon(t_o^+) + q_n \frac{d}{dt}\varepsilon(t_o^+) \tag{h}$$

Use this expression together with the given values of the controlled variable and its first derivative, and the value already established for the initial condition on the dependent variable, to solve for the initial condition sought.

(a) If stress is the dependent variable, expression (h) yields that

$$\frac{d}{dt}\sigma(t_o^+) = \frac{q_{n-1}}{p_n}\varepsilon(t_o^+) + \frac{q_n}{p_n}\frac{d}{dt}\varepsilon(t_o^+) - \frac{p_{n-1}}{p_n}\sigma(t_o^+) \tag{i}$$

Every term on the right-hand side of this expression is known. This means that we could, in principle, use this expression directly to establish the correct initial condition on the first derivative of stress. For convenience, we replace the initial condition on stress by its value in terms of applied strain, collect terms (multiplying and dividing the first term on the right-hand side by the p_n), and arrive at

$$\frac{d}{dt}\sigma(t_o^+) = \frac{q_{n-1}p_n - q_n p_{n-1}}{p_n^2}\varepsilon(t_o^+) + \frac{q_n}{p_n}\frac{d}{dt}\varepsilon(t_o^+) \tag{j}$$

(b) If strain is the dependent variable, the roles of stress and strain are reversed. Then, using (j) and inverting the roles of the p's and q's yield

$$\frac{d}{dt}\varepsilon(t_o^+) = \frac{p_{n-1}q_n - p_n q_{n-1}}{q_n^2}\sigma(t_o^+) + \frac{p_n}{q_n}\frac{d}{dt}\sigma(t_o^+) \tag{k}$$

To establish the proper initial conditions for higher-order derivatives of the dependent variable, one proceeds in the same fashion. For instance, to establish the initial conditions for the second derivative, we integrate $n - 2$ times between t_o^- and t_o^+ and, after some algebra, arrive at the following expressions.

When stress is the dependent variable,

$$\frac{d^2}{dt^2}\sigma(t_o^+) = \frac{q_{n-2}}{p_n}\varepsilon(t_o^+) + \frac{q_{n-1}}{p_n}\frac{d}{dt}\varepsilon(t_o^+) + \frac{q_n}{p_n}\frac{d}{dt}\varepsilon(t_o^+)$$
$$- \frac{p_{n-2}}{p_n}\sigma(t_o^+) - \frac{p_{n-1}}{p_n}\frac{d^1}{dt^1}\sigma(t_o^+) \tag{l}$$

Similarly, when strain is the dependent variable,

$$\frac{d^2}{dt^2}\varepsilon(t_o^+) = \frac{p_{n-2}}{q_n}\sigma(t_o^+) + \frac{p_{n-1}}{q_n}\frac{d}{dt}\sigma(t_o^+) + \frac{p_n}{q_n}\frac{d}{dt}\sigma(t_o^+)$$
$$- \frac{q_{n-2}}{q_n}\varepsilon(t_o^+) - \frac{q_{n-1}}{q_n}\frac{d^1}{dt^1}\varepsilon(t_o^+)$$

(m)

Example 3.2   Use the above rules to decipher the significance of the coefficients of the rheological model of Example 3.1: $p_o\sigma(t) + p_1\partial_t\sigma(t) = q_1\partial_t\varepsilon(t)$ and to establish the correct initial conditions, if the model is subjected to a step-stress history $\sigma(t) = \sigma_o H(t)$.
Solution:
On the basis of the coefficients present in P and Q, this model represents a fluid, as $q_o = 0$, and, because $m = 1 = n$, it will exhibit instantaneous elastic response with glassy compliance $C_g = \frac{p_1}{q_1}$ and its proper initial condition should be $\varepsilon(0^+) = \frac{p_1}{q_1}\sigma_o$.

Example 3.3   Given the following sets of non-zero coefficients of the stress and strain operators P and Q of a set of rheological models, identify which model represents a fluid, which a solid, and which will exhibit instantaneous elastic response.
(a) p_o and q_o
(b) p_o and q_1
(c) p_o, p_1, p_2 and q_o, q_1, q_2
(d) p_o, p_1, p_2, p_3 and q_1, q_2, q_3
   Solution
(a) This model represents a solid because $q_o \neq 0$ and has instantaneous elastic response because $m = n$.
(b) This model represents a fluid, because $q_o = 0$ and does not have instantaneous elastic response because $m < n$.
(c) This model represents a solid because $q_o \neq 0$ and has instantaneous elastic response because $m = n$.
(d) This model represents a fluid because $q_o = 0$ and also exhibits instantaneous elastic response because $m = n$.

3.4   Construction of Rheological Models

As mentioned before, the usefulness of mechanical constitutive equations in differential operator form lies in that they make it easy to construct the corresponding equations for arbitrary combinations of basic elements. The procedure to accomplish this may be summarized in three steps as follows:

1. Identify which elements in the system are acting in parallel with each other, and which ones are in series, and write the constitutive equations for the components of each group of rheological elements accordingly.

   (a) For elements in parallel, write their constitutive equations in stress–strain form, $\sigma_i = (Q_i/P_i)\varepsilon_i$ and note that all elements in parallel with each other experience the same strain and that the total stress in the collection is equal to the sum of the stresses in the individual elements. This is a consequence of force balance—as shown in the free body diagram in Fig. 3.6.

   Thus, for subcollections in parallel, write

   $$\sigma = \sum \sigma_i = \sum \frac{Q_i}{P_i} \varepsilon_i = \sum \frac{Q_i}{P_i} \varepsilon \tag{3.15a}$$

   For sets of elements in series write the constitutive equations in strain–stress form, $\varepsilon_i = (P_i/Q_i)\,\sigma_i$, and note that elements in series with each other will carry the same stress, but the total strain in the collection is equal to the sum of the strains in the individual elements. This is a consequence of force balance, as shown in the free body diagram in Fig. 3.7.

   Thus, for a sub-collection in series, write:

   $$\varepsilon = \sum \frac{P_i}{Q_i} \sigma_i = \sum \frac{P_i}{Q_i} \sigma \tag{3.15b}$$

2. Identify whether the overall assembly corresponds to a system in parallel or to one in series and convert the equations of the subcollections accordingly.
   (a) To stress–strain form ($\sigma_i = [Q_i/P_i]\varepsilon$), if the overall assembly is in parallel.
   (b) To strain–stress form ($\varepsilon_i = [P_i/Q_i]\sigma$), if the overall assembly is in series.
3. Add the responses (stress or strain) of all the subcollections of elements to define the constitutive equation of the assembly.

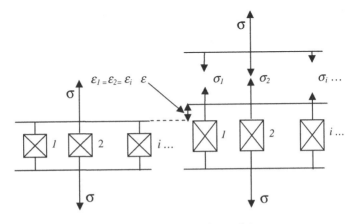

Fig. 3.6 Force balance for a set of rheological units in parallel

Fig. 3.7 Force balance for a set of rheological units in series

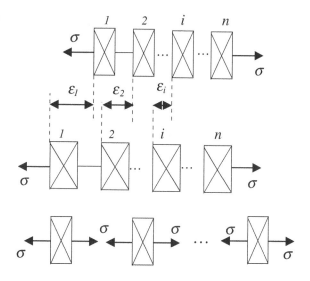

Naturally, the constitutive equations for isolated springs and dashpots would be written directly in either stress–strain or strain–stress form, depending on which form is adequate for the problem at hand.

3.5   Simple Rheological Models

The simplest, non-trivial combination of basic rheological elements consists of one spring and one dashpot, either in series or in parallel with each other. The former arrangement is known as a Maxwell model and the latter as a Kelvin–Voigt model. Both models are widely used in theoretical studies because of their simplicity and not necessarily because they provide good representations of real viscoelastic behavior. As will be seen shortly, the Kelvin arrangement exhibits the behavior of a solid and the Maxwell unit behaves like a fluid. These models are alternatively referred to as the Kelvin, or Kelvin–Voigt solid, and the Maxwell or Maxwell–Wiechert fluid, respectively.

3.5.1   Kelvin–Voigt Solid

This model consists of one spring and one dashpot connected in parallel, as depicted in Fig. 3.8. In what follows, we derive the constitutive equation of the model and then apply it to obtain the response of the model to strain and stress loading; in doing so, the corresponding relaxation and creep compliance functions are developed.

Fig. 3.8 Kelvin–Voigt
model and force balance

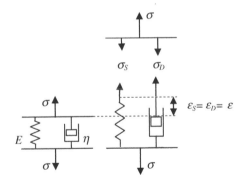

3.5.1.1 Constitutive Equation

We apply the operator procedure following the three suggested steps:
1. We first identify two "subcollections" of elements (in this case, the spring and the dashpot) and write their individual constitutive equations as follows:
   For the spring, $\sigma_S = E \cdot \varepsilon_S$
   For the dashpot, $\sigma_D = \eta \cdot \partial_t \varepsilon_D$
2. The overall Kelvin model is an arrangement in parallel, so we leave the equations of each subset in stress–strain form, using that the strain is the same in each unit, that is: $\varepsilon_s = \varepsilon_d = \varepsilon$.
   For the spring, $\sigma_S = E \cdot \varepsilon$
   For the dashpot, $\sigma_D = \eta \partial_t \cdot \varepsilon$
3. We now add the responses (stresses) of the separate (subsets of) elements and collect terms to write

$$\sigma = \sigma_S + \sigma_D = E \cdot \varepsilon + \eta \partial_t \varepsilon = (E + \eta \partial_t)\varepsilon \qquad (a)$$

In unencumbered form, to identify the stress and strain operators P and Q

$$\sigma = (E + \eta \partial_t)\varepsilon \qquad (3.16a)$$

As seen from this expression, $m = 0 < 1 = n$; $p_o = 1$, $p_1 = 0$; $q_o = E$; and $q_1 = \eta$. We then conclude that the Kelvin unit has no instantaneous response ($m < n$); behaves like a solid ($q_o \neq 0$); and admits the initial condition $\sigma(t_o^+) = \varepsilon(t_o^+)\delta(t - t_o^+)$, if in strain control—this behavior is consistent with what would be expected of the spring and dashpot combination in parallel, where the dashpot would lock at high strain rates, and, $\varepsilon(t_o^+) = \frac{p_1}{q_1}\sigma(t_o^+) = \frac{0}{\eta}\sigma(t_o^+) \equiv 0$, if stress were the controlled variable.

We next evaluate the response of the Kelvin model to general stress and strain loading histories of the type shown in Fig. 3.1, featuring a step discontinuity at $t = t_o$, and apply the solutions to the special cases of the creep and stress recovery experiments, examined earlier in connection with the individual spring and dashpot.

3.5.1.2 Response to Strain Loading

In this case, the controlled variable is strain, so that $\varepsilon(t) = f(t)H(t - t_o)$; the rheological Eq. (3.16a) is not a differential equation in stress. To assess the response of the model, we simply carry out the indicated operations, using the properties of the Heaviside and delta functions, presented in Appendix A,[5] to get

$$\sigma = Ef(t)H(t - t_o) + \eta \left[H(t - t_o)\frac{d}{dt}f(t) + f(t_o)\delta(t - t_o) \right] \tag{a}$$

This expression can be used to establish the relaxation modulus of the Kelvin model. In this case, $f(t) \equiv \varepsilon_o$, $df/dt = 0$, and, by the definition of the relaxation modulus as the stress response to a constant strain of magnitude ε_o, measured per unit of applied strain: $M(t) = \sigma(t)/\varepsilon_o$, (a) yields

$$M(t - t_o) = EH(t - t_o) + \eta\delta(t - t_o) \tag{3.17a}$$

Or, more simply:

$$M(t) = EH(t) + \eta\delta(t) \tag{3.17b}$$

With this, the response of the model to the stress recovery test, $\varepsilon(t) \equiv \varepsilon_o H(t - t_o) - \varepsilon_o H(t - t_1)$, shown conceptually in Fig. 3.9a, may be obtained directly, as

$$\sigma(t) = \{[EH(t - t_o) + \eta\delta(t - t_o)] - [EH(t - t_1) + \eta\delta(t - t_1)]\}\varepsilon_o \tag{b}$$

3.5.1.3 Response to Stress Loading

In this case, the controlled variable is stress, and the rheological Eq. (3.16a) is a general differential equation of first order, for which an integrating factor can be found by the method presented in Appendix A. To this end, we rearrange the differential Eq. (3.16a) and cast it in the standard form of the linear differential equation of first order:

$$\frac{d\varepsilon}{dt} + \frac{E}{\eta}\varepsilon = \frac{1}{\eta}\sigma, \quad t > 0 \tag{3.16b}$$

Following the procedure discussed in Appendix A, and using the notation $\tau_c \equiv \eta/E$, obtain an integrating factor $u = e^{(1/\tau_c)\int_t ds} = e^{t/\tau_c}$ and apply it to the differential equation and, changing the dummy variable of integration on the right-hand side, rewrite it as

---

[5] Specifically, that $[f(t)H(t)]' = f(t)'H(t) + f(t)\delta(t)$ and that $f(t)\delta(t - t_o) = f(t_o)\delta(t - t_o)$.

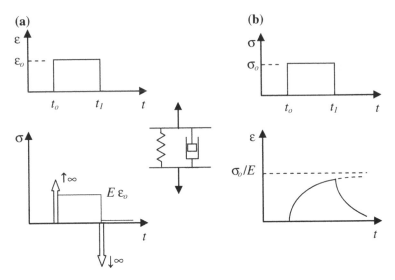

Fig. 3.9 Response of the Kelvin model to stress and strain recovery experiments. **a** Stress recovery response. **b** Creep recovery response

$$d\left[e^{t/\tau_c}\varepsilon(t)\right] = e^{s/\tau_c} \cdot \frac{f(s)}{\eta} H(s - t_o) \cdot ds \qquad \text{(c)}$$

Integrating this expression from 0^- to t, using that $\varepsilon(0^+) = 0$, rearranging, and inserting the exponential e^{-t/τ_c} inside the integral[6] produce the solution sought, as

$$\varepsilon(t) = H(t - t_o) \int_{t_o}^{t} e^{-(t-s)/\tau_c} \cdot \frac{f(s)}{\eta} \cdot ds \qquad \text{(d)}$$

In particular, according to this expression, the response of the model to a step stress, or creep loading, $f(t)H(t - t_o) \equiv \sigma(t) = \sigma_o H(t - t_o)$, noting that $H(0 - t_o) = 0$, would be

$$\varepsilon(t) = H(t - t_o) \int_{t_o}^{t} e^{-(t-s)/\tau_c} \cdot \frac{\sigma_o}{\eta} \cdot ds = \frac{1}{E}\left[1 - e^{-\frac{t-t_o}{\tau_c}}\right] H(t - t_o)\sigma_o \qquad \text{(e)}$$

Consequently, the creep compliance of the Kelvin model, defined as the strain response measured per unit stress, turns out to be

$$C(t - t_o) = \frac{1}{E}\left[1 - e^{-\frac{t-t_o}{\tau_c}}\right], \quad t > 0 \qquad \text{(3.18a)}$$

---

[6] This can be done because t is not the variable of integration.

Or, simply

$$C(t) = \frac{1}{E}\left[1 - e^{-t/\tau_c}\right], \quad t > 0 \tag{3.18b}$$

The parameter $\tau_c \equiv \eta/E$, introduced here, has the dimensional units of time and is referred to as creep or retardation time—the creep time of the Kelvin model—because it is associated with a creep function.

As will be evident from the treatment that follows, when the springs and dashpots are linear, that is, when their own constitutive equations are linear and the higher-order derivatives of the stresses and strains entering their differential equations are equal, the rheological models have equivalent hereditary integral forms (see, for instance, [2, 3].

An important alternate form of the general response presented in (d) may be obtained in terms of the creep compliance. This is done integrating expression (d) by parts, setting $t_o = 0$, for simplicity, so as not to carry the factor $H(t - t_o)$; using $u \equiv \sigma$ and $dv \equiv e^{-(t-s)/\tau_c}ds$:

$$\varepsilon(t) = \frac{1}{E}\left[\sigma(t) - \sigma(0^+)e^{-t/\tau_c}\right] - \int_{0^+}^{t} e^{-(t-s)/\tau_c} \cdot \frac{d}{ds}\sigma(s) \cdot ds \tag{f}$$

Adding and subtracting $\sigma(0^+)/E$ to the right-hand side of this expression and taking the sum $[\sigma(t) - \sigma(0^+)]/E$ inside the integral—changing the sign of the integral to account for the introduction of this term—result in

$$\varepsilon(t) = \frac{1}{E}\left[1 - e^{-t/\tau_c}\right]\sigma(0^+) + \int_{0^+}^{t} \frac{1}{E}\left[1 - e^{-(t-s)/\tau}\right] \cdot \frac{d}{ds}\sigma(s) \cdot ds \tag{g}$$

Using the creep compliance (3.18a), this expression may be cast in integral form, as

$$\varepsilon(t) = C(t)\sigma(0^+) + \int_{0^+}^{t} C(t - s) \cdot \frac{d}{ds}\sigma(s) \cdot ds \tag{3.19}$$

Integrating by parts again, and collecting terms, (3.19) can be expressed in the hereditary integral form involving the derivative of the creep compliance of the model:

$$\varepsilon(t) = C(0)\sigma(t) - \int_{0^+}^{t} \frac{d}{ds}C(t - s) \cdot \sigma(s) \cdot ds \tag{3.20}$$

The response of the Kelvin model to the creep or strain recovery experiment described by $\sigma(t) \equiv \sigma_o H(t - t_o) - \sigma_o H(t - t_1)$ is shown in Fig. 3.9b and may be obtained using (3.19) or (3.20). Using (3.20), for instance, the response, $\varepsilon_1(t)$, to the first part of the load, $\sigma_o H(t - t_o)$, is

$$\varepsilon_1(t) = C(0)\sigma_o H(t - t_o) - \int_{0^+}^{t} \frac{d}{ds} C(t - s) \cdot \sigma_o H(s - t_o) \cdot ds \qquad (h)$$

The integral is evaluated using the following property of the unit step function (c.f. Appendix A):

$$\int_{0^+}^{t} H(s - t_o) \cdot f(s) \cdot ds = H(t - t_o) \int_{t_o}^{t} f(s) \cdot ds \qquad (i)$$

and the fact that the integral of the derivative of a function $(d/ds)C(t - s)$ is the function $C(t - s)$ itself. Proceeding thus, and canceling terms, the expression for $\varepsilon_1(t)$ becomes

$$\varepsilon_1(t) = [C(t - t_o)]\sigma_o H(t - t_o) \qquad (j)$$

The response of the Kelvin model to a general creep recovery experiment is, by linearity, the sum of its individual responses to the loading and unloading parts; hence,

$$\varepsilon(t) = C(t - t_o)\sigma_o H(t - t_o) - C(t - t_1)\sigma_o H(t - t_1) \qquad (k)$$

Or, explicitly, in terms of the creep compliance (3.18a),

$$\varepsilon(t) = \frac{1}{E}\left[1 - e^{-(t-t_o)/\tau_c}\right]\sigma_o H(t - t_o) - \frac{1}{E}\left[1 - e^{-(t-t_1)/\tau_c}\right]\sigma_o H(t - t_1) \qquad (l)$$

Example 3.4 Find the steady-state response of a Kelvin unit having spring and viscosity parameters G and η, respectively, to a sinusoidal strain history $\varepsilon(t) = \varepsilon_o\sin(\omega t)$
Solution:
    Use of the loading history and the corresponding constitutive equation listed in (3.16a) leads to $\sigma(t) = G\varepsilon_o \sin \omega t + \eta\omega\varepsilon_o \cos \omega t$, which, not having any transient terms in it, is the steady-state response of the Kelvin unit to the cyclic strain history.

3.5.2 Maxwell–Wiechert Fluid

The Maxwell model consists of a single spring and a single dashpot connected in series, as depicted in Fig. 3.10. We derive the constitutive equation of the model first and then apply it to evaluate the response of the model to general strain and stress loading.

Fig. 3.10 Maxwell–
Wiechert model and force
balance

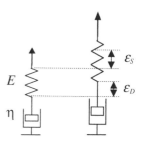

3.5.2.1 Constitutive Equation

The constitutive equation of the Maxwell model is established following the three-step procedure. Accordingly,

1. We first identify two "subcollections" of elements (in this case, the single spring and the dashpot) and write down the corresponding constitutive equations.
   For the spring: $\sigma_S = E \cdot \varepsilon_S$
   For the dashpot: $\sigma_D = \eta \cdot \partial_t \varepsilon_D$
2. he overall Maxwell model is an arrangement in series; for this reason, we rewrite the equations of each unit in its strain–stress form and use that the stress is the same in each individual unit: $\sigma_s = \sigma_D = \sigma$
   For the spring, $\varepsilon_S = \frac{1}{E} \cdot \sigma$
   For the dashpot, $\varepsilon_D = \frac{1}{\eta \partial_t} \cdot \sigma$
3. We now add the responses (strains) of the separate subcollections of elements and use operator algebra to obtain

$$\varepsilon = \varepsilon_S + \varepsilon_D = \frac{1}{E} \cdot \sigma + \frac{1}{\eta \partial_t} \sigma = \frac{(E + \eta \partial_t)}{E \eta \partial_t} \sigma \tag{a}$$

Clearing off fractions—multiplying throughout by $E\eta\partial_t$ to remove the operator from the denominator on the right-hand side—produces the constitutive equation in the standard form ($P \cdot \sigma = Q \cdot \varepsilon$):

$$(E + \eta \partial_t)\sigma = (E\eta \partial_t)\varepsilon \tag{3.21a}$$

As seen from this expression: $m = 1 = n$, $p_o = E$, $p_1 = \eta$, and $q_o = 0$ and $q_1 = E \cdot \eta$. This indicates that the Maxwell model has instantaneous response ($m = n$); represents a fluid ($q_o = 0$); and admits initial conditions of the form $\sigma(t_o^+) = E\varepsilon(t_o^+)$ under applied strain; and $\varepsilon(t_o^+) = \frac{1}{E}\sigma(t_o^+)$, if the controlled variable is stress.

3.5.2.2 Response to Strain Loading

In this case, the controlled variable is strain, so that $\varepsilon(t) = f(t)H(t - t_o)$; we use the rheological Eq. (3.21a), which is an ordinary linear differential equation of first order. To solve it by the procedure in Appendix A, we introduce, for convenience, the time parameter $\tau_r \equiv \eta/E$ and rewrite the constitutive equation, as

$$\frac{d\sigma}{dt} + \frac{1}{\tau_r}\sigma = E\frac{d}{dt'}f(t) \tag{3.21b}$$

In accordance with the solution procedure in Appendix A, we find $u = e^{\int dt/\tau_r} = e^{t/\tau_r}$, as an integrating factor for the equation and, changing the dummy variable on the right-hand side, write it in total differential form, as

$$d\left[e^{t/\tau_r}\sigma(t)\right] = e^{s/\tau_r}E\left[H(s - t_o)\frac{df}{ds}ds + f(s)\delta(s - t_o)ds\right] \tag{b}$$

Upon integration between 0^- and t, using the properties of the impulse and unit functions:

$$e^{t/\tau_r}\sigma(t) - \sigma(0^-) = H(t - t_o)\int_{t_o}^t Ee^{s/\tau_r}\frac{df}{dt'}ds + \int_0^t Ee^{\frac{s}{\tau_r}}f(s)\delta(s - t_o)ds \tag{c}$$

Using that $\sigma(0^-) = 0$, and $f(t_o) = \varepsilon(t_o^+)$, multiplying throughout by e^{-t/τ_r}, taking this factor inside the integral, and regrouping:

$$\sigma(t) = e^{-(t-t_o)/\tau_r}E\varepsilon(t_o^+) + \int_{t_o}^t Ee^{-(t-s)/\tau_r}\frac{d\varepsilon}{ds}ds \tag{d}$$

Using this to evaluate the response of the model to a step-strain load, $\varepsilon(t) = \varepsilon_o H(t)$, noting that, in this case, $d\varepsilon/dt = 0$, in the interval of integration, yields

$$\sigma(t) = \left[Ee^{-(t-t_o)/\tau_r}\right]\varepsilon_o, \quad t > t_o \tag{e}$$

From this, the relaxation modulus, defined as the stress response measured per unit of applied strain, is found to be

$$M(t - t_o) = Ee^{-(t-t_o)/\tau_r}, \quad t > t_o \tag{3.22a}$$

Or, more simply:

$$M(t) = Ee^{-t/\tau_r}, \quad t > 0 \tag{3.22b}$$

Setting $t_o = 0$ in (d) and combining it with (3.22b) produces the following hereditary integral form of the stress–strain law of the Maxwell model:

$$\sigma(t) = M(t)\varepsilon(0^+) + \int_{0^+}^t M(t - s)\frac{d}{ds}\varepsilon(s)ds \tag{3.23a}$$

A second hereditary integral form of the constitutive equation can be derived from this expression. Indeed, integrating it by parts, once, using $u = M(t - s)$ and $dv = d\varepsilon/ds$ and collecting terms:

$$\sigma(t) = M(0)\varepsilon(t) - \int_{0^+}^{t} \frac{d}{dt} M(t - s)\varepsilon(s)ds \qquad (3.23b)$$

In this case, the time parameter, τ_r, is called a relaxation time because it is associated with a relaxation function. The relaxation time of a Maxwell unit represents the time it takes the stress to decay by about 63 %. This is so, because the relaxation modulus, evaluated at $t = \tau_r$ (or $t - t_o = \tau_r$), is $e^{-1} \cdot E \approx 0.37E$.

The response of the Maxwell model to the general stress recovery experiment described by $\varepsilon(t) \equiv \varepsilon_o H(t - t_o) - \varepsilon_o H(t - t_1)$ is shown in Fig. 3.11a and may be obtained using one of the expressions in (d), (e), or (3.23a, b). Using (d) and the properties of the unit step and impulse functions, and collecting terms:

$$\sigma(t) = E\left[e^{-(t-t_o)/\tau_r}H(t - t_o) - e^{-(t-t_1)/\tau_r}H(t - t_1)\right]\varepsilon_o \qquad (f)$$

3.5.2.3 Response to Stress Loading

In this case, the controlled variable is stress, so that $\sigma(t) = f(t)H(t - t_o)$, and the differential equation is recast as

$$\frac{d}{dt}\varepsilon(t) = \frac{1}{E}\frac{d\sigma}{dt} + \frac{1}{\eta}\sigma \qquad (g)$$

Before substituting the applied stress, we integrate this expression between 0^- and t, using that $\varepsilon(0^-) = 0$ and $\sigma(0^+) = 0$, and collect terms to write

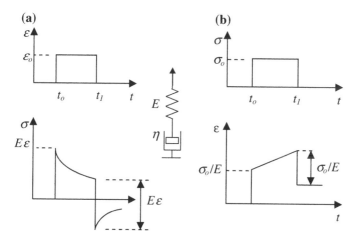

Fig. 3.11 Response of the Maxwell model to stress and strain recovery experiments. **a** Stress recovery response. **b** Creep recovery response

$$\varepsilon(t) = \frac{1}{E}\sigma(t) + \frac{1}{\eta}\int_{0^+}^{t}\sigma(s)ds \tag{h}$$

This expression may be cast in hereditary integral form through integration by parts and some algebraic manipulations. Integrating it by parts once, choosing $u = \sigma$, and $dv = ds$; performing the required operations, bearing in mind that the loading, $\sigma(t)$, has a discontinuity at t_o, and collecting terms, leads to

$$\varepsilon(t) = \left[\frac{1}{E} + \frac{t}{\eta}\right]\sigma(t) - \int_{t_o}^{t}\left(\frac{s}{\eta}\right)\frac{d\sigma}{ds}ds \tag{i}$$

Adding and subtracting the term $(1/E)\sigma(t_o^+)$ and taking the term in brackets inside the integral lead to

$$\varepsilon(t) = \int_{t_o}^{t}\left[\frac{1}{E} + \frac{t-s}{\eta}\right]\frac{d\sigma}{ds}ds \tag{j}$$

Splitting the integral from t_o^- to t_o^+ and from t_o^+ to t and using that $\sigma(t) = f(t)H(t - t_o)$

$$\varepsilon(t) = \left[\frac{1}{E} + \frac{t-t_o}{\eta}\right]f(t_o^+) + \int_{t_o^+}^{t}\left[\frac{1}{E} + \frac{t-s}{\eta}\right]\frac{df}{ds}ds, \quad t > t_o \tag{k}$$

The response of the model to a constant stress load $f(t) \equiv \sigma_o$, that is, $\sigma(t) = \sigma_o H(t - t_o)$ can be easily established from this expression, noting that the derivative of the loading history vanishes inside the interval of integration, as

$$\varepsilon(t) = \left[\frac{1}{E} + \frac{t-t_o}{\eta}\right]H(t - t_o)\sigma_o \tag{l}$$

The result, divided by the applied stress is, by definition, the creep compliance of the Maxwell model:

$$C(t - t_o) = \left[\frac{1}{E} + \frac{t-t_o}{\eta}\right]H(t - t_o) \tag{3.24a}$$

Or, more simply:

$$C(t) = \left[\frac{1}{E} + \frac{t}{\eta}\right]H(t) \tag{3.24b}$$

The hereditary integral form of the strain–stress constitutive equation of the Maxwell model, which we set out to find, may be obtained taking (3.24a) into (k). For clarity, however, we set $t_o = 0$, to write

$$\varepsilon(t) = C(t)\sigma(0^+) + \int_{0^+}^{t} C(t - s)\frac{d\sigma}{ds}\,ds \tag{3.25}$$

A second form of hereditary integral constitutive equation for the Maxwell model, which features the derivative of the creep compliance under the integral sign, is readily obtained from this last relation through integration by parts. Selecting $C(t - s) = u$, and $(d\varepsilon/ds)ds = dv$, and proceeding thus, produces that

$$\varepsilon(t) = C(0)\sigma(t) - \int_{0^+}^{t} \frac{d}{ds} C(t - s)\sigma(s)\,ds \tag{3.26}$$

The response of the Maxwell model to the general creep recovery experiment described by $\sigma(t) \equiv \sigma_o H(t - t_o) - \sigma_o H(t - t_1)$ is shown in Fig. 3.11b and may be obtained using (3.25), and the creep compliance, (3.24b), noting that $d\sigma/ds \equiv 0$ in the integration interval; thus,

$$\varepsilon(t) = \left\{ \left[\frac{1}{E} + \frac{(t - t_o)}{\eta}\right] H(t - t_o) - \left[\frac{1}{E} + \frac{(t - t_1)}{\eta}\right] H(t - t_1) \right\}\sigma_o \tag{m}$$

3.6   Generalized Models

The concepts introduced earlier are used here to develop the constitutive equations for two models involving multiple Kelvin units in series and multiple Maxwell units in parallel. For obvious reasons, the first type of arrangement is called a generalized Kelvin model, while the second type of arrangement is known as a generalized Maxwell model.

3.6.1   Generalized Kelvin Model

A generalized Kelvin model or Kelvin–Voigt model is a collection of Kelvin units in series, plus an isolated spring—or an isolated dashpot, as shown in Fig. 3.12.
In accordance with the operator method, the constitutive equation of this model may be established as the sum of the strain–stress, series form of the equations of the individual Kelvin elements. In detail:
1. The subcollections of the overall model are the individual Kelvin elements. The ith subset in the arrangement has a constitutive equation given by (3.16a), as

$$\sigma_i = (E_i + \eta_i \partial_t)\varepsilon_i \tag{a.1}$$

Fig. 3.12 Generalized
Kelvin–Voigt model

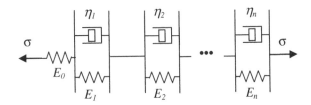

for a spring and dashpot in parallel; and

$$\sigma_o = (E_o)\varepsilon_o \tag{a.2}$$

for the isolated spring
2. which in strain–stress form become

$$\varepsilon_i = \frac{1}{(E_i + \eta_i\partial_t)}\sigma_i \tag{b.1}$$

for the Kelvin unit, and

$$\varepsilon_o = (1/E_o)\sigma_o \tag{b.2}$$

for the isolated spring.
3. The constitutive equation for the whole assembly is established by adding the contributions of all the Kelvin–Voigt "subsets," recalling that $\sigma_i = \sigma$:

$$\varepsilon = \sum_{i=1}^{n} \varepsilon_i = \left[\frac{1}{E_o} + \sum_{i=1}^{n}\frac{1}{(E_i + \eta_i\partial_t)}\right]\sigma \tag{3.27a}$$

Applying the minimum common multiple of the denominators of the operators; setting $\eta_o = 0$, to include the isolated spring as a degenerate Kelvin unit, switching the order of the members of the equation:

$$\sum_{i=o}^{n}\prod_{\substack{j=0 \\ j \neq i}}^{n}(E_i + \eta_i\partial_t)\sigma = \prod_{i=0}^{n}(E_i + \eta_i\partial_t)\varepsilon \tag{3.27b}$$

The linear operator form, $P \cdot \sigma = Q \cdot \varepsilon$, is obtained by performing the indicated operations and collecting like terms. The result is

$$\left[p_o + p_1\frac{d^1}{dt^1} + \cdots + p_m\frac{d^m}{dt^m}\right]\sigma = \left[q_o + q_1\frac{d^1}{dt^1} + \cdots + q_n\frac{d^n}{dt^n}\right]\varepsilon \tag{3.28}$$

Examination of the general constitutive equation (3.27a) or (3.28)[7] reveals the following:

(a) If none of the spring and dashpot coefficients are zero, the order, m, of the higher-order derivative of the stress would be one less than that of the strain ($m = n - 1 < n$). In the context of the analysis of the order of the derivatives of P and Q, this means that a generalized Kelvin model with no degenerate units would have no initial elastic response ($m < n$).

(b) If the arrangement has—at least—one isolated spring (that is, one Kelvin unit with zero viscosity), the operator Q will contain a non-zero constant operator, $q_o \neq 0$; the rheological model will represent a solid.

(c) If the arrangement has one Kelvin unit with no spring, the operator Q will contain a zero constant operator, $q_o = 0$; the rheological model will represent a fluid.

Because the generalized Kelvin model is an arrangement in series, its creep compliance may be obtained directly as the sum of the creep compliance functions of each individual Kelvin unit. Thus, using (3.18b), (3.27a) and that $C_i(t) = \varepsilon_i(t)/\sigma_o$:

$$C(t) = \sum C_i(t) = \frac{1}{E_o} + \sum_{i=1}^{n} \frac{1}{E_i}\left[1 - e^{-t/\tau_{c_i}}\right]; \quad \tau_{c_i} \equiv \frac{\eta_i}{E_i} \tag{3.29}$$

In this expression, $1/E_o$ is the instantaneous or glassy compliance and is denoted by C_g. Also, in molecular dynamic studies, the finite set of pairs (τ_i, $1/E_i$), including the pair (0, C_g), is called a discrete retardation spectrum; the notation $L_i \equiv 1/E_i$ is typically used. With this in mind, in the limit of infinitely many Kelvin units, the summation becomes an integral and the result is a continuous creep or retardation spectrum [4].

$$C(t) = C_g + \int_0^{\infty} L(\tau)\left[1 - e^{-t/\tau}\right]d\tau \tag{3.30}$$

3.6.2   Generalized Maxwell Model

A collection of Maxwell units in parallel, as depicted in Fig. 3.13, is referred to as a generalized Maxwell model, or generalized Maxwell–Wiechert model. Per the operator approach, the corresponding mechanical constitutive equation may be constructed as the sum of the constitutive equations in stress–strain form of all its Maxwell elements.

In detail:

1. The overall assembly may be considered as a collection of Maxwell "elements," in parallel, rather than start at the individual spring and dashpot sets in

---

[7] Because of the explicit summation and multiplication symbols present in (3.1b), it is easier to use that expression to discern the order of the constitutive equation and the nature of its coefficients.

Fig. 3.13 Generalized
Maxwell–Wiechert model

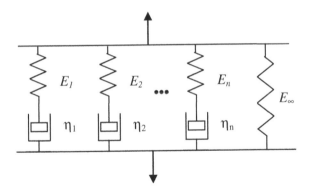

series. We use (3.21a) as the basic constitutive equation and write it for a
generic Maxwell "element":

$$(E_i \eta_i \partial_t) \varepsilon_i = (E_i + \eta_i \partial_t) \sigma_i \tag{c.1}$$

with

$$\sigma_o = (E_\infty) \varepsilon_o \tag{c.2}$$

for the isolated spring.

2. Since the overall arrangement is in parallel, we express it in stress–strain form:

$$\sigma_i = \frac{(E_i \eta_i \partial_t)}{(E_i + \eta_i \partial_t)} \varepsilon_i \tag{d.1}$$

and:

$$\sigma_o = (E_\infty) \varepsilon_o \tag{d.2}$$

for the isolated spring.

3. Adding the contributions of all the elements in the model and using that $\varepsilon_i = \varepsilon$:

$$\sigma = \sum \sigma_i = \left[E_\infty + \sum_{i=1}^{n} \frac{(E_i \eta_i \partial_t)}{(E_i + \eta_i \partial_t)} \right] \varepsilon \tag{3.31a}$$

Multiplying throughout by the minimum common multiple of the denominators
of the operators on the right side, and rearranging:

$$\left[\prod_{j=1}^{n}(E_j + \eta_j \partial_t)\right]\sigma = E_\infty \left[\prod_{j=1}^{n}(E_j + \eta_j \partial_t)\right]\varepsilon$$

$$+ \left[\sum_{i=1}^{n}(E_i \eta_i \partial_t)\prod_{j=1}^{n}(E_j + \eta_j \partial_t)\right]\varepsilon \qquad (3.31b)$$
$$j \neq i$$

After performing the indicated operations, and simplifying, the explicit operator equation, $P \cdot \sigma = Q \cdot \varepsilon$, is seen to be of the same form as for the generalized Kelvin model. In this case, however, examination of the general constitutive equation (3.31a) reveals that

(a) The order of the highest order derivative of the stress would be the same as that of the strain; irrespective of whether the system has an isolated spring or not. This means that a generalized Maxwell model will always exhibits instantaneous response.

(b) If the arrangement has at least one isolated spring, the operator Q will contain a non-zero constant, $q_o \neq 0$; and the rheological model will represent a solid.

(c) If the arrangement has one Maxwell unit without a spring, the operator Q will contain a zero constant, $q_o = 0$; and the rheological model will represent a fluid.

Because the generalized Maxwell model is an arrangement in parallel, its relaxation function is the sum of the relaxation functions of each Maxwell unit. Thus, using (3.22b), (3.31a) and that $M(t) = \sigma(t)/\varepsilon_o$:

$$M(t) = E_\infty + \sum M_i(t) = E_\infty + \sum E_i e^{-t/\tau_{r_i}}; \quad \tau_{r_i} \equiv \frac{\eta_i}{E_i} \qquad (3.32)$$

This finite sum is called a discrete relaxation spectrum. As the number of Maxwell units increases without limit, the summation becomes an integral, giving rise to the continuous relaxation spectrum:

$$M(t) = E_\infty + \int_0^\infty E(\tau)e^{-t/\tau}d\tau \qquad (3.33)$$

3.7 Composite Models

Composite models are those formed by judicious combinations of springs and dashpots, which not only are more complex than the Maxwell and Kelvin–Voigt units, but are capable of reproducing more realistic viscoelastic behavior. Clearly, the choices are very many, and what follows restricts attention to models with three parameters, representing, respectively, solid and fluid behavior.

Fig. 3.14 Standard linear
solid. **a** From generalized
Kelvin model. **b** From
generalized Maxwell model

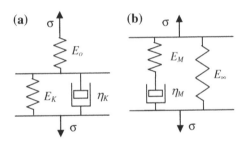

3.7.1   Standard Linear Solid

The two equivalent versions of this model are shown in Fig. 3.14. One of them
consists of a Kelvin unit in series with a spring. It is a degenerate form of a
generalized Kelvin–Voigt model. The other one is composed of a Maxwell element in parallel with a spring, which is a degenerate form of the generalized
Maxwell model. As will be shown, the three-parameter model in Fig. 3.14 exhibits
solid behavior and is thus also referred to as the standard linear (viscoelastic) solid.

The constitutive equation of the standard linear viscoelastic solid is derived
considering it as a degenerate case of the generalized Maxwell model, depicted in
part (a) of Fig. 3.14. The mathematical equivalence between the two models is
demonstrated later, as part of the examples.

For a Kelvin unit in series with a spring, $\sigma_K = [E_K + \eta_K \partial_t]\varepsilon_K$ and $\sigma_S = [E_o]\varepsilon_S$,
and the strains are additive: $\varepsilon = \varepsilon_K + \varepsilon_S$; therefore, by the operator method, the
differential equation of the standard linear solid shown in Fig. 3.14a is given by

$$\varepsilon = \left[\frac{1}{E_K + \eta_K \partial_t} + \frac{1}{E_o} \right] \sigma \tag{3.34}$$

As may be readily verified, this equation could have been written directly as a
particular case of the generalized Kelvin model given in (3.27a). After clearing off
fractions, using operator algebra and rearranging:

$$(E_o + E_K)\sigma + \eta_K \frac{d\sigma}{dt} = (E_o E_K)\varepsilon + (E_o \eta_K) \frac{d\varepsilon}{dt} \tag{3.35a}$$

This equation may be cast in general form:

$$p_o \sigma + p_1 \frac{d\sigma}{dt} = q_o \varepsilon + q_1 \frac{d\varepsilon}{dt} \tag{3.35b}$$

If the following notation is used,

$$p_o \equiv E_o + E_K; \; p_1 \equiv \eta_K; \; q_o \equiv E_o E_K; \; q_1 \equiv E_o \eta_K \tag{3.35c}$$

These relationships show that not all model parameters, p_i and q_i, in the generic equation of the model are independent.

Examination of (3.35a) indicates that, because $m = 1 = n$, the model exhibits glassy response, with modulus $M_o = q_1/p_1 \equiv M_g$, and will admit initial conditions of the form: $p_o\sigma(t_o) = q_o\varepsilon(t_o)$. In addition, as asserted before, the model is a solid, because $q_o \neq 0$, and has long-term modulus $M_\infty = q_o/p_o = E_o\,E_K/(E_o + E_K)$. Also, the derivative terms in its constitutive equation indicate the model captures creep response. These characteristics of the standard linear solid will be demonstrated by evaluating its response to the stress recovery and creep recovery tests. In so doing, it should be noted that the constitutive equation is symmetric or balanced in the stress and the strain. This means that the form of its solution to a strain history is of the same form as its response to a stress history of the same type and vice versa.

3.7.1.1 Response to Strain Loading

In this case, the loading is of the form $\varepsilon(t) = f(t)H(t - t_o)$; the response of the model is obtained as the general solution to the differential Eq. (3.35b), which, for clarity, is rewritten as

$$\frac{d\sigma}{dt} + \frac{1}{\tau}\sigma = \frac{q_o}{p_1}\varepsilon + \frac{q_1}{p_1}\frac{d\varepsilon}{dt}; \quad \tau \equiv \frac{p_1}{p_o} \tag{e}$$

Or, after applying the integrating factor $e^{t/\tau}$ and casting the equation in differential form:

$$d\left(e^{\frac{t}{\tau}}\sigma\right) = \frac{q_o}{p_1}e^{\frac{t}{\tau}}f(t')H(t' - t_o)dt' + \frac{q_1}{p_1}H(t' - t_o)e^{\frac{t'}{\tau}}\frac{df}{dt'}ds + \frac{q_1}{p_1}e^{\frac{t'}{\tau}}f(t)\delta(t' - t_o)dt' \tag{f}$$

Integrating between 0^- and t; using that $\sigma(0^-) = 0$; and recalling the properties of the unit and impulse functions:

$$e^{t/\tau}\sigma(t) = H(t - t_o)\left[\frac{q_o}{p_1}\int_{t_o}^{t}e^{t'/\tau}f(t')dt' + \frac{q_1}{p_1}\int_{t_o}^{t}e^{t'/\tau}\frac{df}{dt'}dt'\right] + \frac{q_1}{p_1}e^{t_o/\tau}f(t_o) \tag{g}$$

Multiplying throughout by $e^{-t/\tau}$; taking the exponential inside the integrals; and collecting terms:

$$\sigma(t) = H(t - t_o)\int_{t_o}^{t}e^{-(t-t')/\tau}\left[\frac{q_o}{p_1}f(t') + \frac{q_1}{p_1}\frac{df}{dt'}\right]dt' + \frac{q_1}{p_1}e^{-(t-t_o)/\tau}f(t_o) \tag{h}$$

This expression may be used to obtain the response of the model to the stress recovery experiment $\varepsilon(t) = \varepsilon_o H(t - t_o) - \varepsilon_o H(t - t_1)$. Because of linearity, the load is split into two parts $\varepsilon_1(t) = f_1(t)H(t - t_o) \equiv \varepsilon_o H(t - t_o)$ and

$\varepsilon_2(t) = f_2(t)\ H(t - t_1) \equiv -\varepsilon_o H(t - t_1)$; the response to each part, calculated separately; and the results added together to obtain the total response. Noting that $f_1(t) \equiv \varepsilon_o$, and $df_1/d_t = 0$; using (h) to evaluate the response to the first part of the load leads to the following response—after performing the indicated integrations, recalling that $\tau \equiv p_1/p_o$, and rearranging

$$\sigma_1(t) = \left\{ \frac{q_o}{p_o} + \left[\frac{q_1}{p_1} - \frac{q_o}{p_o} \right] e^{-(t-t_o)/\tau} \right\} H(t - t_o)\varepsilon_o \tag{i}$$

From this, the relaxation modulus of the standard linear solid follows, as

$$M(t - t_o) = \frac{q_o}{p_o} + \left[\frac{q_1}{p_1} - \frac{q_o}{p_o} \right] e^{-(t-t_o)/\tau}; \quad \tau \equiv \frac{p_1}{p_o} \tag{3.36a}$$

Or:

$$M(t) = \frac{q_o}{p_o} + \left[\frac{q_1}{p_1} - \frac{q_o}{p_o} \right] e^{-t/\tau}; \quad \tau \equiv \frac{p_1}{p_o} \tag{3.36b}$$

The response, $\sigma_2(t)$, to the second part of the loading would be identical to $\sigma_1(t)$, but of opposite sign. Therefore,

$$\sigma(t) = \sigma_1(t) + \sigma_2(t) = \left\{ \frac{q_o}{p_o} \left[1 - e^{-(t-t_o)/\tau} \right] + \frac{q_1}{p_1} e^{-(t-t_o)/\tau} \right\} H(t - t_o)\sigma_o\varepsilon_o$$
$$- \left\{ \frac{q_o}{p_o} \left[1 - e^{-(t-t_1)/\tau} \right] + \frac{q_1}{p_1} e^{-(t-t_1)/\tau} \right\} H(t - t_1)\varepsilon_o \tag{j}$$

Or, in terms of its physical parameters,

$$\sigma(t) = \left\{ M_\infty \left[1 - e^{-(t-t_o)/\tau} \right] + M_g e^{-(t-t_o)/\tau} \right\} H(t - t_o)\varepsilon_o$$
$$- \left\{ M_\infty \left[1 - e^{-\frac{t-t_1}{\tau}} \right] + M_g e^{-\frac{t-t_1}{\tau}} \right\} H(t - t_1)\varepsilon_o; \quad \tau \equiv \frac{\eta_K}{E_o + E_K} \tag{k}$$

This solution is shown schematically in Fig. 3.15. As the figure indicates, the response of the standard linear solid features all aspects of the stress response of a

Fig. 3.15 Stress recovery response of three-parameter, standard linear solid

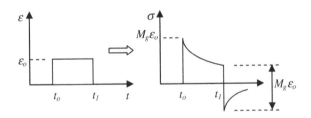

linear viscoelastic solid; it has instantaneous elastic response upon loading and unloading; relaxation under constant strain; and delayed recovery after unloading.

3.7.1.2 Response to Stress Loading

In this case, the loading is of the form $\sigma(t) = f(t)H(t - t_o)$, and strain becomes the controlled variable. Because of the symmetry of the constitutive equation (e) in stress and strain, all the arguments used for the case of strain loading may be repeated interchanging the roles of stress and strain, as well as interchanging the parameters p_o and p_1 with q_o and q_1. This leads to the following relations:

$$\frac{d\varepsilon}{dt} + \frac{1}{\lambda}\varepsilon = \frac{p_o}{q_1}\sigma + \frac{p_1}{q_1}\frac{d\sigma}{dt}; \quad \lambda \equiv \frac{q_1}{q_o} \tag{1}$$

$$C(t - t_o) = \frac{p_o}{q_o} + \left[\frac{p_1}{q_1} - \frac{p_o}{q_o}\right]e^{-(t-t_o)/\lambda}; \quad \lambda \equiv \frac{q_1}{q_o} \tag{3.37a}$$

$$C(t) = \frac{p_o}{q_o} + \left[\frac{p_1}{q_1} - \frac{p_o}{q_o}\right]e^{-t/\lambda}; \quad \lambda \equiv \frac{q_1}{q_o} \tag{3.37b}$$

And, lastly, to

$$\varepsilon(t) = \left\{\frac{p_o}{q_o}\left[1 - e^{-(t-t_o)/\lambda}\right] + \frac{p_1}{q_1}e^{-(t-t_o)/\lambda}\right\}H(t - t_o)\sigma_o$$
$$- \left\{\frac{p_o}{q_o}\left[1 - e^{-(t-t_1)/\lambda}\right] + \frac{p_1}{q_1}e^{-(t-t_1)/\lambda}\right\}H(t - t_1)\sigma_o \tag{m}$$

And, in terms of the model's physical parameters,

$$\varepsilon(t) = \left\{\frac{1}{M_\infty}\left[1 - e^{-(t-t_o)/\lambda}\right] + \frac{1}{M_g}e^{-(t-t_o)/\lambda}\right\}H(t - t_o)\sigma_o$$
$$- \left\{\frac{1}{M_\infty}\left[1 - e^{-(t-t_1)/\lambda}\right] + \frac{1}{M_g}e^{-(t-t_1)/\lambda}\right\}H(t - t_1)\sigma_o; \quad \lambda \equiv \frac{\eta_K}{E_K} \tag{n}$$

As depicted in Fig. 3.16, the creep response of the standard linear solid exhibits all behavioral characteristics of the creep response of viscoelastic solids.

Example 3.5 Show that the model of Fig. 3.14b, also called the Zener model, is mathematically equivalent to that in Fig. 3.14a.
Solution:
Since this model is a particular case of the generalized Maxwell model, its constitutive equation is given by (3.31a), with $n = 1$; thus, $\sigma = \left[E_\infty + \frac{(E_M \eta_M \partial_t)}{(E_M + \eta_M \partial_t)}\right]\varepsilon$
which upon rearrangement may be recast as

Fig. 3.16 Creep response of
standard linear solid.
a Elastic, impact response.
b Creep. **c** Elastic recovery.
d Delayed recovery

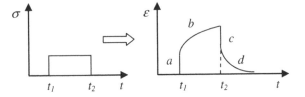

$$E_M \sigma + \eta_M \frac{d\sigma}{dt} = E_M E_\infty \varepsilon + \eta_M (E_M + E_\infty) \frac{d\varepsilon}{dt} \qquad (3.35d)$$

This expression is mathematically equivalent to (3.35a), which proves the point. As may also be seen, equating the corresponding coefficients of the two equations reveals that the parameters of the two versions of the standard linear solid are related as follows:

$$\eta_K = \eta_M; \quad E_o = E_M + E_\infty; \quad \frac{1}{E_K} = \frac{1}{E_M} + \frac{1}{E_\infty}. \qquad (o)$$

3.7.2 Three-Parameter Fluid

There are two equivalent versions of this model; they are shown in Fig. 3.17. As seen in the figure, one version of the model consists of a Kelvin unit in series with a dashpot. It is a degenerate form of a generalized Kelvin–Voigt model. The other one is composed of a Maxwell element in parallel with a dashpot, which is a special case of the generalized Maxwell model. As will be shown shortly, the three-parameter model exhibits fluid behavior and is thus referred to as the three-parameter fluid.

As in the case of the three-parameter solid, the constitutive equation of each version of the three-parameter fluid shown Fig. 3.17 may be derived either from the equation of the generalized Kelvin model or from the equation of the Maxwell model. For instance, the equation of the model in part (a) of the figure results from equation (3.27a), using $n = 2$, $E_1 = 0$, $\eta_1 = \eta_o$, and removing the term $1/E_o$, which had been introduced to account for an isolated spring. This leads to $\varepsilon = \left[\frac{1}{\eta_o \partial_t} + \frac{1}{(E_K + \eta_K \partial_t)} \right] \sigma$. After clearing off fractions, using operator algebra, and rearranging

$$E_K \sigma + (\eta_0 + \eta_K) \frac{d\sigma}{dt} = E_K \eta_o \frac{d\varepsilon}{dt} + \eta_o \eta_K \frac{d^2\varepsilon}{dt^2} \qquad (3.38)$$

Accordingly, $m = 1 < 2 = n$, so the model does not exhibit impact response. Also, $q_o = 0$, which indicates the model represents a fluid, as claimed earlier.

Fig. 3.17 Three-parameter
fluid. **a** From generalized
Kelvin model. **b** From
generalized Maxwell model

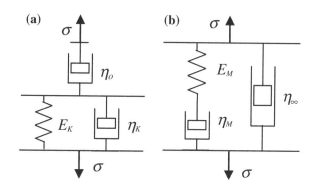

3.8 Problems

P.3.1 Determine the initial conditions for the model of Example 2, if the controlled variable is stress and the loading is defined by $\sigma(t) = \sigma_o H(t)$.

Answer: $\varepsilon(0^+) = (p_1/q_1)\sigma_o$

Hint: Use expression (d) of Sect. 3.3.4, with $t_o^+ = 0^+$ and $\sigma(0^+) = \sigma_o$, to get the desired result. Note that since $m = 1 = n$, this model has instantaneous elasticity with compliance p_1/q_1, from which the result follows.

P.3.2 Obtain the response of a Kelvin unit to the constant strain rate history $\varepsilon(t) = Rt$.

Answer: $\sigma(t) = ERt + \eta R$

Hint: The easiest way to solve this problem is by direct application of (3.16a) to the given strain history. Doing this produces the result.

P.3.3 Find the steady-state response of a Maxwell unit with spring and viscosity parameters G and η, respectively, to a sinusoidal strain history $\varepsilon(t) = \varepsilon_o \sin(\omega t)$

Answer: $\sigma(t) = G\left[\frac{(\omega\tau)^2}{1+(\omega\tau)^2}\sin(\omega\tau) + \frac{\omega\tau}{1+(\omega\tau)^2}\cos(\omega\tau)\right]\varepsilon_o$

Hint: Take the strain history $f(t) = \varepsilon_o \sin(\omega t)$ and $E = G$, $\tau = \eta/G$ into Eq. (3.21b) and use integration by parts, twice, to resolve the integral. The full solution should be

$\sigma(t) = G\left[\frac{(\omega\tau)^2}{1+(\omega\tau)^2}\sin(\omega\tau) + \frac{\omega\tau}{1+(\omega\tau)^2}\cos(\omega\tau)\right]\varepsilon_o - \frac{\omega\tau}{1+(\omega\tau)^2}e^{-t/\tau}\varepsilon_o$. From which the

steady-state solution follows dropping the exponential, transient term. This problem is solved more efficiently by the methods of Chap. 4.

P3.4 Show that the model in Fig. 3.14b is mathematically equivalent to that in Fig. 3.14a.

Hint: Use either the operator method or Eq. (3.31a) for a generalized Maxwell model, with $n = 1$, $E_1 = E_M$, and $\eta_1 = \eta_M$, to write $\sigma = \left[E_\infty + \frac{E_M\eta_M\partial_t}{E_M+\eta_M\partial_t}\right]\varepsilon$ and resolve the expression to arrive at $[E_M + \eta_M\partial_t]\sigma = [E_\infty E_M + E_M + \eta_M\partial_t]\varepsilon$; which, being of the same form as (3.35a), proves the point.

Fig. 3.18 Problem 3.5

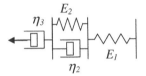

P.3.5 Derive the constitutive equation for the four-parameter model shown in Fig. 3.18, treating it as a degenerate generalized Kelvin model. Use physical arguments regarding how the springs and dashpots of the arrangement would behave under a constant stress to show that the system would respond like a fluid and that it provides for initial elastic response. This arrangement is known as the four-parameter or Burgers fluid.

Answer: $P\sigma = Q\varepsilon$, with: $P \equiv E_1 E_2 + (E_1\eta_2 + E_2\eta_3 + E_1\eta_3)\frac{d}{dt} + \eta_2\eta_3\frac{d^2}{dt^2}$, and: $Q \equiv E_1 E_2\eta_3\frac{d}{dt} + (E_1\eta_2\eta_3)\frac{d^2}{dt^2}$. Since $m = 2 = n$, this model exhibits instantaneous elastic response with impact, or glassy modulus $M_g = q_2/p_2 = E_1$. Also, since $q_o = 0$, the model represents a fluid. On physical grounds alone, under instantaneous strain, the dashpots will lock, but the isolated spring, will respond elastically. In the limit of extremely long time under load, the isolated dashpot will extend like a fluid would.

Hint: use constitutive equation (3.27a) for the generalized Kelvin model with $E_3 = 0$ and $\eta_1 = 0$, to arrive at the result.

P.3.6 Develop the differential equation for the rheological model depicted in Fig. 3.19 adjacent sketch, show that it behaves as a solid, and determine its instantaneous and long-term moduli.

Answer: $\left[(E_1 + E_2) + \eta_1\frac{d}{dt}\right]\sigma = \left[(E_2 E_3 + E_3 E_1 + E_1 E_2) + (E_2 + E_3)\eta_1\frac{d}{dt}\right]\varepsilon$.

In this case, $m = 1 = n$; the model exhibits instantaneous elastic response with glassy modulus $M_g = q_1/p_1 = E_2 + E_3$. Also, since $q_o \neq 0$, the model represents a solid with long-term modulus, $M_\infty = \frac{q_o}{p_o} = \frac{E_2 E_3 + E_3 E_1 + E_1 E_2}{E_1 + E_2}$.

Hint: Note that this model is composed of a standard linear solid in parallel with an isolated spring and add the corresponding constitutive equations in stress–strain form, $\sigma_{std} = \left[\frac{E_1 E_2 + E_2\eta_1\partial_t}{E_1 + E_2 + \eta_1\partial_t}\right]\varepsilon$, and $\sigma_{spring} = E_3\varepsilon$, respectively, and arrive at the desired result.

P.3.7 The simple exponential function $M(t) = Ee^{-\alpha t}$ was fitted to the results of a shear relaxation test of a polymer. Use the Laplace transform to obtain the corresponding creep compliance and explain why the relaxation function given models fluid response.

Fig. 3.19 Problem 3.6

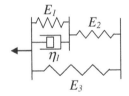

Answer: $C(t) = \frac{1}{E} + \frac{\alpha}{E}t$

Hint: Use the relationship $\bar{C}(s) = \frac{1}{s^2 \bar{M}(s)}$ and the Laplace transform $\bar{M}(s) = E\frac{1}{s+\alpha}$ of the relaxation modulus to arrive at the expression $\bar{C}(s) = \frac{1}{E}\left(\frac{1}{s}\right) + \frac{\alpha}{E}\left(\frac{1}{s^2}\right)$. The inverse Laplace transform of this yields the desired result. The relaxation function given will model fluid response because it has zero long-term modulus, which is a characteristic of fluid behavior. In addition, the corresponding creep compliance indicates that the model used for this polymer will exhibit unlimited flow under a sustained constant stress (see the behavior of a Maxwell model).

P.3.8 The three parameters of a standard linear solid model were fitted to the results of a shear test of a rubber. Using the Laplace transform, determine the corresponding shear relaxation modulus.

Answer: $M(t) = \frac{q_o}{p_o} + \left[\frac{q_1}{p_1} - \frac{q_o}{p_o}\right]e^{-t/\tau}; \quad \tau \equiv \frac{p_1}{p_o}$

Hint: Use the relationship $\bar{M} = \frac{\bar{Q}}{s\bar{P}}$ and the Laplace transform of the differential operators $P = p_o + p_1\partial_t$ and $Q = q_o + q_1\partial_t$ to write $\bar{M} = \frac{q_o + q_1 s}{s(p_o + p_1 s)}$. Factor p_1 out of the denominator; introduce the relaxation time $\tau = p_1/p_o$; and use partial fraction expansion (see Appendix A), to arrive at $\bar{M} = \frac{q_o/p_o}{s} + \frac{q_1/p_1 - q_o/p_o}{(s+1/\tau)}$, from which the desired result follows by inversion to physical space.

P.3.9 A viscoelastic material of Kelvin type obeys the law $\sigma(t) = E\varepsilon(t) + \eta\frac{d}{dt}\varepsilon(t)$. Find its steady-state response to the cyclic strain history $\varepsilon = \varepsilon_o \sin(\omega t)$.

Answer: $\sigma(t) = E\varepsilon_o \sin(\omega t) + \eta\omega\varepsilon_o \cos(\omega t)$.

Hint: Insert the given strain history into the analytical model and perform the needed operations to reach the result.

P.3.10 Repeat Problem P.3.9 for a viscoelastic material of Maxwell type whose stress–strain law is described by $\frac{d}{dt}\sigma(t) + \frac{1}{\tau}\sigma(t) = E\frac{d}{dt}\varepsilon(t)$.

Answer: $\sigma(t) = E \cdot \varepsilon_o\left\{\frac{(\omega t)^2}{1+(\omega t)^2}\sin(\omega t) + \frac{\omega t}{1+(\omega t)^2}\cos(\omega t)\right\}$

Hint: Insert the cyclic strain history into the differential equation of the model, using the integrating factor $u = e^{t/\tau}$ and perform integration by parts, twice, to arrive at the result $\sigma(t) = E \cdot \varepsilon_o\left\{\frac{(\omega t)^2}{1+(\omega t)^2}\sin(\omega t) + \frac{\omega t}{1+(\omega t)^2}\cos(\omega t)\right\} - \frac{E\varepsilon_o}{1+(\omega t)^2}e^{-t/\tau}$. After a sufficiently long time, the last term becomes negligible, and the steady-state solution listed is reached.

References

1. D.L. Kreider, R.G. Kuller, D.R. Ostberg, *Elementary Differential Equations* (Addison-Wesley, Reading, 1968), pp. 66–69
2. R.M. Christensen, *Theory of Viscoelasticity*, 2nd edn. (Dover, NY, 2003), pp. 14–20
3. A.D. Drosdov, *Finite Elasticity and Viscoelasticity* (World Scientific, Singapore, 1996), pp. 250–255
4. J.D. Ferry, *Viscoelastic Properties of Polymers*, 3rd edn. (Wiley, NY, 1980), pp. 60–67

Constitutive Equations for Steady-State Oscillations

4

Abstract

Although the viscoelastic constitutive equations in either integral or differential form apply in general, irrespective of the type of loading, or the point in time at which the response is sought, it is possible to derive from them constitutive equations of a form especially well suited to steady-state oscillations. This chapter uses complex algebra to transform the integral and differential constitutive equations of viscoelasticity, defined in the time domain, into algebraic expressions in the complex plane. The chapter also examines the relationships between the material property functions defined in the time domain, and their complex-variable counterparts, and examines the problem of energy dissipation during steady-state oscillations, important in the design of mounts for vibratory equipment, among others.

Keywords

Steady state · Complex · Real · Imaginary · Harmonic · Excitation · Frequency · Storage · Loss · Constitutive · Euler · Fourier · Argand · Dirichlet · Laplace · Newton · Transform · Modulus · Compliance · Amplitude · Integral · Differential · Energy

4.1 Introduction

It is often necessary to evaluate the steady-state response of a viscoelastic material to harmonic loading. Although the constitutive equations in either integral or differential form (c.f. Chaps. 2, 3) apply irrespective of the type of loading and the point in time at which the response is sought, it is possible to derive from them constitutive equations of a form especially well suited to steady-state oscillatory

D. Gutierrez-Lemini, *Engineering Viscoelasticity*, DOI: 10.1007/978-1-4614-8139-3_4,   93
© Springer Science+Business Media New York 2014

conditions. Such equations, expressed in terms of complex quantities, enter the problem of harmonic oscillations in a natural way.

This chapter uses Euler's formula: $e^{j\theta} = \cos(\theta) + j\sin(\theta)$, and complex algebra[1] to represent the steady-state harmonic excitations imposed on a viscoelastic material, and their steady-state response to such loads. As will be seen, this transforms the integral and differential constitutive equations, defined in the time domain, into simple algebraic expressions in the complex plane.[2] As it turns out, the resulting equations in the complex plane are of exactly the same forms as the corresponding Laplace-transformed equations and hence are of the same forms as their elastic counterparts. This similarity is due to the fact that the constitutive equations in complex-variable form may be obtained through an integral transformation, called the Fourier transform, of the general constitutive equations, which is equivalent to the Laplace transform for functions typically occurring in viscoelasticity (see Appendix A).

The chapter also examines the relationships between the time-domain material property functions and their complex-variable counterparts. This treatment is followed by an evaluation of the energy dissipated per cycle. There, a simple analysis shows that the energy dissipated by a viscoelastic material in a cycle is given by an expression that resembles that for energy storage in linear elastic materials; in that it is proportional to the product of some adequate measure of modulus and squared strain; or compliance and squared stress.

Finally, although there are several excellent books on viscoelasticity dealing with the present subject, we list only three as references: Ferry's seminal book on properties of polymers [1], Flügge's classical text on viscoelasticity [2], and the more recent treatment by Wineman and Rajagopal [3].

4.2   Steady-State Constitutive Equations from Integral Constitutive Equations

The aim here is to develop constitutive equations which apply specifically to steady-state oscillatory loads, where the controlled variable, say strain, is of the form $\varepsilon(t) = \varepsilon_o\sin(\omega t)$, or $\varepsilon(t) = \varepsilon_o\cos(\omega t)$; and where both, ε_o and ω, representing, respectively, the amplitude of the excitation and its frequency, are real numbers. Noting that Euler's exponential form of a complex number can be used to represent a sine or a cosine function of its argument, we introduce the complex strain $\varepsilon^*(j\omega t)$ (c.f. Appendix A):

$$\varepsilon^*(j\omega t) \equiv \varepsilon_o e^{j\omega t} = \varepsilon_o[\cos(\omega t) + j\sin(\omega t)] \tag{4.1}$$

---

[1] In the main body of the text, the symbol j is used to denote the imaginary unit: $j \equiv \sqrt{-1}$.

[2] Because the frequency of the excitations becomes the independent variable in this transformation, the resulting equations are said to be defined in the frequency domain.

where the real part of the complex excitation, $Re[\varepsilon^*(j\omega t)]$, would represent excitations of cosine type; and the imaginary part, $Im[\varepsilon^*(j\omega t)]$, would stand for excitations of sine type.

The compex stress, σ^*, is obtained by inserting (4.1) into (2.32a), setting the
lower limit to $-\infty$, in accordance with the closed-cycle condition for steady-state
response to oscillatory loads. Carrying out the differentiation under the integral
sign, introducing the change of variable $u = t - \tau$, and regrouping, using that
$\varepsilon^*(j\omega t) \equiv \varepsilon_o e^{j\omega t}$, leads to:

$$\sigma^*(j\omega t) = \varepsilon_o j\omega \int_{-\infty}^{t} M(t - \tau)e^{j\omega \tau}d\tau \equiv \left[j\omega \int_0^\infty M(u)e^{-j\omega u}du \right]\varepsilon^*(j\omega t) \qquad (a)$$

Or:

$$\sigma^*(j\omega t) = \left[j\omega \int_0^\infty M(u)e^{-j\omega u}du \right]\varepsilon^*(j\omega t) \equiv M^*(j\omega)\varepsilon^*(j\omega t) \qquad (4.2)$$

This expression shows that, just as in an ideally linear elastic material, where the
stress response is directly proportional to the applied strain—and vice versa—the
steady-state complex stress in a viscoelastic substance subjected to harmonic
excitation is directly proportional to the complex strain. As discussed in Appendix
A, the factor of proportionality—in fact, a function of frequency—is termed the
complex modulus, M^*, and is defined by (4.2), as:

$$M^*(j\omega) \equiv j\omega \int_0^\infty M(t)e^{-j\omega t}dt \qquad (4.3)$$

The integral in (4.3) is the Fourier transform of the relaxation modulus, $M(t)$. In
general, the Fourier transform of an arbitrary function $f(t)$ will exist if the function
is piecewise continuous, has a finite number of finite discontinuities, and is absolutely integrable—i.e., the integral of its absolute value is bounded.[3] As discussed
in Appendix A, the Fourier transform and the inverse Fourier transform, referred to
as the Fourier transform pair, are, respectively, defined as:

$$\mathcal{F}\{f(t)\} \equiv \bar{f}(j\omega) = \int_0^\infty e^{-j\omega t}f(t)dt \qquad (4.4a)$$

$$f(t) = \frac{1}{\pi} \int_0^\infty e^{j\omega t}\bar{f}(j\omega)d\omega \qquad (4.4b)$$

---

[3] These are the so-called Dirichlet conditions.

In this manner, the complex modulus, M^*, is the $j\omega$-multiplied Fourier transform of the relaxation function: $M^* = j\omega\mathcal{F}\{M(t)\} \equiv j\omega\bar{M}(j\omega)$. The relaxation function $M(t)$ may be recovered from (4.3) dividing through by $j\omega$ and applying the inverse Fourier transform [4]:

$$M(t) = \frac{1}{\pi}\int_0^\infty \frac{M^*(j\omega)}{j\omega}e^{j\omega t}d\omega \tag{4.5}$$

Example 4.1 Obtain the analytical form of the complex modulus of an isotropic viscoelastic solid with shear relaxation modulus given by the following finite sum of exponentials:

$$G(t) = G_e + \sum_{k=1}^n G_k e^{-t/\tau_k}$$

Solution:
The complex modulus is obtained by putting this expression into (4.3) and carrying out the indicated integrations. Thus:

$$G^*(j\omega) \equiv j\omega\int_0^\infty \left[G_e + \sum_{k=1}^n G_k e^{-t/\tau_k}\right]e^{-j\omega t}dt$$

$$= j\omega\left[G_e\int_0^\infty e^{-j\omega t}dt + \sum_{k=1}^n G_k\int_0^\infty e^{-\left(\frac{1+j\omega\tau_k}{\tau_k}\right)t}dt\right]$$

$$G^*(j\omega) = G_e + \sum_{k=1}^n G_k\frac{j\omega\tau_k}{1+j\omega\tau_k} = G_e + \sum_{k=1}^n G_k\frac{(\omega\tau_k)^2}{1+(\omega\tau_k)^2} + j\sum_{k=1}^n G_k\frac{(\omega\tau_k)}{1+(\omega\tau_k)^2}$$

It is common practice to express the complex modulus in Cartesian form; that is in terms of its real and imaginary components, which are denoted, respectively, as M' and M'':

$$M^*(j\omega) = M'(\omega) + jM''(\omega) \tag{4.6}$$

As with any other complex number, the graphical representation of M^* in the Argand plane, which is shown in Fig. 4.1, leads to the polar form of M^* in terms of its amplitude, $\|M^*\|$, and phase angle, δ_M (see Appendix A):

$$M^*(j\omega) = \|M^*\|e^{j\delta_M} \tag{4.7a}$$

Fig. 4.1 Graphical
representation of the complex
modulus in the *Argand Plane*

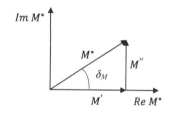

$$\|M^*\| = \sqrt{(M')^2 + (M'')^2} \tag{4.7b}$$

$$\tan \delta_M = M''/M' \tag{4.7c}$$

$$M' = \|M^*\| \cos \delta_M, \quad M'' = \|M^*\| \sin \delta_M \tag{4.7d}$$

For reasons that will become clear later, M' and M'' are also frequently referred to as the storage and loss modulus, respectively; and $\tan(\delta_M)$, as the loss tangent.

The following important alternate form of (4.2), results from combining it with (4.1) and (4.7 a):

$$\sigma^*(j\omega t) = \|M^*\| \varepsilon_o e^{j(\omega t + \delta_M)} \equiv \sigma_o e^{j(\omega t + \delta_M)}; \sigma_o \equiv \|M^*\| \varepsilon_o \tag{4.8}$$

This expression defines, quite naturally, the amplitude of the modulus, M^*, as the ratio σ_o/ε_o, of the stress amplitude to the strain amplitude—as for linear elastic materials. In addition, this expression and (4.1) state that to a complex strain $\varepsilon^*(j\omega t) = \varepsilon_o e^{j\omega t}$ there corresponds a complex stress $\sigma^*(j\omega t) = \sigma_o e^{j(\omega t + \delta_M)}$; which is of the same frequency as the controlled strain, but out of phase with it by δ_M radians.

The algebraic relationships between the complex modulus M^* and its real and imaginary components, M' and M'', result from taking (4.7a) into (4.3); using Euler's formula to split the integral into its real and imaginary parts; separating the equilibrium modulus, so that $M_t(t) = M(t) - M_e$ and grouping:

$$M'(j\omega) = M_e + \omega \int_0^\infty M_t(t) \sin(\omega t) dt \tag{4.9a}$$

$$M''(j\omega) = \omega \int_0^\infty M_t(t) \cos(\omega t) dt \tag{4.9b}$$

These are the Fourier sine and cosine transforms, respectively, of the relaxation modulus, M. The corresponding inverse transforms are given by:

$$M_t(t) = \frac{2}{\pi} \int_0^\infty \frac{M'(j\omega) - M_e}{\omega} \sin(\omega t) d\omega \tag{4.10a}$$

$$M_t(t) = \frac{2}{\pi} \int_0^\infty \frac{M''(j\omega)}{\omega} \sin(\omega t) d\omega \tag{4.10b}$$

These expressions indicate that the Cartesian components, M' and M'', of the complex modulus M^* are interdependent, since the relaxation modulus M, may be obtained from either component.

Example 4.2 Determine the storage and loss moduli, and the loss tangent of a material with relaxation modulus $E(t) = Ee^{-t/\tau}$.
Solution:
The complex modulus of the material may be obtained directly from (4.3). After carrying out the integration, multiplying and dividing the result by $(1 - j\omega)$, and collecting real and imaginary parts, the result is $M^* = E\frac{j\omega\tau}{1+j\omega\tau} \equiv E\frac{(\omega\tau)^2 + j(\omega\tau)}{1+()^2}$, with the following Cartesian components: $M' = E\frac{(\omega\tau)^2}{1+(\omega\tau)^2}$; $M'' = E\frac{\omega\tau}{1+(\omega\tau)^2}$. The loss tangent is obtained from these components and (4.7a), as: $\tan\delta = \frac{M''}{MI} = \frac{1}{\omega\tau}$

The strain–stress constitutive equations applicable to steady-state conditions can be established by reversing the roles of stress and strain in the previous derivations. In this case, similar to (4.1), the complex form of the given excitation, $\sigma(t) = \sigma_o \cos(\omega t)$, or $\sigma(t) = \sigma_o \sin(\omega t)$, would be:

$$\sigma^*(j\omega t) = \sigma_o e^{j\omega t} = \sigma_o[\cos(\omega t) + j\sin(\omega t)] \tag{4.11}$$

Inserting this into (2.33a) and proceeding as before, leads to the strain–stress equation:

$$\varepsilon^*(j\omega t) = C^*(j\omega)\sigma^*(j\omega t) \tag{4.12}$$

With the complex compliance C^*:

$$C^*(j\omega) \equiv j\omega \int_0^\infty C(t)e^{-j\omega t} dt \tag{4.13}$$

In addition, combining (4.2) and (4.12) produce that:

$$M^*(j\omega)C^*(j\omega) = 1 \tag{4.14}$$

This shows that the complex modulus and complex compliance are inverses of each other: $M^* = 1/C^*$.

Moreover, in view of (4.7a) and (4.13):

$$C^*(j\omega) = \frac{1}{M^*(j\omega)} = \frac{1}{\|M^*(j\omega)\|} e^{-j\delta_M} \equiv \|C^*(j\omega)\| e^{-j\delta_M} \equiv \|C^*(j\omega)\| e^{j\delta_C} \quad (4.15)$$

From this expression, it then follows that:

$$\|C^*(j\omega)\| = \frac{1}{\|M^*(j\omega)\|} \quad (4.16)$$

$$\delta_C = -\delta_M \quad (4.17)$$

That is, the amplitudes of the complex modulus and complex compliance are reciprocals of each other—as for linear elastic materials; and the phase angle, δ_C, of the complex compliance is the negative of the phase angle, δ_M, of the complex modulus. Because of this last result, a negative sign is typically used in front of the imaginary part, C'', of the Cartesian form of the complex compliance, and the same phase angle $\delta \equiv \delta_M$ used for the modulus and compliance. Then, corresponding to (4.6), the complex compliance is usually written as:

$$C^*(j\omega) = C'(j\omega) - jC''(j\omega) \quad (4.18)$$

The analogous of (4.7a) results from this expression, the definition of magnitude of a complex number, and the relationship between its components; thus:

$$\|C^*\|(j\omega) = \|C^*\| e^{-j\delta} \quad (4.19a)$$

$$\|C^*\| = \sqrt{(C')^2 + (c'')^2} \quad (4.19b)$$

$$\tan \delta = C''/C' \quad (4.19c)$$

$$C' = \|C^*\| \cos \delta, \ C'' = \|C^*\| \sin \delta \quad (4.19d)$$

Likewise, combining (4.12) and (4.15) with the complex stress $\sigma^*(t) = \sigma_o \, e^{j\omega t}$, given in (4.11), yields the following result, in analogy with (4.8):

$$\varepsilon^*(j\omega t) = \|C^*\| \sigma_o e^{j(\omega t - \delta)} \equiv \varepsilon_o e^{j(\omega t - \delta)}; \quad \varepsilon_o \equiv \|C^*\| \sigma_o \quad (4.20)$$

Just as in the case of the complex modulus, this expression defines the amplitude of the complex compliance, $\|C^*\|$, as the ratio ε_o/σ_o, of the amplitudes of the strain and stress, as for linear elastic materials.

Other relationships involving the creep compliance, such as their Fourier sine and cosine transforms are entirely similar to those for the relaxation modulus. Specifically, using $C_t(t) = C(t) - C_e$:

$$C'(j\omega) = C_e + \omega \int_0^\infty C_t(t) \sin(\omega t) dt \qquad (4.21a)$$

$$C''(j\omega) = -\omega \int_0^\infty C_t(t) \cos(\omega t) dt \qquad (4.21b)$$

$$C_t(t) = \frac{2}{\pi} \int_0^\infty \frac{C'(j\omega) - C_e}{\omega} \sin(\omega t) d\omega \qquad (4.22a)$$

$$C_t(t) = -\frac{2}{\pi} \int_0^\infty \frac{C''(j\omega)}{\omega} \sin(\omega t) d\omega \qquad (4.22b)$$

In practice, all functions involved in the response of viscoelastic materials—such as excitations, stress or strain response, and material properties—are of Dirichlet type; and hence, their Fourier transforms exist. It is then permissible to apply the Fourier transform to viscoelastic relationships, such as constitutive equation. In particular then, the Fourier transform of Eq. (2.33a) and (2.33b), for instance, would lead to (c.f. Appendix A):

$$\mathcal{F}\{\sigma(t)\} = \mathcal{F}\left\{ \int_{-\infty}^t M(t - \tau) \frac{d}{d\tau} \varepsilon(\tau) d\tau \right\} \Rightarrow \bar{\sigma}(j\omega) = j\omega \bar{M}(j\omega) \bar{\varepsilon}(j\omega) \qquad (4.23)$$

$$\mathcal{F}\{\varepsilon(t)\} = \mathcal{F}\left\{ \int_{-\infty}^t C(t - \tau) \frac{d}{d\tau} \sigma(\tau) d\tau \right\} \Rightarrow \bar{\varepsilon}(j\omega) = j\omega \bar{C}(j\omega) \bar{\sigma}(j\omega) \qquad (4.24)$$

Combining these expressions one may arrive at the following additional result:

$$\bar{M}(j\omega) \bar{C}(j\omega) = \frac{1}{(j\omega)^2} \qquad (4.25)$$

Expressions (4.23) to (4.25) are of exactly the same form as (2.29–2.31), derived in Chap. 2, taking the Laplace transform of constitutive equations (2.32). As explained in Appendix A, there is a close relationship between the Laplace and Fourier transforms; and it is to emphasize this fact, that the term $1/(j\omega)^2$, in (4.25), has been left as such, instead of as $-1/\omega^2$.

As emphasized in Chap. 1, except for very long observation times, the response of a viscoelastic solid is indistinguishable from that of a viscoelastic fluid. Consequently, the behavior of viscoelastic fluids under cyclic loading can be evaluated using either one of the complex property functions, M^* or C^*, derived earlier. Quite often, however, in reference to Newton's constitutive equation for a viscous fluid: $\sigma = \eta \cdot d\varepsilon/dt$, the complex viscosity, $\eta^* = \eta' - j\eta''$, is used for this purpose. The relationship between the complex modulus and complex viscosity can be easily established introducing complex quantities, as before. Hence, if $\varepsilon^*(j\omega t) = \varepsilon_o e^{j\omega t}$, and $\sigma^*(j\omega t) \equiv \eta^* \, d\varepsilon^*/dt = (j\omega)\eta^* \varepsilon^*$. Then, since $\sigma^*(j\omega t) \equiv M^* \varepsilon^*(j\omega t)$, it follows that:

$$\eta^* \equiv \eta' - j\eta'' = \frac{M^*}{j\omega} \tag{4.26}$$

The Cartesian components of the complex viscosity follow from this and (4.6), as:

$$\eta' = \frac{M''}{\omega}; \quad \eta'' = \frac{M'}{\omega} \tag{4.27}$$

This expression shows that the phase relationships for the complex viscosity and complex modulus are exactly the opposite of one another. In addition, (4.26) and (4.3) define the complex viscosity as the Fourier transform of the relaxation modulus:

$$\eta^* = \frac{M^*}{j\omega} = \int_0^\infty e^{-j\omega t} M(t)dt \tag{4.28}$$

In the limit of very slow frequency excitations ($\omega \to 0$), this expression leads to what is termed—for this reason—the zero shear-rate viscosity, η_o, as:

$$\eta^*(0) \equiv \eta_o = \int_0^\infty M(t)dt \tag{4.29}$$

The zero shear-rate viscosity enters the compliance function of viscoelastic fluids, in the term t/η_o.

Example 4.3 Determine the steady-state response of the viscoelastic substance of Example 4.2 to the harmonic excitation: $\varepsilon(t) = \varepsilon_o \cos(\omega t)$.
Solution:

The steady-state response may be computed with the constitutive equation $\sigma^* = M^* \varepsilon^*$, and the excitation in complex form $\varepsilon^*(j\omega t) = \varepsilon_o e^{j\omega t} = \varepsilon_o[\cos(\omega t) + j\sin(\omega t)]$. Since the applied strain corresponds to the real part of the complex strain, the response will be the real part of the resulting complex stress; that is $\sigma(t) = Re\{\sigma^*(j\omega t)\}$. *Putting* $G(t) = Ge^{-t/\tau}$ in (4.3) produces the complex modulus, after integrating and reordering, as:

$$G^*(j\omega) = j\omega \int_0^\infty e^{-j\omega t} G e^{-t/\tau} dt = \frac{j\omega}{1+j\omega\tau} G \equiv G \frac{(\omega\tau)^2 + j(\omega\tau)}{1+(\omega\tau)^2}.$$

Taking the complex modulus and the complex applied strain into (4.2) yields the complex stress as:

$$\sigma^*(j\omega t) = G^*(j\omega)\varepsilon^*(j\omega t) = \frac{G\varepsilon_o}{1+(\omega\tau)^2} [(\omega\tau)^2 + j(\omega\tau)][\cos(\omega t) + j\sin(\omega t)].$$

Or, in full:

$$\sigma^*(j\omega t) = \frac{G\varepsilon_o}{1+(\omega\tau)^2} \{ [(\omega\tau)^2 \cos(\omega t) - (\omega\tau)\sin(\omega t)] + j[(\omega\tau)\cos(\omega t) + (\omega\tau)^2 \sin(\omega t)] \}.$$

The steady-state response to the actual applied loading, $\varepsilon(t) = Re\{\varepsilon^*\}$, then becomes: $\sigma(t) = Re\{\sigma^*(j\omega t)\} = \frac{G\varepsilon_o}{1+(\omega\tau)^2} [(\omega\tau)^2 \cos(\omega t) - (\omega\tau)\sin(\omega t)]$

It is worth pointing out that because of the linearity of the Fourier transform, the steady-state response of a viscoelastic system to a sum of excitations is equal to the sum of the steady-state responses to the individual excitations. The simplest way to proceed in this case is to establish the steady-state response of the system to each cosine or sine component of the excitation and add the corresponding real or imaginary part of the complex responses.

Example 4.4 Determine the steady-state response of a viscoelastic material having shear relaxation modulus $G(t)$, to the harmonic excitation: $\varepsilon(t) = \varepsilon_{1o}\sin(\omega_1 t) + \varepsilon_{2o}\cos(\omega_2 t)$.
Solution:
The solution may be readily obtained as the sum of the steady-state response to the two strain histories, $\varepsilon_1^*(j\omega_1 t) = \varepsilon_{1o}e^{j\omega_1 t}$ and $\varepsilon_2^*(j\omega_2 t) = \varepsilon_{2o}e^{j\omega_2 t}$, representing the actual excitations $\varepsilon_1(t) = \varepsilon_{1o}sin(\omega_1 t)$ and $\varepsilon_2(t) = \varepsilon_{2o}cos(\omega_2 t)$. Using complex algebra, let $M^* = M' + jM''$, and calculate the steady-state responses $\sigma_1^* = M^*(j\omega_1)\varepsilon_1^*(j\omega_1 t)$ and $\sigma_2^* = M^*(j\omega_2)\varepsilon_2^*(j\omega_2 t)$. Then, take the imaginary part of σ_1^*—because of the sine function—and the real part of σ_2^*—because of the cosine function—to arrive at the result:

$$\sigma(t) = [G'(\omega_1)\sin(\omega_1 t) + G''(\omega_1)\cos(\omega_1 t)\varepsilon_{1o}]$$
$$+ [G'(\omega_2)\cos(\omega_2 t) - G''(\omega_2)\sin(\omega_2 t)\varepsilon_{2o}]$$

4.3 Steady-State Constitutive Equations from Differential Constitutive Equations

As indicated in the introduction, the constitutive equations for steady-state conditions can also be derived from constitutive equations in differential form. This is done in the same manner as before, by:

- Introducing a complex excitation, $\varepsilon^* = \varepsilon_o e^{j\omega t}$, say, to represent the controlled variable

- Invoking the closed-cycle condition to take the response as harmonic too, with the same frequency as the excitation, but out of phase with it: $\sigma^* = e^{j(\omega t + \delta_M)}$.
- Taking the excitation and response into differential constitutive Eq. (3.9), repeated here, for reference:

$$p_o\sigma + p_1\frac{d^1}{dt^1}\sigma + \cdots + p_m\frac{d^m}{dt^m}\sigma = q_o\varepsilon + q_1\frac{d^1}{dt^1}\varepsilon + \cdots + q_n\frac{d^n}{dt^n}\varepsilon \qquad \text{(a)}$$

Proceeding as indicated, performing the required operations and collecting terms, yields:

$$\left[p_o + p_1(j\omega)^1 + \cdots + p_m(j\omega)^m\right]\sigma^* = \left[q_o + q_1(j\omega)^1 + \cdots + q_n(j\omega)^n\right]\varepsilon^* \qquad \text{(4.30a)}$$

More succinctly:

$$P^*\sigma^* = Q^*\varepsilon^* \qquad \text{(4.30b)}$$

With the definitions:

$$P^* \equiv p_o + p_1(j\omega)^1 + \cdots + p_m(j\omega)^m \qquad \text{(4.31a)}$$

$$Q^* \equiv q_o + q_1(j\omega)^1 + \cdots + q_n(j\omega)^n \qquad \text{(4.31b)}$$

Since (4.30b) is an algebraic relation involving complex numbers, it may be rewritten in either of the following forms:

$$\sigma^* = \frac{Q^*}{P^*}\varepsilon^* \qquad \text{(4.32a)}$$

$$\varepsilon^* = \frac{P^*}{Q^*}\sigma^* \qquad \text{(4.32b)}$$

Proceeding as in the previous section yields the complex material property functions, M^* and C^*, and the relationship between them:

$$\sigma^* \equiv M^*\varepsilon^* \Rightarrow M^* = \frac{Q^*}{P^*} \qquad \text{(4.33a)}$$

$$\varepsilon^* \equiv C^*\sigma^* \Rightarrow C^* = \frac{P^*}{Q^*} \qquad \text{(4.33b)}$$

$$M^*C^* = 1 \qquad \text{(4.34)}$$

The practical implication of expressions (4.33) is that to determine the complex modulus or the complex compliance of any given rheological model, it suffices to evaluate P^* and Q^*, in accordance with (4.31), and use complex algebra to remove the complex quantities from the resulting denominator. The following example illustrates this point.

Example 4.5 Derive the material property functions of a Maxwell fluid for steady-state conditions.

Solution:

Use (3.21b): $\frac{d\sigma}{dt} + \frac{1}{\tau_r}\sigma = E\frac{d}{dt}f(t)$, and identify: $p_o = (1/\tau_r)$, $p_1 = 1$, $q_o = 0$, $q_1 = E$. With them, obtain $P^* = (1/\tau_r)(1 + j\omega\tau_r)$, and $Q^* = j\omega E$. Then, apply Eq. (4.33a) to establish $M^* = \frac{Ej\omega\tau_r}{(1+j\omega\tau_r)} \equiv \frac{E[(\omega\tau_r)^2+j(\omega\tau_r)]}{1+(\omega\tau_r)^2}$. The complex compliance, C^* may be determined directly from (4.33b), as: $C^* = \frac{1+j\omega\tau}{E_oj\omega\tau} \equiv \frac{1}{E_oj\omega\tau} + \frac{1}{E_o} \equiv \frac{1}{E_o} - j\frac{1/E_o}{\omega\tau}$.

It is worth pointing out that in accordance with (4.3) and (4.13) the complex material property functions of rheological operators may be obtained directly from their relaxation or creep compliance functions.

Example 4.6 Use the relaxation modulus of the standard linear solid to derive its complex modulus M^*.

Solution:

The relaxation modulus of the standard or three-parameter solid was obtained in (3.36b), as: $M(t) = \frac{q_o}{p_o} + \left[\frac{q_1}{p_1} - \frac{q_o}{p_o}\right]e^{-t/\tau_r}$; $\tau_r \equiv \frac{p_1}{p_o}$; where the index r has been appended to the relaxation time, for clarity. Taking this into (4.3) and performing the integrations yields: $M^* = \frac{q_o}{p_o} + \frac{\left[\frac{q_1}{p_1}-\frac{q_o}{p_o}\right]\left[(\omega\tau_r)^2+j\omega\tau_r\right]}{1+(\omega\tau_r)^2}$.

Alternatively, using that $q_o/p_o \equiv M_\infty \equiv M_e$, $q_1/p_1 \equiv M_o \equiv M_g$ (Sects. 3.3.2, 3.3.4):

$$M^* = M_\infty + \frac{[M_o - M_\infty]\left[(\omega\tau_r)^2 + j\omega\tau_r\right]}{1 + (\omega\tau_r)^2}$$

$$= \frac{\left[M_\infty + M_o(\omega\tau_r)^2\right]}{1 + (\omega\tau_r)^2} + j\frac{[M_o - M_\infty]\omega\tau_r}{1 + (\omega\tau_r)^2}$$

In similar fashion, using (3.37b): $C(t) = \frac{p_o}{q_o} + \left[\frac{p_1}{q_1} - \frac{p_o}{q_o}\right]e^{-t/\tau_c}$; $\tau_c \equiv \frac{q_1}{q_o}$ and applying (4.13), leads, after some algebra to: $C^* = \frac{[C_\infty+C_o(\omega\tau_c)^2]}{1+(\omega\tau_c)^2} + j\frac{[C_o-C_\infty]\omega\tau_c}{1+(\omega\tau_c)^2}$

4.4   Limiting Behavior of Complex Property Functions

Since the complex modulus and compliance are related to their time-domain counterparts through expressions such as (4.3), it is reasonable to expect that their asymptotic values $M^*(\infty)$ and $M(\infty)$, or $M^*(0)$ and $M(0)$, etc., are related in some manner. That such values are related may be seen from a physical perspective. For instance, since the glassy response ($t = 0$) is elastic and $M(0)$ is a constant, and since ω is proportional to $1/t$, one should expect that $M^*(\infty) = M(0)$.

To fix ideas, we obtain the relationships between the asymptotic values of the moduli and use (4.3) for that effect. Before applying the limiting process, we integrate the equation by parts:

$$M^*(j\omega) = M(0) + \int_0^\infty e^{-j\omega t} \frac{\partial}{\partial t} M(t) dt \qquad \text{(a)}$$

Taking the limit of this expression as $\omega \to \infty$ confirms the physical argument made before:

$$\lim_{\omega \to \infty} M^*(j\omega) = M(0) \qquad (4.35)$$

Or, using the Cartesian components and equating real and imaginary parts, respectively:

$$M'(\infty) = M(0); \quad M''(\infty) = 0 \qquad (4.36)$$

To obtain the limit as $\omega \to 0$, we introduce the change of scale $\eta = \omega t$, and write:

$$M^*(j\omega) = M(0) + \int_0^\infty e^{-j\eta} \frac{\partial}{\partial \eta} M(\frac{\eta}{\omega}) d\eta \qquad \text{(b)}$$

From which, follows:

$$\lim_{\eta \to 0} M^*(j\omega) = M(0) + \int_0^\infty \frac{\partial}{\partial \eta} M(\frac{\eta}{\omega}) d\eta = M(0) + M(\infty) - M(0) \qquad \text{(c)}$$

Finally, since $\omega \to 0$ as $\eta \to 0$:

$$\lim_{\omega \to 0} M^*(j\omega) = M(\infty) \qquad (4.37)$$

Or, equivalently, in terms of the real and imaginary components:

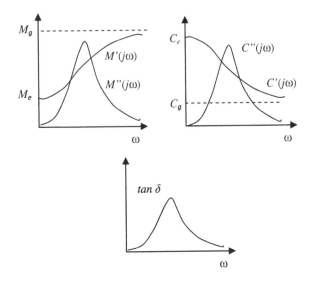

Fig. 4.2 Real and imaginary components of the complex property functions of a viscoelastic solid. $M^*(\omega \to 0) = M(t \to \infty) \equiv M_\infty \equiv M_e$; $and\ M^*(\omega \to \infty) = M(t \to 0) \equiv M_o \equiv M_g$. Therefore: $M'(0) = M_e, M''(0) = 0$, and $M'(\infty) = M_g$, $M''(\infty) = 0$. Also: $C^*(\omega \to 0) = 1/M_e = C'(0) = C_e$; thus: $C''(0) = 0$, $and\ C^*(\omega \to \infty) = 1/M_g = C'(\infty) = C_g$; and so, $C''(\infty) = 0$

$$M'(0) = M(\infty); \quad M''(0) = 0 \tag{4.38}$$

Starting with Eq. (4.13) and using arguments entirely similar as with the complex modulus, the following relationships are obtained describing the limiting behavior of the complex compliance. The analogs of (4.35) to (4.38) are, thus:

$$\lim_{\omega \to \infty} C^*(j\omega) = C(0) \tag{4.39}$$

$$C'(\infty) = C(0); \quad C''(\infty) = 0 \tag{4.40}$$

$$\lim_{\omega \to 0} C^*(j\omega) = C(\infty) \tag{4.41}$$

$$C'(0) = C(\infty); \quad C''(0) = 0 \tag{4.42}$$

Expressions (4.35) to (4.42) indicate that the glassy and equilibrium complex moduli and compliances are elastic—as they should be—in agreement with the properties of the relaxation and creep functions, discussed in Chap. 2. Typical graphs for the real and imaginary components of the complex property functions, including the loss tangent are as shown in Fig. 4.2 for a typical viscoelastic solid.

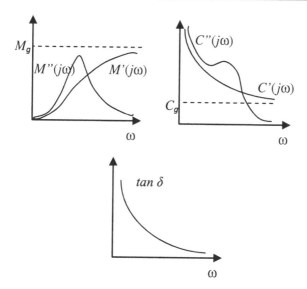

Fig. 4.3 Real and imaginary components of the complex property functions of a typical viscoelastic fluid $M^*(\omega \to 0) = M(t \to \infty) = 0$; and $M^*(\omega \to \infty) = M(t \to 0) \equiv M_0 \equiv M_g$. Therefore $M'(0) = 0$; $M''(0) = 0$ and $M'(\infty) = M_g$ and $M''(\infty) = 0$. $C^*(\omega \to 0) = C(t \to \infty) \to \infty$; thus $C' = (0) \infty$; $C''(0) = \infty$ and $C^*(\omega \to \infty) = C(t \to 0) = 1/M_g \equiv C_g = C'(\infty)$; $C''(\infty) = 0$

Regarding viscoelastic fluids, we know, from Chaps. 1 and 2, that the nature of their response to loads is in general indistinguishable from that of viscoelastic solids, except in the long run, because viscoelastic fluids have zero equilibrium modulus. For this reason, in particular, $\tan \delta(0) = M''/M' \to \infty$. Consequently, the graphical forms of the complex material property functions of a typical viscoelastic fluid are as shown in Fig. 4.3.

Quite naturally, the graphical representations of the complex material property functions of a rheological model depend on the specific arrangements of its elements. Because the mechanical response of the standard linear solid to loads exhibits all the characteristics of a viscoelastic solid, it follows that its complex property functions should resemble those shown in Fig. 4.2, whereas those of a fluid should be as in Fig. 4.3.

The complex modulus and compliance of the Maxwell fluid are depicted in Fig. 4.4. The representations in the figure are constructed taking the indicated limits of the results of Example 4.5.

The complex modulus and compliance of a Kelvin solid are displayed in Fig. 4.5, which was constructed using the results of Problem P.4.3.

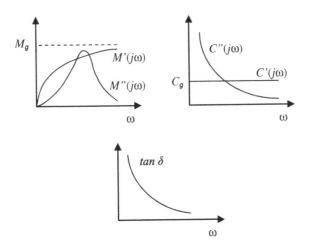

Fig. 4.4 Real and imaginary components of the complex properties of a Maxwell fluid $M^*(\omega \to 0) = 0; M^*(\omega \to \infty) = M_g; C^*(\omega \to 0) = -j\infty; C^*(\omega \to \infty) = 1/M_o \equiv 1/M_g$ Therefore :$M'(0) = 0, M''(0) = 0; M'(\infty) = M_g, M''(\infty) = 0, C'(0) = 0, C''(0) = \infty;$ $C'(\infty) = C_g, C''(\infty) = 0;$ with : $C_g \equiv 1/M_g$

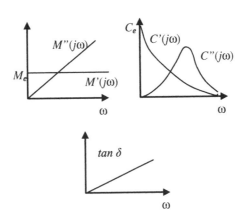

Fig. 4.5 Real and imaginary components of the complex properties of a Kelvin Solid $M^*(\omega \to 0) = M_\infty \equiv M_e; M^*(\omega \to \infty) = j\infty; C^*(\omega \to 0) = 1/M_\infty \equiv C_e; \quad C^*(\omega \to \infty) = 0 \to M'(0) = M_e, M''(0) = 0, M'(\infty) = M''(\infty) = \infty C'(0) = C_e, C''(0) = 0, C'(\infty) = C''(\infty) = 0$

4.5   Energy Dissipation

In Chap. 1, we stated the physical expectation that viscoelastic substances should have some capacity to store energy and some to dissipate it. This section proves that statement by means of a simple experiment.

The ability of a material to store or dissipate energy may be ascertained subjecting a one-dimensional specimen of the material to a full load-unload cycle. Without loss of generality, we consider a uniaxial specimen subjected to the excitation $\Delta l(t) = \Delta l_o \sin(\omega t)$, which divided by the original length of the specimen produces the strain $\varepsilon(t) = \varepsilon_o \sin(\omega t)$. The test is then carried out for a complete cycle, so that $t_d = 2\pi/\omega$. Using Eq. (2.46):

$$W_V|_0^{t_d} = (1/V)W|_0^t = \int_0^{t_d} \sigma \frac{d\varepsilon}{ds}ds \tag{4.43}$$

and noting that the stress corresponding to the applied strain is $\sigma(t) = \sigma_o \sin(\omega t + \delta)$, leads to:

$$W_V|_0^{t_d} = \sigma_o \varepsilon_o \int_0^{2\pi/\omega} \sin(\omega t + \delta)\frac{d}{dt}\sin(\omega t)dt = \sigma_o \varepsilon_o \omega \sin \delta \int_0^{2\pi/\omega} (1 + \cos 2\omega t)dt \tag{a}$$

which integrates to:

$$W_V|_0^{t_d} = \pi \sigma_o \varepsilon_o \sin \delta \tag{4.44}$$

This result indicates that maximum dissipation occurs when $\delta = \pi/2$; which according to the discussions in Chaps. 1, 2 and 3, corresponds to fluid behavior. By contrast, zero or no dissipation, corresponding to solid, elastic response, occurs at $\delta = 0$.

Using (4.8): $\sigma_o = \|M^*\| \varepsilon_o$, and noting that $\| M^* \| \sin \delta = M''$, puts (4.44) in the form:

$$W_V|_0^P = \pi \cdot \varepsilon_o^2 \cdot M''(j\omega) \tag{4.45a}$$

In words: energy dissipation depends on M''; which, for this reason, is termed the loss modulus. By contrast, zero dissipation, and hence fully recoverable elastic response occurs at $\delta = 0$, when: $M^* = M'$. For this reason, M' is called the storage modulus.

Alternatively, substituting $_o = \|C^*\|\sigma_o$, in accordance with (4.20), into (4.45a), and noting that, per the definition in (4.19): $\|C^*\| \sin \delta = C''$

$$W_V|_0^P = \pi \cdot \sigma_o^2 C''(j\omega) \tag{4.45b}$$

The same remarks can be made about C' and C'', relative to energy storage or dissipation, respectively, as were made in relation to M' and M''. Thus, C' is the storage compliance, and C'', the loss compliance. Furthermore, since the total energy dissipated in a cycle has to be positive, in accordance with the closed-cycle condition, expressions (4.45a) show that $M''(j\omega) > 0$ and $C''(j\omega) > 0$. This last inequality is one reason why the complex compliance is written with a negative sign in front of C''.

Example 4.7 Derive an expression for the maximum energy per unit volume dissipated per cycle by a material with constitutive equation of the standard linear solid type.

Solution:

Use the results of Example 4.6: $C^* = \frac{\left[C_\infty + C_o(\omega\tau_c)^2\right]}{1+(\omega\tau_c)^2} + j\frac{[C_o - C_\infty]\omega\tau_c}{1+(\omega\tau_c)^2}$, together with

(4.45b) to write: $\frac{W}{V} = \pi\sigma_o^2 \frac{[C_o - C_\infty]\omega\tau_c}{1+(\omega\tau_c)^2}$; where, $C_\infty \equiv p_o/q_o$, $C_o \equiv p_1/q_1$, and

$\tau_c \equiv q_1/q_o$.

4.6 Problems

P.4.1. Generalize the result of Example 4.2 to the case when the relaxation function is given by a finite sum of exponentials, such as that in Example 4.1.

Answer:

$$G'(j\omega) = G_e + \sum_{i=1}^{n} G_i \frac{(\omega_i\tau_i)^2}{1 + (\omega_i\tau_i)^2}; \quad G''(j\omega) = \sum_{i=1}^{n} G_i \frac{(\omega_i\tau_i)}{1 + (\omega_i\tau_i)^2}$$

Hint: Use the results of Example 1 and follow the approach in Example 2.

P.4.2. Verify the result of Example 4.3 by direct integration of the hereditary integral.

Hint: Start with the integral constitutive equation (2.33a), and apply integration by parts twice to derive the steady-state response.

P.4.3. Derive the material property functions of a Kelvin solid appropriate to steady-state conditions.

Answer:

$$M^* = M_\infty(1 + j\omega\tau_K); \quad C^* = \left(\frac{1}{M_\infty}\right)\frac{(1 + j\omega\tau_K)}{1 + (\omega\tau_K)^2}; \tau_K \equiv \frac{q_1}{p_1}; M_\infty \equiv \frac{q_o}{p_o}$$

Hint: Write (3.16a) in operator form: $p_o\sigma = (q_o + q_1\partial_t)\varepsilon$, and obtain $P^* = p_o$, and $Q^* = q_o + j\omega q_1 = q_o(1 + j\omega\tau_K)$; where $\tau_K \equiv q_1/q_o$ is the relaxation time of the model. Use this and (4.33a) to get $M^* = \frac{q_o}{p_o}(1 + j\omega\tau_K) = M_\infty(1 + j\omega\tau_K)$.

Establish the complex compliance by inverting this relation: $C^* = \frac{1}{M_\infty} \left[\frac{1+j\omega\tau}{1+(\omega\tau)^2} \right]$

P.4.4. Derive the expression for the maximum energy dissipated per cycle by a material with relaxation function of Prony-Dirichlet type, like that in P.4.1, under oscillations of amplitude ε_o.

Answer:

$$\frac{W}{V} = \pi\varepsilon_o^2 \sum_{k=1}^{n} G_k \frac{\omega_k\tau_k}{1+(\omega_k\tau_k)^2}$$

Hint: Use the results of P.4.1 together with (4.45a) to arrive at the desired result.

P.4.5. Calculate the maximum energy dissipated per cycle by a bar of a standard linear solid under a uniaxial cyclic strain of amplitude ε_o?

Answer: $\pi\varepsilon_o^2 \frac{[M_o-M_\infty]\omega\tau_r}{1+(\omega\tau_r)^2}$; where: $M_\infty \equiv q_o/p_o$, $M_o \equiv q_1/p_1$, and $\tau_r \equiv p_1/p_o$.

Hint: Use the results of Example 4.6, that: $M^* = \frac{[M_\infty+M_o(\omega\tau_r)^2]}{1+(\omega\tau_r)^2} + j\frac{[M_o-M_\infty]\omega\tau_r}{1+(\omega\tau_r)^2}$, together with (4.45a) to arrive at the quoted result.

P.4.6. Obtain the complex viscosity of the standard linear solid of Problem P.4.5.

Answer:

$$\eta^*(j\omega) = \frac{[M_o - M_\infty]\tau_r}{1+(\omega\tau_r)^2} - j\frac{\left[M_\infty + M_o(\omega\tau_r)^2\right]}{\omega\left[1+(\omega\tau_r)^2\right]}$$

Hint: Insert the expression for M^* derived in Example 4.6 into Eq. (4.26) and (4.27)for the complex viscosity: $\eta^*(j\omega) = \frac{M^*}{j\omega} = \frac{M''}{\omega} - j\frac{M'}{\omega}$, to arrive at the result.

P.4.7. Calculate the complex compliance for a viscoelastic solid whose creep compliance is $C(t) = C_e - \sum_{i=1}^{n} C_i e^{-t/\tau_i}$.

Answer:

$$C^*(j\omega) = C_e - \sum_{i=1}^{n} C_i \frac{(\omega_i\tau_i)^2}{1+(\omega_i\tau_i)^2} - j\sum_{i=1}^{n} C_i \frac{\omega_i\tau_i}{1+(\omega_i\tau_i)^2}$$

Hint: Insert $C(t)$ into (4.13), performing the indicated integrations and simplifying to obtain the result.

P.4.8. Calculate the work done per cycle of oscillation under a stress $\sigma(t) = \sigma_o \cdot sin(\omega t)$, for a bar made of the material of Problem P.4.7.

Answer:

$$\frac{W}{V} = \pi\sigma_o^2 \sum_{i=1}^{n} C_i \frac{\omega_i\tau_i}{1+(\omega_i\tau_i)^2}$$

Hint: Use (4.45b): $\frac{W}{V} = \pi \cdot \sigma_o^2 \cdot C''(j\omega)$ and the result: $C''(j\omega) = \sum_{i=1}^{n} C_i \frac{\omega_i \tau_i}{1+(\omega_i \tau_i)^2}$,

from Problem P.4.7, to arrive at the desired result.

P.4.9. Derive a general expression for the steady-state response of a linear viscoelastic material to a cyclic stress $\sigma(t) = \sum_{k=1}^{N} \sigma_{ok} \sin(\omega_k t)$.

Answer:

$$\varepsilon(t) = \sum_{k=1}^{N} \sigma_{ok}[C'(j\omega_k)\sin(\omega_k t) - C''(j\omega_k)\cos(\omega_k t)]$$

Hint: The solution may be readily obtained as the sum of the steady-state response to each of the stress histories, $\sigma_k^*(j\omega_k t) = \sigma_{ok}e^{j\omega_k t}$. Proceeding as in Example 4.4, let $C_k^* \equiv C^*(j\omega_k) \equiv C'(j\omega_k) - C''(j\omega_k) \equiv jC_k' - jC_k''$, calculate the steady-state strain responses $\varepsilon_k^* = C_k^*\varepsilon_k^*$ and add them together. Since all components of the excitation are sine functions, the solution sought corresponds to the imaginary part of the total complex strain thus obtained.

P.4.10. Assuming that $p \equiv 2\pi/\omega_1$ and $\omega_k \equiv k\omega_1$, for $k = 1, 2, \ldots, N$, calculate the work done in the period, p, by the stress history of Problem P.4.9 acting on a viscoelastic bar having creep compliance $C(t)$.

Answer:

$$W_V\big|_t^{t+p} = \sum_{k=1}^{N} \pi \sigma_{ok}^2 C''(j\omega_k), \quad p \equiv 2\pi/\omega_1, \quad \omega_k \equiv k\omega_1, \quad j \equiv \sqrt{-1}$$

Hint: Take the stress history $\sigma(t) = \sum_{k=1}^{N} \sigma_{ok} \sin(\omega_k t)$ given in Problem P.4.9, and the derivative $\frac{d}{dt}\varepsilon(t) = \sum_{k=1}^{N} \sigma_{ok}\omega_k[C'(j\omega_k)\cos(\omega_k t) + C''(j\omega_k)\sin(\omega_k t)]$ of the strain response calculated there, into (4.43). Perform the indicated algebra, and use the orthogonality of the sinusoidal functions over an interval spanning the fundamental period (see Appendix A) to arrive at the result.

References

1. J.D. Ferry, Viscoelastic Properties of Polymers (Wiley, New York, 1980), pp. 11–14
2. W. Flügge, Viscoelasticity, 2nd edn (Springer, New York, 1975), pp. 95–120
3. A.S. Wineman, K.R. Rajagopal, Mechanical Response of Polymers, an Introduction (Cambridge University Press, Cambridge, 2000), pp. 115–147
4. R. M. Christensen, Theory of viscoelasticity, 2nd edn (Dover, New York, 1982), pp. 21–26

Structural Mechanics

5

Abstract

This chapter is devoted to structural mechanics, developing the theories of bending, torsion, and buckling of straight bars, and presenting a detailed account of vibration of single-degree-of-freedom viscoelastic systems, including vibration isolation. A balanced treatment is given to stress–strain equations of integral and differential types, and to stress–strain relations in complex-variable form, which are applicable to steady-state response to oscillatory loading. All equations in this chapter are developed from first principles, without presuming previous knowledge of the subject matter being presented. This approach is followed for two reasons: first, because it is necessary for readers without a formal training in mechanics of materials; and secondly, because it provides the reader—even one with formal training in classical engineering—with a method to follow when the use of popular shortcuts, like the integral transform techniques, might be questionable or unclear.

Keywords

Static · Bending · Torsion · Buckling · Navier · Hereditary · Integral · Differential · Steady-state · Euler · Critical · Creep · Spring · Correspondence · Mass · Free · Forced · Vibration · Amplification · Transmissibility

5.1 Introduction

Previous chapters have examined the various forms available of the stress–strain laws for uniaxial conditions, using a straight bar of viscoelastic material as the physical specimen to which these laws applied. This chapter expands the one-dimensional applications to examine the quasi-static[1] bending, torsion, and buckling of structural members made of homogeneous, linear, isotropic viscoelastic materials. Free and forced vibrations of one-dimensional viscoelastic spring-mass systems are also examined in detail, including the important topic of vibration isolation. As will be explained, the case of free vibrations is considerably more complicated than that of forced vibration, to which the methods of Chap. 4 apply directly. For free vibrations, then, only the solution outline is given. For completeness of presentation as well as for the benefit of readers without formal training in mechanics of materials, all topics are treated in great detail.

As will be seen, the relations derived for viscoelastic materials have the same forms as their elastic counterparts. The kinematic relations and the equilibrium equations are established for materials whose stress–strain law is in integral as well as in differential operator form. The approach used to derive the equilibrium equations is as follows (c.f. Chaps. 2 and 3):

1. Establish the equilibrium equation relating the pertinent mechanical element,[2] F, to the corresponding stress, σ:

$$F(x,t) = A(x)\sigma(x,t) \tag{a}$$

2. Introduce the stress–strain law:
   a. For materials of integral type, $\sigma(x,t) = E(t-s) * d\varepsilon(s)$ (c.f. Chap. 2).[3] Thus,

$$F(x,t) = A(x)E(t-s) * d\varepsilon(s) \tag{b}$$

   b. For materials of differential type apply operator $P[\cdot]$ to both sides of the equilibrium equation, and use that $P[\sigma] = Q[\varepsilon]$, to write (c.f. Chap. 3)

$$P[F(x,t)] = A(x)Q[\varepsilon] \tag{c}$$

3. Use the pertinent kinematic relation to replace the generalized strain, ε, with the corresponding generalized displacement to arrive at the desired form.

---

[1] That is, such that the external stimuli vary sufficiently slowly that no significant inertial effects develop.

[2] The mechanical elements that may act on a bar's cross section are two bending moments, one torsional moment, one axial force, and two shear forces.

[3] In this chapter, to avoid confusion, the letter M is used to denote the moment of the external forces, while E will denote relaxation modulus. As in previous chapters, C will still denote compliance.

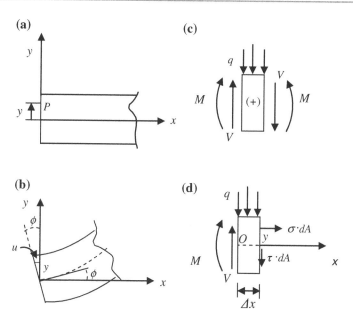

Fig. 5.1 Bending of a straight beam. Kinematics of deformation. **a** Initial geometry. **b** Deformed geometry. **c** Sign convention. **d** Force resultant at x

5.2 Bending

This section examines the mechanical response of initially straight viscoelastic beams subjected to transverse loads only.[4] The derivations assume that the lateral deflection and in-plane rotation of the beam are small; and that, in accordance to Navier's assumptions, plane sections initially normal to the axis of the beam remain plane and normal to the beam's axis during deformation [1].

Regarding geometry, the axis of the beam is taken along the x-axis; the y- and z-axes are assumed principal centroidal axes of inertia of the cross section; the y-axis directed upward, and the z-axis defined by the right-hand-screw rule, as shown in Fig. 5.1. To keep the presentation simple, all transverse loads will be assumed to act in the x–y plane, along the y-direction.

As can be seen from Fig. 5.1b, the axial displacement, $u(x,y,z,t)$, of the bar depends only on the axial, x, and transverse, y, coordinates. For this reason, in the sequel, we omit the out-of-plane coordinate, z. Hence,

$$u(x, y, z, t) \equiv u(x, y, t) = -y\tan\phi, \forall z \in A(x) \qquad (5.1)$$

---

[4] Generalization to beam-columns, which include axial loads, is straightforward and is left as an exercise.

For small deflections, $v(x,t)$, and rotations, φ, of the bar's axis: tan $\varphi \approx \varphi = \partial v/\partial x$; thus,

$$u(x,y,t) \approx -y\phi = -y\frac{\partial}{\partial x}v(x,t), \forall z \in A(x) \tag{5.2}$$

Introducing the axial strain in the bar—i.e., the change in length per unit original length—leads to:

$$\varepsilon(x,y,t) = \frac{\partial}{\partial x}u(x,y,t) = -y\frac{\partial^2}{\partial x^2}v(x,t) \equiv -yv''(x,t), \forall z \in A(x) \tag{5.3}$$

Figure 5.1d shows the internal stresses σ and τ which develop on a generic cross section to balance the externally applied loads. Taking moments with respect to point O, located on the generic section at station x, we write

$$M(x,t) + q(x,t)dx \cdot \frac{dx}{2} + \int_A y\sigma(x,y,t)dA + \left[\int_A \tau(x,y,t)dA\right] \cdot dx = 0 \tag{a}$$

In the limit as $dx \to 0$, this expression produces the resultant moment acting on the cross section at a generic axial station, x, as

$$M(x,t) = -\int_A \sigma(x,y,t)y dA \tag{5.4}$$

The following relations result from force and moment equilibrium taken between any two stations separated an infinitesimal distance Δx along a beam's axis, as depicted in Fig. 5.2, and do not contain material properties.

Indeed,

a. Moment equilibrium about the bottom left-hand corner of the elemental length of beam leads to

$$[V + \Delta V]\Delta x - [M + \Delta M] + M + [q\Delta x]\frac{\Delta x}{2} = 0 \tag{b}$$

Fig. 5.2 Force and moment equilibrium of a differential beam element

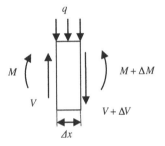

Neglecting terms of order higher than the first,

$$V(x,t) = \frac{\partial}{\partial x} M(x,t) \tag{c}$$

Just like for elastic solids, this expression is independent of material properties.
b. Equilibrium of forces requires that $V - q\Delta x - [V + \Delta V] = 0$, which, neglecting higher-order terms yields:

$$q(x,t) = -\frac{\partial}{\partial x} V(x,t) \tag{d}$$

As for elastic materials, this expression is also independent of material properties.

5.2.1 Hereditary Integral Models

Introducing the strain–stress form $\varepsilon(t) = C(t - \tau) * d\sigma(\tau)$ of the constitutive equation into (5.3), and convolving the result with the relaxation modulus, leads to (c.f. Chap. 2)[5]

$$\sigma(x,y,t) = -yE(t - \tau) * \partial v''(x,\tau) \tag{5.5}$$

Combining (5.4) and (5.5), and separating the spatial and time operations produces

$$M(x,t) = \left[\int_A y^2 dA \right] E(t - \tau) * \partial v''(x,\tau) \tag{d}$$

Or, more succinctly,

$$M(x,t) = I(x)E(t - \tau) * \partial v''(x,\tau) \tag{5.6}$$

This is the relaxation integral analogue of the elastic expression $M = I_z \cdot E \cdot v''$, in which $I_z(x)$ represents the second moment of area—also called area moment of inertia—of the beam's cross section at station x.

To solve for the deflection function, $v(x,t)$, of the beam, (5.6) is convolved with the tensile compliance function, $C(t)$, of the beam's material (c.f. Chap. 2). Proceeding thus, and rearranging,

---

[5] Note that the same result may be reached by taking the kinematic relation (5.3) into the stress–strain constitutive equation $\sigma(t) = E(t - \tau) * d\varepsilon(\tau)$.

$$\frac{\partial^2 v}{\partial x^2} = \frac{1}{I} C(t - \tau) * \partial M(\tau) \tag{5.7a}$$

This expression is also often written in terms of the rotation angle, $\varphi = \partial v / \partial x$, as

$$\frac{\partial \varphi}{\partial x} = \frac{1}{I} C(t - \tau) * dM(\tau) \tag{5.7b}$$

The relation between the normal stress and the external moment is obtained combining (5.5) and (5.6). This is accomplished by first rewriting (5.5), as

$$E(t - \tau) * \partial v''(x, \tau) = -\frac{\sigma(x, y, t)}{y} \tag{e}$$

Inserting this result on the right-hand side of (5.6), and rearranging, leads to

$$\sigma(x, y, t) = -\frac{M(x, t)}{I(x)} y; \quad \forall z \in A(x) \tag{5.8}$$

This expression is the same as that for an elastic beam in bending; as it should be, since the bending stress is independent of material constitution.

5.2.2 Differential Operator Models

Proceeding as suggested in the introduction, we apply the operator Q to the strain–displacement relation (5.3) and use that $P[\sigma] = Q[\varepsilon]$ to write

$$P[\sigma(x, y, t)] = Q[\varepsilon(x, y, \tau)] \equiv -yQ[v''(x, t)]; \forall z \in A(x) \tag{5.9}$$

Applying the operator P to expression (5.4), combining the result with (5.9), and separating the spatial and temporal parts, leads to the counterparts of (5.6) and (5.7a, b), respectively,

$$P[M(x, t)] = I(x)Q[v''(x, t)] \tag{5.10}$$

$$Q[v''(x, t)] = \frac{P[M(x, t)]}{I(x)} \tag{5.11a}$$

$$Q\left[\frac{\partial}{\partial x} \varphi(x, t)\right] = \frac{P[M(x, t)]}{I(x)} \tag{5.11b}$$

Fig. 5.3 Example 5.1

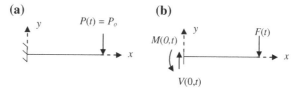

To derive the counterpart of (5.8), first rewrite (5.9) as $Q[v''(x,t)] = -\frac{P[\sigma(x,y,t)]}{y}$. Then, combine the result with (5.10) to get $P[\sigma(x,y,t)] = -\frac{P[M(x,t)]}{I(x)}y$. Applying the operator P^{-1} to both sides of this expression leads to the elastic form in (5.8).

Example 5.1 A cantilevered beam of length L and uniform cross section with moment of inertia I is made of a viscoelastic material with tensile relaxation modulus $E(t)$ and is loaded as shown in Fig. 5.2. Determine the deflection at its tip. Solution:

As indicated in Fig. 5.3, force and moment equilibrium—as in the statics of elastic solids—produce the reactions $V(0,t) = F(t)$ and $M(0,t) = -LF(t)$, on the beam's support. Using these results and the sign convention, shown in Fig. 5.1, produces the moment $M(x,t) = -LF(t) + F(t)x$, at station x. Taking this into (5.7a), integrating the resulting differential equation twice using that $v(0,t) = v'(0,t) = 0$, and inserting $F(t)$, yields, after some algebra, $v(x,t) = -\frac{L^3}{3I}\left[\frac{3}{2}\left(\frac{x}{L}\right)^2 - \frac{1}{2}\left(\frac{x}{L}\right)^3\right]C(t-\tau) * dF(\tau)$. The value of this expression at $x = L$, gives the result $v(L,t) = -\frac{L^3}{3I}C(t-\tau) * dF(\tau)$.

5.2.3 Models for Steady-State Oscillations

The interest here is on the steady-state lateral deflections of straight viscoelastic beams of uniform cross section, which are made of materials of either integral or differential type and are subjected to cyclically varying loads of frequency, ω. To develop the appropriate relations, we take the complex forms $M^*(x, j\omega t) = M_o(x)e^{j\omega t}$ and $v^*(x, j\omega t) = v_o(x)e^{j(\omega t+\delta)}$, of the cyclic moment and the bending deflection, respectively, and insert them into (5.7a) for materials of hereditary integral type, and (5.11a) for materials of differential type, proceeding as in Chap. 4.

For materials of integral type, convolve (5.7a) with the tensile relaxation modulus E, say, to cast it in the form: $M(t) = IE(t-\tau) * \partial v''(x,\tau)$. Then, replace M and v with their complex forms and proceed as in Chap. 4 to get[6]

---

[6] Here $E^*(j\omega) \equiv (j\omega)\int_0^\infty e^{-j\omega t}E(t)dt \equiv (j\omega)\mathcal{F}[E(t)]$; where $\mathcal{F}[E(t)]$ is the Fourier transform of $E(t)$.

Fig. 5.4 Torsion of a
straight bar of circular cross
section. Kinematics of
deformation

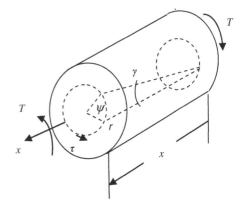

$$M^*(j\omega t) = IE^*(j\omega)\frac{d^2}{dx^2}v^*(x, j\omega t) \tag{5.12a}$$

In similar fashion, inserting M^* and v^* into (5.11a), and rearranging the terms, leads to[7]

$$M^*(j\omega t) = I\frac{Q(j\omega)}{P(j\omega)}\frac{\partial^2}{\partial x^2}v^*(x, j\omega t) = IE^*(j\omega)\frac{\partial^2}{\partial x^2}v^*(x, j\omega t) \tag{5.12b}$$

The expression on the far right uses (4.33) with $E^* = \frac{Q^*}{P^*}$, as the corresponding modulus.

5.3 Torsion

The torsional response of an initially straight bar of circular cross section is considered in what follows. Because of mechanical and geometric symmetry, a straight circular bar subjected to pure torsion deforms in such a way that initially plane cross sections, which are normal to the bar's axis before deformation, rotate without distortion and remain plane and normal to the bar's axis after deformation [2]. Therefore, any straight line initially parallel to the axis of the bar and located at any radius from the bar's axis will remain straight during deformation. As a consequence, the shear strain, γ, at any radial distance, r, from the center of the cross section of the bar must be proportional to r, as indicated in Fig. 5.4.

The kinematics of deformation in this case leads to the following relation between the angle of twist, ψ, and the shear strain, γ:

$$\gamma(r, \varphi, t) \cdot x = r \cdot \psi(x, t) \tag{5.13}$$

---

[7] As shown in Chap. 4, $Q(j\omega)/P(j\omega) = (j\omega) \cdot \mathcal{F}[E(t)] \equiv E^*(j\omega)$.

Differentiating this expression with respect to x, and introducing the angle of twist per unit length, θ, as

$$\theta(x,t) \equiv \frac{d\psi}{dx} \tag{5.14}$$

We arrive at the relation:

$$\gamma(r,\varphi,x,t) = r\frac{d\psi}{dx} \equiv r\theta(x,t) \tag{5.15}$$

Moment equilibrium at axial station, z requires that

$$T(x,t) = \int_A r\tau(r,x,t)dA \tag{5.16}$$

Next, the kinematic and equilibrium relations in (5.15) and (5.16) will be combined with the constitutive equations in hereditary and differential operator forms.

5.3.1 Hereditary Integral Models

Use (5.13) and the convolution form of the constitutive equations for a linear homogeneous viscoelastic material with shear relaxation modulus $G(t)$: $\tau = G * d\gamma$, to write (c.f. Chap. 2):

$$\tau(r,xt) = G(t-\tau) * \partial\gamma(r,x,\tau) = rG(t-\tau) * \partial\theta(x,\tau) \tag{5.17}$$

Combining this with the equilibrium Eq. (5.16) yields

$$T(x,t) = [\int_A r^2 dA]G(t-\tau) * \partial\theta(x,\tau) \equiv J(x)G(t-\tau) * \partial\theta(x,\tau) \tag{5.18}$$

This is the viscoelastic equivalent of the elastic relation $T = J \cdot G \cdot \theta$; and, just as in the elastic case, $J(x)$ represents the polar moment of inertia of the cross section at station, z.

If the twist per unit length is the prescribed loading, (5.18) may be inverted by convolving it with the shear creep compliance function C_G to read:

$$\theta(x,t) \equiv \frac{\partial}{\partial z}\psi(x,t) = \frac{1}{J(x)}C_G(t-\tau) * dT(x,\tau) \tag{5.19a}$$

This expression is also often written in terms of the angle of twist, ψ, using (5.14), as

$$\frac{\partial \psi}{\partial x} = \frac{1}{J(x)} C_G(t - \tau) * \partial T(x, \tau) \tag{5.19b}$$

The relationship between shear stress and external torque is obtained by replacing the expression $G * d\theta = \tau/r$, obtained from (5.18), into (5.19a). The result is identical to that for the torsion of an elastic bar:

$$\tau(r, x, t) = \frac{T(r, t)r}{J(x)} \tag{5.20}$$

As with the bending of straight beams, the stress is independent of material constitution.

5.3.2 Differential Operator Models

The stress–strain relations governing the torsion of bars of material of differential type are derived using that $P[\tau] = Q[\gamma]$ with expression (5.15). Thus,

$$P[\tau(r, x, t)] = Q[\gamma(r, x, \tau)] \equiv Q[r\theta(x, t)] \equiv rQ[\theta(x, t)] \tag{5.21}$$

The relationship between torque and angle of twist per unit length is obtained by applying operator $P[\cdot]$ to both sides of (5.16), and using (5.21):

$$P[T(z, t)] = \int_A rP[\tau(r, x, t)]dA \equiv \int_A r^2 Q[\theta(x, t)]dA \tag{a}$$

Performing the integration, leads to the final form in terms of the second polar moment of area—polar moment of inertia—J of the cross section at axial station x.

$$P[T(x, t)] = J(x)Q[\theta(x, t)] \tag{5.22a}$$

And, in terms of the total twist, ψ:

$$P[T(x, t)] = J(x)Q[\frac{\partial}{\partial x}\psi(x, t)] \tag{5.22b}$$

Finally, expression (5.20), which shows that the shear stress is independent of material properties, may be derived taking $Q[\theta] = P[\tau]/r$, from (5.21), into (5.22a) and rearranging to get $P[\tau] = \frac{P[T]}{J(x)}r$, which, after integration—i.e., after application of the inverse operator P^{-1} to both sides, yields (5.20).

Example 5.2 A straight bar with polar moment of inertia J, and length L, is subjected to torsional moments $\pm M(t)$ at its ends. Derive an expression for the total angle of twist if the bar's material is a standard linear solid. Determine the rotation after t_o units of time if the applied torque is the step function $M(t) = M_o \cdot H(t)$.

Solution:

The solution sought may be obtained solving (5.22a) for $\theta = \psi/L$. To do this, first set $Q[\psi] = LP[M]/J$, and then insert the operators $P = p_o + p_1 \partial_t$ and $Q = q_o + q_1 \partial_t$ for the material at hand (c.f. Chap. 3), as well as the controlled variable, M. This leads to $\frac{d\psi}{dt} + \frac{1}{\tau}\psi = \frac{L}{J}\left[\frac{p_o}{q_1}M + \frac{p_1}{q_1}\frac{dM}{dt}\right]$, with $\tau = q_1/q_o$. This first-order differential equation has an integrating factor $u = e^{t/\tau}$, and thus, $d\left[\psi e^{t/\tau}\right] = \frac{Le^{t/\tau}}{J}\left[\frac{p_o}{q_1}M + \frac{p_1}{q_1}\frac{dM}{dt}\right]dt'$ [c.f. Appendix A]. Assuming $\psi(0) = 0$, $\psi(t)e^{t/\tau} = \frac{L}{J}\int_0^t e^{t'/\tau}\left[\frac{p_o}{q_1}M(t') + \frac{p_1}{q_1}\frac{dM}{dt'}\right]dt'$. Using $M(t) = M_o H(t)$, at $t = t_o$, leads to $\psi(t_o) = \frac{M_o L}{J}\left[\frac{p_o}{q_0} + \left(\frac{p_1}{q_1} - \frac{p_o}{q_0}\right)e^{-t_o/\tau}\right]$.

5.3.3 Models for Steady-State Oscillations

The interest here is on the steady-state torsional deflections of straight viscoelastic beams of uniform circular cross section, made of materials of either integral or differential type, subjected to cyclically varying loads of frequency, ω. Following the same procedure as was used to develop the steady-state equations for beam bending, represent the complex torque and rotation as $T^*(x, j\omega t) = T_o(x)e^{j\omega t}$ and $\psi^*(x, j\omega t) = \psi_o(x)e^{j(\omega t+\delta)}$, respectively, and insert them into (5.19a) for hereditary integral type materials, and (5.22a, b) for materials of differential type, and proceed as in Chap. 4.

For materials of integral type, convolve (5.19b) with the shear relaxation modulus G, say, to cast it in the form: $T(t) = JG(t - \tau) * d\frac{d\psi}{dx}(x, \tau)$. Then, replace T and ψ with their complex forms and proceed as in Chap. 4 to get

$$T^*(j\omega t) = JG^*(j\omega)\frac{\partial}{\partial x}\psi^*(x, j\omega t) \qquad (5.23a)$$

In similar fashion, inserting T^* and ψ^* into (5.22b), and rearranging the terms, leads to

$$T^*(j\omega t) = J\frac{Q(j\omega)}{P(j\omega)}\frac{\partial}{\partial x}\psi^*(x, j\omega t) = JG^*(j\omega)\frac{\partial}{\partial x}\psi^*(x, j\omega t) \qquad (5.23b)$$

The expression on the far right uses (4.33a), with $G^* = \frac{Q^*}{P^*}$, as the corresponding modulus.

5.4   Column Buckling

This section discusses the stability of an initially straight axially compressed viscoelastic bar of constant cross section. The axial load is assumed independent of position along the bar's axis. As will be shown, under constant sustained loading, a viscoelastic column may remain stable indefinitely, become instantaneously unstable, or turn unstable only after enough time has elapsed since the load was applied—a phenomenon referred to as creep buckling. A bar made of an elastic material, by contrast, can only either remain stable, or become unstable instantly.

The critical load of a compressed bar is that at which its initial configuration ceases to be stable. An unstable configuration is an equilibrium configuration that may differ only very slightly from the initial, static equilibrium configuration. For this reason, to determine the critical load of a column, one must find the load under which equilibrium is possible in a configuration that is lightly different from the undistorted one [3].

The pin-ended column of Fig. 5.5 will be used to develop the subject matter. In so doing, it will be assumed that the axial load is applied in some time-dependent manner until it reaches its target value, N_o, which will be held constant thereafter. Using the free-body diagrams in the figure to establish equilibrium in the slightly bent configuration yields

$$N(t) \cdot v(x,t) = M(x,t)$$

$$v(0,t) = 0 \tag{5.24}$$

$$v(l,t) = 0$$

Fig. 5.5 Equilibrium of pin-ended column in slightly bent configuration. **a** Initial configuration. **b** Equilibrium in a slightly deformed state

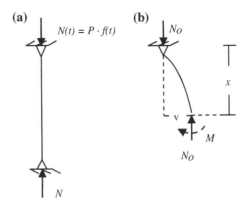

Fig. 5.6 Temporal
dependence of axial load and
lateral deflection of an
initially straight viscoelastic
column

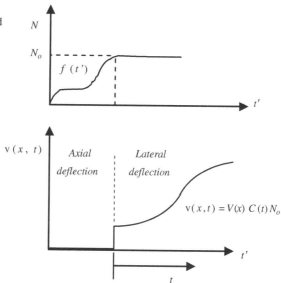

The response of the bar prior to the load reaching its target value is immaterial
for the present discussion. Hence, the origin of the time scale is shifted to the time
when the target load N_o is reached. Under these conditions, the lateral deflection is
controlled by the material's creep function, so that $v(x,t) = V(x)C(t)N_o$, as indicated in Fig. 5.6 (c.f. Chaps. 2 and 3), in which $V(x)$ represents the magnitude of
the lateral deflection $v(x,t)$.

5.4.1   Hereditary Integral Models

In this case, using that $v(x,t) = V(x)C(t)N_o$ with (5.6) and (5.24) leads to

$$N_o \cdot C(t) \cdot N_oV(x) - I(x)E(t - \tau) * dC(\tau)N_oV''(x) = 0$$
$$v(0, t) = 0$$
$$v(l, t) = 0 \tag{5.25}$$

After canceling out the common term N_o, using that $E*dC = H(t)$ (c.f. Chap. 2),
and rearranging, this take the form:

$$N_oC(t) - I(x)\frac{V''(x)}{V(x)} = 0 \tag{5.26}$$

The only way in which the algebraic sum of two functions of different variables
can be equal to zero is either if both functions are zero—a trivial case—or if both

functions equal the same constant. Calling this constant β^2 leads to the following characteristic and evolution equations, respectively:

$$
\begin{aligned}
V''(x) + (1/I)\beta^2 V(x) &= 0 \\
V(0) &= 0 \\
V(l) &= 0
\end{aligned}
\tag{5.27}
$$

$$
N_o \cdot C(t) = \beta^2
\tag{5.28}
$$

Introducing the notation $\lambda^2 = \beta^2/I$, the non-trivial solution of (5.27) is: $V(x) = A_1 sin(\lambda x) + A_2 sin(\lambda x)$, which, in view of the boundary conditions, yields the following Eigen-values of the equation:

$$
\lambda \equiv \frac{\beta}{\sqrt{I}} = \frac{n\pi}{l}, \quad n \ge 1
\tag{5.29a}
$$

$$
\beta = \sqrt{I}\frac{n\pi}{l}, \quad n \ge 1
\tag{5.29b}
$$

The critical loads are obtained combining (5.26), (5.28), and (5.29), replacing N_o with N_{cr}, for consistency with common usage, as

$$
N_{cr}(t) = \frac{n^2\pi^2 I}{l^2 C(t)}; \quad n \ge 1
\tag{5.30}
$$

From this, using the relations $E(0) \equiv E_g = 1/C(0)$, and $E(\infty) \equiv E_e = 1/C(\infty)$, the following asymptotic expressions may be obtained (c.f. Chap. 2):

$$
\lim_{t \to 0} N_{cr}(t) \equiv N_{crg} = \frac{n^2\pi^2}{l^2}IE(0) \equiv \frac{n^2\pi^2}{l^2}IE_g; \quad n \ge 1
\tag{5.31}
$$

$$
\lim_{t \to \infty} N_{cr}(t) = N_{cre} = \frac{n^2\pi^2}{l^2}IE(\infty) \equiv \frac{n^2\pi^2}{l^2}IE_e; \quad n \ge 1
\tag{5.32}
$$

These are the "short-term" or glassy, N_{crg}, and long-term or equilibrium, N_{cre}, Euler loads. Quite often, the critical Euler load of interest corresponds to $n = 1$. The physical meaning of the above results is that as follows:

a. A viscoelastic column will be asymptotically stable if the applied load does not exceed its long-term critical Euler load, N_{cre}

b. A viscoelastic column will become asymptotically unstable if the applied load is larger than its long-term critical Euler load, N_{cre}

c. A viscoelastic column will be instantaneously unstable if the applied load is larger than its short-term critical Euler load, N_{crg}.

In addition, if the viscoelastic column does not buckle instantly after reaching the target load, the presence of the compliance function, $C(t)$, in (5.30) allows one to estimate the time, measured from that point on, that it would take the column to buckle.

Example 5.3 Find the instantaneous and long-term buckling loads of a straight column 100 cm long, and rectangular cross section 15 cm by 10 cm, if it is made of a viscoelastic material with modulus $E(t) = 3.5 + 8.5e^{-t/0.25}$ MPa.
Solution:
    The critical load depends directly on the second moment of area and is a minimum for the smallest value of the latter. The minimum centroidal moment of inertia for the column is $I = \frac{1}{12}bh^3 = 15 \cdot 1,000/12 = 1,250$ cm^4 $= 1.250 \cdot 10^{-5}$ m^4. The long-term modulus of the material is 3.5 MPa; and its instantaneous modulus, $3.5 + 8.5 = 12$ MPa. Inserting these values in (5.31) and (5.32), and setting $n = 1$, produces the critical loads: $N_{crg} \approx 1.48$ *kN* and $N_{cre} \approx 0.43$ *kN*.

5.4.2   Differential Operator Models

When the constitutive equations are given in differential operator form, applying operator P to both sides of (5.24), and invoking (5.10) produces

$$P[N(t)v(x,t)] = P[M(x,t)] \equiv IQ[v''(x,t)]$$

Proceeding as before, using $N(t) = N_o$, $v(x,t) = V(x)C(t)$, and rearranging, produces the same characteristic Eq. (5.27), and Eigen-values (5.29), as for materials of integral type. In this case, the evolution equation, equivalent to (5.28), takes the form:

$$\frac{N_o}{Q[C(t)]/P[C(t)]} = \beta^2 \tag{5.33}$$

The critical load is obtained combining (5.29) and (5.33) to give

$$N_{cr} = \frac{n^2\pi^2}{L^2}IQ[C(t)]/P[C(t)] \tag{5.34}$$

As seen from this last expression, the limiting values of the critical load depend on the specific form of the differential operators.
    As discussed in Chap. 3, for rheological models whose operators P and Q are of the same order, $Q[C(t)]|_{t=0} = q_nC(0)$ and $P[C(t)]|_{t=0} = p_nC(0)$. Likewise, for models with non-zero long-term modulus, $Q[C(t)]|_{t=\infty} = q_oC(\infty)$, and $P[C(t)]|_{t=\infty} = p_oC(\infty)$. Taking these values into (5.33), using the notation $E_g \equiv q_n/p_n$, for the impact or glassy modulus, and $E_e \equiv q_o/p_o$ for the long-term, or equilibrium modulus, produces the same expressions as (5.31) and (5.32).

Fig. 5.7 Example 5.4

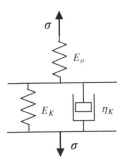

Example 5.4 Find the instantaneous and long-term buckling loads for a straight column of length L, and minimum moment of inertia, I, made of a viscoelastic material of standard linear solid type, with parameters as shown in Fig. 5.7.
Solution:
The solution is given by (5.31) and (5.32), which require the glassy and equilibrium moduli. These are obtained from the differential operators $P = (E_o + E_K) + \eta_K \partial_t$, and $Q = (E_o E_K) + E_o \eta_K \partial_t$ [see Eq. (3.35a)], and the relationships $E_g \equiv q_n/p_n = E_o$; and $E_e \equiv q_o/p_o = (E_o E_K)/(E_o + E_K)$. Hence, $N_{crg} = \pi^2 E_o I/L^2$, and $N_{cre} = \pi^2 \frac{E_o E_K}{E_o + E_K} \left[\frac{I}{L^2}\right]$.

5.5   Viscoelastic Springs

Analytical models of viscoelastic springs find many important practical applications, both under quasi-static and dynamic conditions. Composite elastomeric bearings, for instance, are made up of alternating layers of elastomers and elastic materials, such as is shown conceptually, in Fig. 5.8. The elastomeric layers are called pads, and the elastic layers, shims. In general, each layer—either pad or shim—may have a different thickness.

Since the elastic materials used for the shims are orders of magnitude stiffer than the elastomers, all deformation in these bearings is essentially due to that of the pads. Idealizing each pad as a massless spring of the appropriate type (axial, shear, bending, or torsion), one may construct a multidimensional structural model of the bearing from a combination of viscoelastic springs connected by rigid interleaves. The same is true of analytical models for more elaborate parts.

Fig. 5.8 Flat-pad composite elastomeric bearing and possible types of external loads

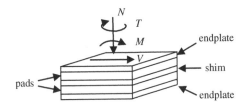

In this section, the equations of static equilibrium are invoked to develop the force–deflection relationships for viscoelastic springs of axial, shear, bending, and

Fig. 5.9 Geometry, material properties, and loading of viscoelastic spring models

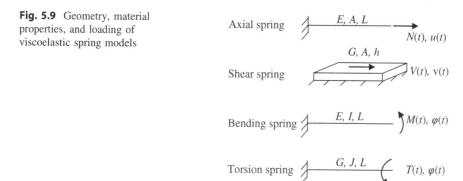

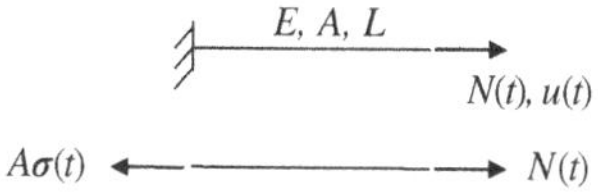

torsion type. For simplicity, straight bars of uniform cross section are used to represent axial, bending, and torsional springs, and a plate to idealize a shear spring, as shown in Fig. 5.9.

Fig. 5.10 Axial spring

5.5.1 Axial Spring

The bar-like idealized model of an axial spring and the corresponding free body diagram are shown in Fig. 5.10.

Use of the figure to establish force equilibrium leads to

$$N(t) = A\sigma(t) \tag{a}$$

Assuming the deflection at the end of the spring to be $u(L,t) = u_o f(t)$ the axial deflection at station x along the bar's axis takes the form: $u(x,t) = (u_o/L) x \cdot f(t)$, producing the axial strain:

$$\varepsilon(x,t) \equiv \partial u(x,t)/\partial x = (u_o/L) \cdot f(t) = u(L,t)/L \tag{b}$$

These expressions will now be used to develop the force–deflection relationship—and stiffness—for an axial spring in terms of its integral, differential, and steady-state constitutive equations (c.f. Chaps. 2, 3 and 4).

a. Hereditary integral springs. In this case, substitution of $\sigma = E*d\varepsilon$ into (a), using (b), leads to

$$N(t) = \frac{A}{L} E(t - \tau) * du(\tau) \equiv K_N(t - \tau) * du(\tau) \tag{5.35a}$$

Hence, the operational form of the spring stiffness becomes:

$$K_N(t - \tau) * d \equiv \frac{A}{L} E(t - \tau) * d \tag{5.35b}$$

b. Differential operator springs. Here, applying the operator P to both sides of (a), using (b) and $P[\sigma] = Q[\varepsilon]$ on the right-hand side produce:

$$P[N(t)] = \frac{A}{L} Q[u(t)] \tag{5.36a}$$

The spring stiffness is only indirectly defined by this expression. Its Laplace transform yields a form more directly comparable to its elastic counterpart:

$$\overline{K}_N(s) \equiv \frac{A\overline{Q}(s)/\overline{P}(s)}{L} \tag{5.36b}$$

c. Springs for steady-state oscillations. Introducing the complex strain $\varepsilon^* = u^*/L$, and complex force, N^*, and integrating the constitutive equation (c.f. Chap. 4), results in the complex modulus E^* and produces the relationship:

$$N^*(j\omega) = \frac{A}{L} E^*(j\omega) u^*(j\omega) \equiv K_N^*(j\omega) u^*(j\omega) \tag{5.37a}$$

This defines the complex stiffness of axial springs made of materials of integral type, as

$$K_N^*(j\omega) \equiv \frac{A E^*(j\omega)}{L} \tag{5.37b}$$

For rheological models, use of N^* and u^* with (5.36a) and (4.33a) leads to

$$K_N^*(j\omega) \equiv \frac{A}{L} \left\{ \frac{Q(j\omega)}{P(j\omega)} \right\} = \frac{A E^*(j\omega)}{L} \tag{5.37c}$$

5.5.2 Shear Spring

The idealized model of a shear force spring is shown in Fig. 5.11. Using it to establish force balance leads to

Fig. 5.11 Shear spring

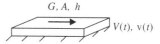

G, A, h

$V(t), v(t)$

$$V(t) = A\tau(t) \tag{c}$$

Assuming the shear deflection at the end of the spring is given by $v(L,t) = v_o \cdot f(t)$, and using the thickness, h, of the spring, yields the shearing strain:

$$\gamma(x, t) = \frac{v(x, t)}{h} \tag{d}$$

Expressions (c) and (d) will be used to develop the force–deflection relationship—and stiffness—for a shear spring in terms of its integral, differential, and steady-state constitutive equations (c.f. Chaps. 2, 3 and 4).

a. Hereditary integral springs. In this case, taking (d) and $\tau = G*d\gamma$ into (c) leads to

$$V(t) = \frac{A}{h} G(t - \tau) * dv(\tau) \equiv K_V(t - \tau) * dv(\tau) \tag{5.38a}$$

With this, the operational form of the spring stiffness becomes:

$$K_V(t - \tau) * d \equiv \frac{A}{h} G(t - \tau) * d \tag{5.38b}$$

b. Differential operator springs. Here, applying the operator P to both sides of (c) and using (d) with $P[\tau] = Q[\gamma]$ on the right-hand side produces

$$P[V(t)] = \frac{A}{h} Q[v(t)] \tag{5.39a}$$

The spring stiffness is only indirectly defined by this expression; but, its Laplace transform yields a form that is directly comparable to its elastic counterpart, as

$$\overline{K}_V(s) \equiv \frac{A\overline{Q}(s)/\overline{P}(s)}{h} \tag{5.39b}$$

c. Springs for steady-state oscillations. Here, we introduce the complex strain $\gamma^* = v^*/h$, and complex force V^*, integrate the constitutive equation (as in Chap. 4), to obtain the complex modulus G^*, and arrive at the relationship:

$$V^*(j\omega) = \frac{A}{h}G^*(j\omega)v^*(j\omega) \equiv K_V^*(j\omega)v^*(j\omega) \tag{5.40a}$$

This defines the complex stiffness of shear springs made of materials of integral type, as

$$K_V^*(j\omega) \equiv \frac{AG^*(j\omega)}{h} \tag{5.40b}$$

For rheological models, use of V^* and v^* with (5.39a) and (4.33a) leads to

$$K_N^*(j\omega) \equiv \frac{A}{L}\left\{\frac{Q(j\omega)}{P(j\omega)}\right\} = \frac{AG^*(j\omega)}{h} \tag{5.40c}$$

5.5.3 Bending Spring

The idealized model of a bending spring is shown in Fig. 5.12. The pertinent relations for this case were derived earlier as Eq. (5.7), for constitutive equation of integral type, and (5.11a, b), for materials of differential type.

a. Hereditary integral springs. The load–deflection relationship in this case is obtained by integrating (5.7b) with respect to position, x, along the axis, noting that M is a function of time only. Setting $\varphi(0) = 0$, this leads to

$$\varphi(t) = \frac{L}{I}C(t - \tau) * dM(\tau) \tag{e}$$

Taking the convolution of this expression with the tensile relaxation modulus and rearranging, the following form is obtained, which resembles the elastic relation:

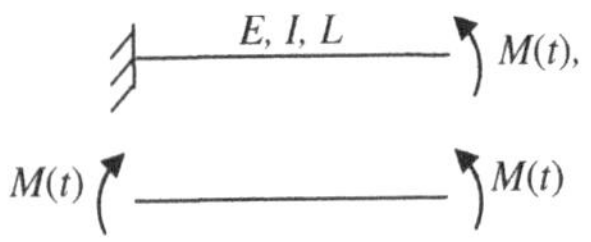

Fig. 5.12 Bending spring

$$M(t) = \frac{I}{L}E(t - \tau) * d\varphi(\tau) \equiv K_M(t - \tau) * d\varphi(\tau) \qquad (5.41a)$$

This defines the spring stiffness in operational form as

$$K_M(t - \tau) * d \equiv \frac{I}{L}E(t - \tau) * d \qquad (5.41b)$$

b. Differential operator springs. The load–deflection equation for a viscoelastic beam of differential type in bending was obtained as (5.11b). Integrating it with respect to x, noting that M and φ are independent of position along the axis, setting $\varphi(0) = 0$, and reordering terms leads to

$$P[M(t)] = \frac{IQ[\varphi(t)]}{L} \qquad (5.42a)$$

The spring stiffness is only indirectly defined by this expression; but its Laplace transform yields a form directly comparable to its elastic counterpart, as

$$\overline{K}_M(s) \equiv \frac{I\overline{Q}(s)/\overline{P}(s)}{L} \qquad (5.42b)$$

c. Springs for steady-state oscillations. In this case, introduce the complex rotation $\varphi^*(j\omega t) \equiv \varphi_o e^{j\omega t}$, integrate (5.7a) with respect to x, and effect the convolution integral with respect to time of as in Chap. 2. Recognizing that now the applied moment is a complex quantity, M^*, and replacing C^* with $1/E^*$, in accordance with (4.14), leads to the form:

$$M^*(j\omega t) = \frac{I}{L}E^*(j\omega)\varphi^*(j\omega t) \equiv K_M^*(j\omega)\varphi^*(j\omega) \qquad (5.43a)$$

This defines the complex stiffness of bending springs of materials of integral type, as

$$K_M^*(j\omega) = \frac{I}{L}E^*(j\omega) \qquad (5.43b)$$

For rheological models, use of M^* and φ^* with (5.42a) and (4.33a) leads to

$$K_N^*(j\omega) \equiv \frac{I}{L}\left\{\frac{Q(j\omega)}{P(j\omega)}\right\} = \frac{IE^*(j\omega)}{L} \qquad (5.43c)$$

5.5.4 Torsion Spring

The idealized bar-like model of a torsion spring is shown in Fig. 5.13. Equations (5.19b) and (5.22b) will be used to derive the force–deflection relationships for viscoelastic springs of integral and differential types, respectively, as well as those applicable to steady-state conditions.

a. Hereditary integral springs. The load–deflection relationship for this case is obtained by integrating (5.19b) with respect to position along the axis. Noting that T is independent of the position coordinate, this leads to

$$\int_0^L \frac{\partial}{\partial x}\psi(x,t)dx \equiv \psi(L,t) - \psi(0,t) = \frac{L}{J}C_G(t-\tau) * dT(\tau) \tag{f}$$

Setting $\psi(0) = 0$, for reference, convolving this result with the shear relaxation modulus, G, and rearranging, produces a form that is reminiscent of the elastic equation $T = \frac{GJ}{L}\psi$:

$$T(t) = \frac{J}{L}G(t-\tau) * d\psi(\tau) \equiv K_T(t-\tau) * d\psi(\tau) \tag{5.44a}$$

This defines the hereditary operator form of the stiffness of a viscoelastic torsion spring, as

$$K_T(t-\tau) * d \equiv \frac{J}{L}G(t-\tau) * d \tag{5.44b}$$

b. Differential operator springs. The load–deflection equation for the viscoelastic torsion spring of differential type is obtained integrating (5.22b) with respect to axial position. This is done noting that T is independent of position, and operators P and Q are functions of time only, which allows interchanging the order of integration and differentiation. Proceeding thus, leads to

$$\int_0^L P[T(t)]dx \equiv P[T(t)]L = \int_0^L JQ\left[\frac{\partial}{\partial x}\psi(x,t)\right]dx \equiv J\{Q[\psi(L,t)] - Q[\psi(0,t)]\}$$

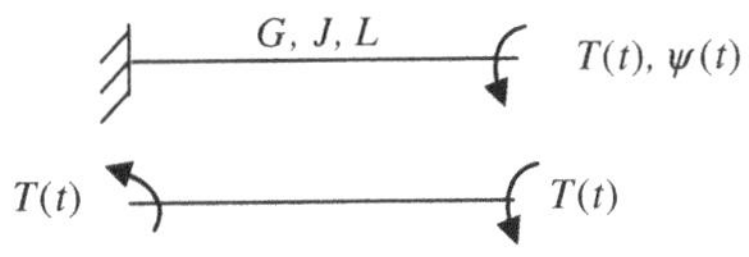

Fig. 5.13 Torsion spring

Setting $\psi(0, t) = 0$, for reference, leads to the following differential operator form of the torque–twist relationship for the viscoelastic torsion spring:

$$P[T(t)] = \frac{JQ[\psi(t)]}{L} \tag{5.45a}$$

Again, as for other springs, the torsional spring stiffness is only indirectly defined by this expression. However, taking its Laplace transform and rearranging yields that

$$\overline{T}(s) = \frac{J\overline{Q}(s)/P(s)}{L}\overline{\psi}(s) \equiv \overline{K}_T(s)\overline{\psi}(s) \tag{g}$$

This defines the Laplace transform of the viscoelastic torsional spring stiffness:

$$\overline{K}_T(s) \equiv \frac{J\overline{Q}(s)/\overline{P}(s)}{L} \tag{5.45b}$$

c. Springs for steady-state oscillations. Here, as with all other types of spring, the approach is to introduce the complex twist $\psi^*(j\omega t) \equiv \psi_o e^{j\omega t}$, and complex torque $T^* = T_o e^{j(\omega t+\delta)}$, into (5.44a) for materials of integral type, and (5.45a), for those of differential types.
Starting with (5.44a), and proceeding as in Chap. 4, leads to[8]:

$$T^*(j\omega t) = \frac{J}{L}E^*(j\omega)\psi^*(j\omega t) \equiv K_T^*(j\omega)\psi^*(j\omega) \tag{5.46a}$$

From this follows the complex stiffness of torsional springs made of materials of integral type:

$$K_T^*(j\omega) = \frac{J}{L}G^*(j\omega) \tag{5.46b}$$

For rheological models, use of T^* and ψ^* with (5.45a) and (4.33) leads to

$$K_T^*(j\omega) \equiv \frac{J}{L}\left\{\frac{Q(j\omega)}{P(j\omega)}\right\} = \frac{J}{L}G^*(j\omega) \tag{5.46c}$$

Example 5.5 The straight viscoelastic beam shown in Fig. 5.14 having length *L* and uniform cross section with second moment of area *I* is subjected to a sinusoidal concentrated load of frequency ω at its midpoint. Determine the

---

[8] Note that, just as $E^*(j\omega t) \equiv (j\omega) \cdot \mathcal{F}[E(t)]$, for the uniaxial modulus in tension, the complex shear modulus is defined by $G^*(j\omega t) \equiv (j\omega) \cdot \mathcal{F}[G(t)]$.

Fig. 5.14 Example 5.5

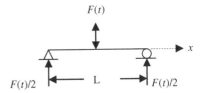

complex stiffness of the beam corresponding to lateral deflections of its midpoint, if the beam's material is of differential type.

Solution:

Take the complex form of the lateral deflection as $v^*(x, j\omega t) = v_o(x)e^{j(\omega t + \theta)}$ into (5.12b) and integrate it twice with respect x, using the boundary condition $v^*(0, j\omega t) = 0$, to arrive at the expression: $IQ(j\omega)/P(j\omega)v^*(x, j\omega t) = \frac{x^3}{12}F^*(j\omega t) + A(j\omega t)x$. By mechanical as well as geometric symmetry: $\frac{d}{dx}v^*(x, j\omega t)\big|_{x=L/2} = 0$, so that $A(j\omega t) = -F^*(j\omega t)L^2/16$. Inserting this result in the general solution, evaluating it at $x = L/2$, and rearranging it, yields the following force–deflections relationship: $F^*(j\omega t) = \frac{48IQ(j\omega)/P(j\omega)}{L^3}v^*(L/2, j\omega t)$. Hence, the complex stiffness for mid-span deflections is $K^*_{L/2}(j\omega t) = \frac{48IQ(j\omega)/P(j\omega)}{L^3}$.

5.6   Elastic–Viscoelastic Correspondence

All expressions derived previously could have been established directly from their elastic counterparts, invoking the correspondence between elastic and viscoelastic relationships introduced in an elementary fashion in Chaps. 2 and 3. According to this version of the so-called elastic–viscoelastic correspondence, replacing each quantity by its transform and each material constant entering an elastic formula by its Carson transform[9] produces the transform of the corresponding viscoelastic relationship. The inverse transformation of the latter yields the viscoelastic relation being sought. By this approach, elastic expressions that do not involve elastic constants are valid, in exactly the same form, for viscoelastic materials.[10] In applying this method to materials whose constitutive equations are of differential type, it is important to remember that $s\overline{M} = \frac{\overline{Q}}{\overline{P}}$ (c.f. Chap. 3).

Example 5.6 Use the correspondence principle and the elastic relationship $EIy'' = M$, for the deflection $y(x)$ of a beam of Young modulus E and second moment of area I, subjected to a bending moment $M(t)$, to establish the pertinent

---

[9] The Carson transform of a function is the transform-variable multiplied transform of the function.

[10] This is true for all materials, irrespective of their constitution.

relationship for a viscoelastic beam. Assume that the constitutive equation of the beam may be given in either integral or differential form.
Solution:

Start by writing the Laplace transform of the given expression as $s\overline{EIy''} = \overline{M}$. Then, reorder it to read: $I\overline{Esy''} = \overline{M}$, which makes it easier to associate with the transform of the product of E and the time derivative of y''. Then, apply the inverse transform to arrive at the desired results:

a. For a material of integral type, the inverse transform gives $IE(t - \tau) * dy''(\tau) = M(t)$, which, as expected, is Eq. (5.6).

b. For a material of differential type, replace $s\overline{E} = \frac{\overline{Q}}{\overline{P}}$ to get $I\frac{\overline{Q}}{\overline{P}}y'' = \overline{M}$; then, rewrite it as $I\overline{Qy''} = \overline{PM}$ and take the inverse Laplace transform to arrive at expression (10): $P[M] = IQ[y'']$.

Example 5.7 Use the elastic–viscoelastic correspondence and the elastic relationship: $P_{cr} = \pi^2 EI/L^2$, for the critical Euler load of a simply supported elastic column, of Young modulus E, length L, and second moment of area, I, to establish the instantaneous and long-term critical loads for a viscoelastic solid of differential type.

Solution:

Write the elastic expression in transform space as $\overline{P}_{cr} = \pi^2 s\overline{EI}/L^2$ and use that $s\overline{E} = \frac{\overline{Q}}{\overline{P}}$. Then, invoke the Initial- and Final-Value theorems for the Laplace transform: $\lim_{s \to \infty} s\overline{f}(s) = \lim_{t \to 0} f(t)$, and $\lim_{s \to 0} s\overline{f}(s) = \lim_{t \to \infty} f(t)$, to arrive at the results (c.f. Appendix A): $P_{crg} = \pi^2 Q(0)/P(0)/L^2 \equiv \pi^2 E_g/L^2$ and $P_{cre} = \pi^2 Q(\infty)/P(\infty)/L^2 \equiv \pi^2 E_e/L^2$.

Example 5.8 Use the elastic–viscoelastic correspondence and the elastic relationship, $v = PL^3/(3EI)$, for the deflection under the point of load application of a cantilever elastic beam that is subjected to a concentrated load at its free end, to establish the deflection of a viscoelastic beam of hereditary integral type. In the elastic expression, E, L, and I represent the Young modulus of the elastic material, and the length and second moment of area of the beam.
Solution:

Invoke the elastic–viscoelastic correspondence to write the transform of the elastic relationship as $\overline{v} = \overline{PL^3}/(3s\overline{EI})$.. Rewrite it to cast it in the more convenient form: $\overline{Esv} = \overline{PL^3}/(3I)$ and take the inverse transform to get $E(t - \tau) * dv(\tau) = \frac{L^3}{3I}P(t)$. Convolving this expression with the tensile compliance function C_E produces the desired result: $v(t) = \frac{L^3}{3I}C_E(t - \tau) * dP(\tau)$.

Fig. 5.15 Viscoelastic
spring-mass system

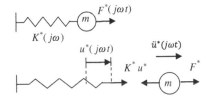

5.7   Mechanical Vibrations

This section examines the mechanical response of single-degree-of-freedom springs-mass systems subjected to excitations which are harmonic functions of time, and in which the springs are made of homogeneous, linear isotropic viscoelastic materials. Attention will be limited to single-degree-of-freedom (SDOF) systems, consisting of a mass element attached to a single massless spring element of one of the types described previously. For simplicity and ease of representation, all derivations are based on the axial spring-mass system depicted in Fig. 5.15 in terms of complex quantities.

Because an axial spring-mass system is used in the derivations, the equation of motion is established from the complex balance of linear momentum, $\sum F_i^* = m\ddot{u}^*$, assuming that $u^* = u_o e^{j(\omega t - \delta)}$. As may be seen from Fig. 5.15, this leads to[11]

$$F^*(j\omega t) - K^*(j\omega)u^*(j\omega t) = -\omega^2 m u^*(j\omega t) \qquad (5.47)$$

The same approach is applicable to other types of spring-mass systems and leads to entirely similar expressions. In particular, it should be clear that the expressions that would result for the shear spring-mass system would be identical to those of the axial case, except that spring height, h, would replace spring length, L, and shear, instead of axial relaxation modulus, would enter the equations.

The expressions for bending and torsional spring-mass systems are likewise similar to those of the axial spring-mass system, except that as follows:
a. Axial deflection is replaced by bending, φ, or torsional rotation, ψ, respectively
b. Axial acceleration, $\ddot{u}^*$, is replaced by angular acceleration, $\ddot{\psi}^*$ or rotational acceleration, $\ddot{\psi}^*$ respectively.
c. The mass element, m, is replaced by bending, I_M, or torsional, J_M, mass moments of inertia, respectively
d. The equation of motion is established from the balance of angular momentum:

---

[11] In general, because the steady-state viscoelastic input and response are out of phase, whenever the complex-controlled variable is, say, $c^* = c_o e^{j\omega t}$, the corresponding response variable is of the form $r^* = r_o e^{j(\omega t + \delta)}$ where the sign of the phase angle is positive if the response is of stress type, and negative, if it is of strain type.

Fig. 5.16 Example 5.9

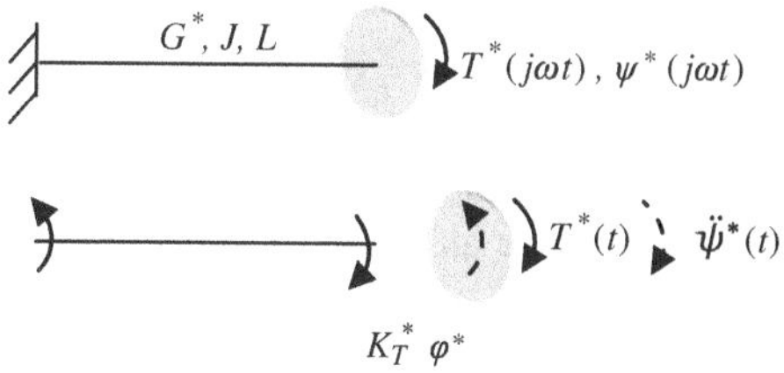

$$\sum M_i^* = I_M \ddot{\varphi}^*, \text{or} \sum T_i^* = J_M \ddot{\varphi}^* \tag{a}$$

Example 5.9 Establish the equation governing the torsional vibrations of the SDOF viscoelastic spring-mass system shown in Fig. 5.16, assuming that the mass is concentrated in a disk with centroidal polar mass moment of inertia of magnitude J_M, and the spring material is of integral type, with shear relaxation modulus G.

Solution:

Take $\psi^*(j\omega t) = \psi_o e^{j\omega t}$ into Newton's second law for angular motion to obtain $\sum T_i^*(j\omega t) = J_M \ddot{\psi}^*$. Noting that $(j)^2 = -1$, leads to $T^* - K_T^* \psi^* = -\omega^2 J_M \psi^*$; or, rearranging: $\omega^2 J_M \psi^* - K_T^* \psi^* = -T^*$

5.7.1 Forced Vibrations

Proceeding as in Chap. 4, we replace the forcing function $F(t)$ and displacement $u(t)$ with the complex quantities $F^*(j\omega t) \equiv F_o e^{j\omega t}$, and $u^* \equiv u_o e^{j(\omega t - \theta)}$. Inserting them in (5.47), rearranging, and canceling the common factor, produces

$$\left[K^*(j\omega) - m\omega^2\right] u_o = F_o e^{j\theta} \tag{5.48}$$

In this case, F_o, m, and ω, are all real quantities, as are the component K's and K'' of K^*. Using this, we factor K' out of the left-hand side of (5.48), use that, according to (4.7c), and the geometric properties of the axial viscoelastic spring:

$$\tan\delta \equiv \frac{E''}{E'} \equiv \frac{AE''/L}{AE'/L} = \frac{K''}{K'} \tag{5.49}$$

and rearrange the result to read:

$$u_o = \frac{F_o/K'}{\left[\left(1 - \beta^2\right) + j\tan\delta\right]} e^{j\theta} \tag{5.50}$$

where β is newly defined, as

$$\beta^2 \equiv \frac{m\omega^2}{K'} \tag{5.51}$$

Since u_o is real, the imaginary part of (5.50) must equal zero. This condition yields the phase angle, θ, in terms of β and $tan\delta \equiv E''/E' = K''/K'$, as

$$\tan\theta = \frac{\tan\delta}{(1-\beta^2)} \tag{e}$$

Using the definition of the tangent function, this expression can be represented as in Fig. 5.17:

Fig. 5.17 Geometric representation of phase angle

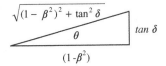

As seen in the figure,

$$\sin\theta = \frac{\tan\delta}{\sqrt{\left(1-\beta^2\right)^2+\tan^2\delta}}; \qquad \cos\theta = \frac{\left(1-\beta^2\right)}{\sqrt{\left(1-\beta^2\right)^2+\tan^2\delta}}; \tag{f}$$

Equating the real parts of (5.50) and using the previous relations for the sine and cosine functions produces the following expression for the displacement amplitude, u_o, as

$$u_o = \frac{F_o/K'}{\sqrt{\left[\left(1-\beta^2\right)^2+\tan^2\delta\right]}} \tag{5.52}$$

Because F_o/K' may be interpreted as a static displacement under the load amplitude F_o, the quantity:

$$A_F \equiv \frac{1}{\sqrt{\left[\left(1-\beta^2\right)^2+\tan^2\delta\right]}} \tag{5.53}$$

is called the "amplification factor," and is one of the three main formulas used in the analyses and design of viscoelastic spring-mass systems. Since $tan\delta \neq 0$, expression (5.53) shows that the amplification factor for viscoelastic materials is never infinite, not even when the frequency ratio, β, equals one.

As mentioned earlier, all the previous expressions are of the same form as those for a viscously damped elastic spring-mass system. In the present case, the frequency ratio, β, is defined by (5.51), and the damping factor, $2\zeta\,\beta$, of the elastic case [4] is replaced by the loss tangent; that is

$$2\zeta\beta = \tan\delta \tag{5.54}$$

In this expression, ζ is the critical damping ratio—i.e., the ratio of the system's actual damping to the critical damping. Critically damped systems do not oscillate. Also, it is common practice to refer to the real part, K', of the complex stiffness, K^*, as "dynamic stiffness" [5]; and, in analogy with elastic systems, to use the quantity:

$$\omega_N \equiv \sqrt{K'/m} \tag{5.55}$$

as a measure of the spring's natural frequency. On this note, we put expression (5.51) in the form:

$$\beta \equiv \omega/\omega_N \tag{5.56}$$

Expression (5.55) is the second formula required in vibration analysis and design of viscoelastic systems.

The third formula used in vibration analysis concerns the maximum force that is transmitted to the support of a spring-mass system. This force may be computed from the expression $F^* = K^* u^*$, using (5.50) and the fact that $u^* = u_o e^{j(\omega t - \theta)}$. Indeed, the magnitude, F^* is $|F^*| = |K^* u^*| \equiv |K^*||u^*|$, which, noting that $|e^{j(\omega t - \theta)}| = 1$, yields that $|F^*| = |K^*||u_o|$. Introducing $K'' = K' \tan\delta$, one may write $|K^*| = K'\sqrt{1 + \tan^2\delta}$, and using this with (5.50) results in:

$$F_o^{max} = \frac{\sqrt{[1 + \tan^2\delta]}}{\sqrt{\left[\left(1 - \beta^2\right)^2 + \tan^2\delta\right]}} F_o \tag{5.57}$$

The magnitude of the ratio of the maximum force, F_o^{max}, transmitted to the support, to the force F_o, that would be transmitted to the support under static conditions, is the transmission ratio, T_R, of the system. The quantity $1 - T_R$ is called transmissibility. The transmission ratio is given by:

$$T_R = \left| \frac{\sqrt{[1 + \tan^2\delta]}}{\sqrt{\left[\left(1 - \beta^2\right)^2 + \tan^2\delta\right]}} \right| \tag{5.58}$$

The expressions for the amplification factor and the transmission ratio for elastic and viscoelastic materials are similar in appearance. However, the damping and forcing frequency for elastic materials are decoupled; whereas for viscoelastic materials, they are not, because *tanδ*, which is responsible for damping in visco-elastic materials, depends on forcing frequency, intrinsically.

The graphs of the amplification factor and the transmission ratio for single-degree-of-freedom rheological models of viscoelastic systems are identical as hose for viscously damped elastic systems. In them, the system's damping is separated from the forcing frequency, through the term $2\zeta(\omega/\omega_N)$. These graphs, like those shown in Fig. 5.18, are strictly valid only for rheological models with one dashpot, like the Maxwell, Kelvin, and standard linear solid and fluid.

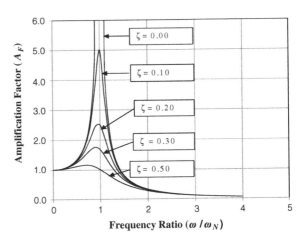

Fig. 5.18 Amplification factor for rheological model with one dashpot

For viscoelastic systems in general, the dependence of *tanδ* on frequency and temperature is as shown in Fig. 5.19 for a lightly damped natural rubber compound. The corresponding amplification factor is presented in Fig. 5.20.

Example 5.10 A rubber compound used in a shock mount application has a loss tangent $tan(\delta) = 0.05$ at a frequency of 6 Hz. What will the maximum expected

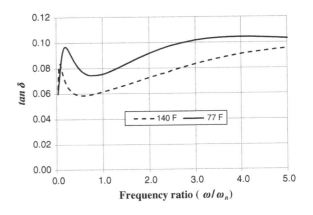

Fig. 5.19 Frequency dependence of the loss tangent of a lightly damped natural rubber

Fig. 5.20 Amplification factor for a lightly damped natural rubber compound

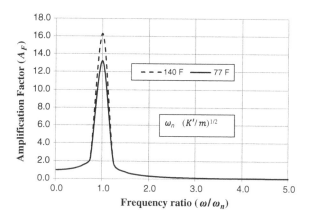

Fig. 5.21 Example 5.11

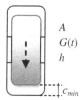

displacement of this shock mount be if it is designed to support a package of weight W_o?

Solution:

According to (5.52), the maximum deflection occurs at resonance, when $\beta = 1$. Since $F_o = W_o$, this expression yields $u_o^{max} = \frac{F_o/K'}{tan\delta} \equiv 20 \cdot W_o/K'$.

Example 5.11 To avoid damage during operation, an electronic instrument is supported inside its container by two shear mounts, as shown in Fig. 5.21.

The shear relaxation modulus of the rubber used for the mounts may be represented by $G(t) = 0.4 + 1.40e^{-t/0.06}$ MPa, with t measured in seconds. The mass of the instrument is 100 kg, and each pad has plan area $A = 25$ cm^2 and thickness $h = 1$ cm. Determine the minimum vertical clearance needed to avoid damage to the instrument at an operating frequency of 6 *Hz*.

Solution:

The vertical clearance needed to avoid the instrument from hitting the bottom wall of its container must be larger than the sum of the long-term shear deflection, v_e, of the mounts plus the maximum dynamic deflection, v_o^{max}, during operation. The long-term shear deflection is given by $v_e = W/K_e$ where $W = 100 \cdot 9.81 \approx 980$ N, $G_e = 0.4$ MPa $= 0.4$ N/mm^2 and $K_e = AG_e/h \equiv 2(25 \cdot 10^2$ mm^2) $(0.4 \cdot$ N/mm^2)/(10 mm) $= 200$ N/mm. With this, $v_e = 980/200$ mm ≈ 4.9 mm. Note that this produces a long-term shearing strain of $0.5/1 = 50$ %, which is well below the capability of typical elastomers. Now the maximum dynamic deflection is given by (5.52). This requires K', $\beta = \omega/\omega_n$, and $tan\delta = K''/K'$. From P.4.1, using $\omega = 2\pi \cdot f = 2\pi \cdot 6 \approx 37.7$ rad/s, one may obtain $G' = G_e + G_t\frac{(\omega\tau)^2}{1+(\omega\tau)^2} \approx$ 1.571 MPa; and $G'' = G_t\frac{\omega\tau}{1+(\omega\tau)^2} \approx 0.518$Mpa. This gives $tan\delta \approx 0.330$. Therefore, $K' = \frac{AG'}{h} \approx 785.5$N/mm. Using (5.55) with this value gives $\omega_n^2 = \frac{K'}{m} = \frac{785.5 \cdot 10^3}{100} \approx 7855\left(\frac{rad}{s}\right)^2$ and $\beta^2 = (\omega/\omega_n)^2 \approx 0.181$. Evaluating (5.52) gives the maximum dynamic deflection: $v_o^{max} = \frac{980/785.5}{\sqrt{(1-0.181)^2+0.33^2}} \approx 1.4$ mm. This and the long-term deflection yield a minimum vertical clearance of $c_{min} \geq 0.63$ cm. In practice, a safety factor of 2 is not unusual; leading to $c_{min} \approx 1.25$ cm.

5.7.2 Free Vibrations

This is the type of motion with which a body—viscoelastic or otherwise—would respond to a momentary perturbation from its initial position of static equilibrium. In such case, the response of the system would be due entirely to inertia effects, and the motion would have to be such that the points where the system is supported (a single point in our spring-mass system) would not move, in addition, all externally applied forces would be identically zero, except for their temporary application to start the motion. In formal terms, free vibrations of a viscoelastic body require zero boundary conditions and zero initial conditions:

$$u(0,t) = 0, F(l,t) = 0, t > 0 \tag{5.59}$$

The method of Chap. 4 developed for steady-state response to oscillatory excitations does not apply here. For the present case, however, the equation of motion $F = m\ddot{u}$ may be written using that $\varepsilon(t) = u(t)/L$, and $F(t) = A\sigma(t)$, in equation (2.40a). Thus,

$$m\ddot{u}(t) + \frac{AM_g}{L}\{u(t) - \Gamma(t-s) * u(s)\} = 0 \tag{5.60}$$

Dividing by m, identifying AM_g/L with the glassy stiffness of the spring-mass system:

$$K_g \equiv AM_g/L \tag{5.61}$$

and introducing the following logical notation for the glassy frequency, ω_g:

$$K_g \equiv m\omega_g^2 \tag{5.62}$$

transforms (5.60) into

$$\ddot{u}(t) + \omega_g^2 u(t) - \Gamma(t-s) * u(s) = 0 \tag{5.63}$$

Unlike the case of forced vibrations, posed by (5.48), Eq. (5.63) is an integral equation in the unknown displacement function $u(t)$. The solution to equations of this type may be sought, for instance, by means of the Laplace transform (see Appendix A).

5.8   Problems

P. 5. 1. Derive the equation for the defection of a straight viscoelastic beam of uniform cross section, subjected to the loads shown in Fig. 5.22.

Fig. 5.22 Problem 5.1: geometry and loading

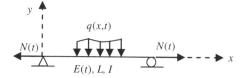

As a matter of terminology, when the axial force is tension, the beam is called a tie, and when it is compression, beam-column.

Answer: $IE(t - \tau) * dv^{IV}(x, \tau) - N(t)v''(x, t) = q(x, t)$

Hint: Establish the equations of force and moment equilibrium of a differential element of the beam located at an arbitrary axial station, x, as indicated in Fig. 5.23.

Fig. 5.23 Problem 5.1: differential beam element

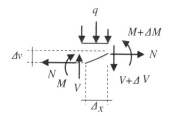

This should yield $\frac{d}{dx}V(x, t) = -q(x, t)$ and $\frac{d}{dx}M(x, t) - N\frac{d}{dx}v(x, t) = V(x, t)$. Combine these two expressions and insert (5.6) into the result to obtain the solution.

P. 5. 2. Determine the instantaneous and long-term deflections of the loaded end of the beam in Example 5.1.

Answer: $v(L, 0) = -\frac{F(0)L^3}{3E_g I}$; $v(L, \infty) = -\frac{F(\infty)L^3}{3E_e I}$

Hint: The impact and equilibrium responses correspond, respectively, to the limits as $t \to 0$, and $t \to \infty$ of the solution: $v(L, t) = -\frac{L^3}{3I}C(t - \tau) * dF(\tau)$, listed in the example. Take the stated limits and use that $C(0) = C_g = 1/E_g$ and $C(\infty) = C_e = 1/E_e$, for the impact or glassy properties and the long-term or equilibrium properties (c.f. Chap. 2). Substitute these values in the load–deflection formulas to arrive at the results.

P. 5. 3. Determine the deflection of the loaded end of the problem given in Example 5.1, for a material of differential type.

Answer: $Q[v(L,t)] = -\frac{L^3}{3I}P[F(t)] \equiv -\frac{F_oL^3}{3I}P[F(t)]$

Hint: Insert the expression $M(x, t) = -LF(t) + F(t)x$, for the moment at station x, into (5.10) and integrate the resulting differential equation twice with respect to x, using the boundary conditions $v(0,t) = v'(0,t) = 0$. Insert the external load, $F(t) = F_o f(t)$, and evaluate the expression at $x = L$, to reach the result.

P. 5. 4. Determine the instantaneous and long-term deflections of the loaded end of the beam in Problem P.5.3, if the beam's material is a standard linear solid.

Answer: $v(L,0) = -\frac{F(0)L^3}{3M_gI}$; $v(L,\infty) = -\frac{F(\infty)L^3}{3M_eI}$;

Hint: The impact and equilibrium responses correspond, respectively, to the limits as $t \to 0$, and $t \to \infty$ of the solution $Q[v(L, t)] = -\frac{L^3}{3I}P[F(t)]$, obtained in P.5.2. Since the material parameters have not been specified, assume the forms: $P = p_o + p_1\partial_t$ and $Q = q_o + q_1\partial_t$. Take the stated limits and use that the impact and long-term moduli for the standard linear solid are, respectively, $M_g = q_1/p_1$ and $M_e = q_o/p_o$ (c.f. Chap. 3). Substitute these values in the load–deflection formulas to get the results.

P. 5. 5. The pipe with uniform circular cross section shown in Fig. 5.24 is made of a viscoelastic material having tensile relaxation modulus $E(t)$ and shear relaxation modulus $G(t) = E(t)/3$. Find the complex stiffness corresponding to vertical deflections of the free end of the pipe, assuming its polar moment of area is J.

Answer: $K^*(j\omega) = 3E^*(j\omega)J\left[\frac{1}{2L_{ab}^3+2L_{bc}^3+9L_{ab}L_{bc}^2}\right]$

Hint: The vertical deflection of point c is the sum of three components. (a) The deflection at point b of cantilevered beam a–b: $v_{cb}^* = F^*(j\omega t)L_{ab}^3/(3E^*I)$. (b) The deflection of segment b–c due to the twist of segment a–b under the torque $L_{bc} \cdot F^*$: $v_{c\psi}^* = L_{bc}\psi \equiv L_{bc}M_T^*L_{ab}/(G^*J)$. (c) The deflection of cantilevered beam b-c due to load $F^*(j\omega t)$: $v_{cc}^* = F^*L_{bc}^3/(3E^*I)$. Add these three quantities together, and using that $I = J/2$, and $G = E/3$, express the result in the form: $F^*(j\omega t) = K^*(j\omega) \cdot v_c^*(j\omega t)$, to obtain K^*.

Fig. 5.24 Problem 5.5

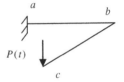

P. 5. 6. The beam of Example 5.1, of length L, and second moment of area I, is made of a viscoelastic material of Kelvin type having spring stiffness E and viscosity parameter η. If the concentrated load at its free end is $F(t) = Rt$, determine the rotation of the point of load application.

Answer: $\varphi(L,t) = -\frac{L^2}{2EI}\left[Rt - R\tau\left(1 - e^{-t/\tau}\right)\right], \tau \equiv \frac{\eta}{E}$

Hint: Use (3.16a): $\sigma = (E + \eta\partial_t)\varepsilon$ to get $P \equiv 1$ and $Q \equiv (E + \eta\partial_t)$, and insert these into (5.11b): $Q\left[\frac{\partial}{\partial x}\varphi(x,t)\right] = \frac{P[M(x,t)]}{I(x)}$, to write $EI\varphi'(x,t) + I\eta\frac{\partial}{\partial t}\varphi'(x,t) = M(x,t)$. Then, evaluate $M(x,t) = F(t)x - F(t)L$, as in Example 5.1, and integrate once with respect to x, to obtain $EI\varphi(x,t) + I\eta\frac{\partial}{\partial t}\varphi(x,t) = F(t)\frac{x^2}{2} - F(t)Lx$. Evaluate this expression at $x = L$ to get $EI\varphi(L,t) + I\eta\frac{\partial}{\partial t}\varphi(L,t) = RL^2t/2$. Divide through by $I\eta$ and define $\tau \equiv \eta/E$ to arrive at the differential equation: $\frac{\partial\varphi}{\partial t} + \frac{1}{\tau}\varphi = RL^2t/2$. Use the integrating factor $u = e^{t/\tau}$; integrate by parts once and simplify to arrive at the result.

P. 5. 7. A beam of length L, second moment of area I, built in at one end, and made of a viscoelastic solid of hereditary integral type, is subjected to a load $F(t) = F_o\sin(\omega t)$ at its free end. Find the steady-state deflection of the beam's tip.

Answer: $v(L,t) = -\frac{F_o L^3}{3I}\left[C'(\omega)\sin(\omega t) - C''(\omega)\cos(\omega t)\right]$

Hint: First, introduce the complex load: $F^*(j\omega t) = F_o e^{j\omega t} \equiv F^*$, and using (5.12a): $M^*(j\omega t) = IE^*(j\omega)\frac{d^2}{dx}v^*(x, j\omega t)$, with $M^*(j\omega t) = F^*x - F^*L$, integrate twice with respect to x, and making use of the boundary conditions: $v'(0,t) = v(0,t) = 0, \forall t$, to arrive at $v^*(L, j\omega t) = -\frac{L^3}{3I}\left(\frac{F^*}{E^*}\right)$. Use that $F^* = F_o e^{j\omega t} = F_o[\cos(\omega t) + j\sin(\omega t)]$ together with (4.18): $1/E^* = C^* = C' - jC''$, perform the indicated operations and collect terms, to arrive at the result.

P. 5. 8. A circular bar of uniform cross section with polar moment inertia, J, and length L, which is made of a viscoelastic solid of integral type, is fixed at one end while its other end is subjected to a cyclic torque $T(t) = T_o\sin(\omega t)$. Find the steady-state rotation of the tip of the bar.

Answer: $\psi(t) = \frac{T_o L}{J}\left[C'\sin(\omega t) - C''\cos(\omega t)\right]$

Hint: Rewrite (5.46a) as $\psi^*(j\omega t) = \frac{L}{J}C^*(j\omega)T^*(j\omega t)$; then, introduce the complex torque $T^*(j\omega t) = T_o e^{j\omega t} = T_o[\cos(\omega t) + j\sin(\omega t)]$, and using (4.14) and (4.18): $1/E^* = C^* = C' - jC''$, perform the indicated operations and simplify. Since the torque actually applied corresponds to the imaginary part of the complex torque, the solution sought is the imaginary part of the result thus obtained.

P. 5. 9. Calculate the energy dissipated per cycle for the bar of Problem P.5.8.

Answer: $W_V = \pi\frac{L}{J}T_o^2 C''(\omega)$

Hint: Apply (2.37) in the form $W|_t^{t+p} = \int_t^{t+p} T(t)\frac{d\psi}{dt}\,dt$, with $p = 2\pi/\omega$ using the rotation $\psi(t) = \frac{T_o L}{J}\left[C'\sin(\omega t) - C''\cos(\omega t)\right]$ calculated in Problem P.5.8, and the applied torque $T(t) = T_o\sin(\omega t)$. Carry out the integration invoking the orthogonality of the sinusoidal functions [see Appendix A] and arrive at the result.

P. 5. 10. A simply supported viscoelastic beam of length L has constant cross section with second moment of area I and supports a concentrated mass M, at its midpoint. If the concentrated mass is acted on by the cyclic load $F(t) = F_o cos(\omega t)$, and the self-weight of the beam is ignored in comparison, calculate the steady-state oscillation of the attached mass, in terms of the creep compliance of the beam's material.

Answer: $v(t) = \frac{F_o L^3}{48I} [C'(\omega) \cos(\omega t) + C''(\omega) \sin(\omega t)]$

Hint: Introduce the complex load: $F^* = F_o e^{j\omega t} = F_o[cos(\omega t) + jsin(\omega t)]$, together with the result of Example 5.5: $K_{L/2}^*(j\omega t) = \frac{48IQ(j\omega)/P(j\omega)}{L^3}$, and the fact that the ratio $Q(j\omega)/P(j\omega) = E^*(j\omega t)$ to get the complex stiffness $K_{L/2}^* = 48IE^*/L^3$. With these, obtain the complex deflection of the concentrated mass as $v^* = F^*/K_{L/2}^*$. Substitute the complex forms of F^* and $K_{L/2}^*$, using (4.14) and (4.18): $1/E^* = C^* = C' - jC''$; perform the indicated operation, collect terms, and, noting that the applied load is the real part of the complex load, select the real part of the result to arrive at the solution sought.

References

1. E. Volterra, J.H. Gaines, *Advanced Strength of Materials* (Prentice-Hall, Inc., Englewood Cliffs, 1971), pp 257–268
2. J.T. Oden, *Mechanics of Elastic Structures* (McGraw-Hill Book Co., New York, 1967), pp 30–33
3. A. Chajes, *Principles of Structural Stability Theory* (Prentice-Hall, Englewood Cliffs, 1974), pp 1–3
4. R. M. Christensen, *Theory of viscoelasticity*, 2nd edn. (Dover, New York, 1982), pp. 21–26
5. A.N. Gent, *Engineering with Rubber* (Hanser, Munich 2001), pp. 73–87

Temperature Effects

6

Abstract

This chapter examines the simultaneous dependence of material property functions on time and on temperature. The time–temperature superposition principle and the concept of time–temperature shifting are introduced first. The dependence of the glass transition temperature both on the time of measurement and on surrounding pressure is examined in detail. The integral and differential constitutive equations are then generalized to include thermal strains and strains due to changes in humidity. Two ways used in practice to represent the material property function that accounts for thermal strains are considered: independent of time and time dependent.

Keywords

Time-temperature superposition · Shift · Time · Frequency · Scale · Master · Thermorheologically simple · WLF · Transition · Thermal · Volume · Pressure · Temperature · Moisture · Hygrothermal

6.1 Introduction

As indicated in Chap. 1, the specific values of some properties, such as the relaxation modulus, creep compliance, or strain capability of viscoelastic materials, depend on the time and the temperature at which the measurements are taken. The discussions up to this point have assumed that all required material property functions were available at the temperature of interest. Temperature was simply thought of as a parameter rather than as an independent variable on which material properties depend and was thus omitted from explicit consideration.

D. Gutierrez-Lemini, *Engineering Viscoelasticity*, DOI: 10.1007/978-1-4614-8139-3_6,   149
© Springer Science+Business Media New York 2014

This chapter examines the dependence of material property functions on time and temperature. The time–temperature superposition principle and the concept of time–temperature shifting, stating that increasing temperature is equivalent to shortening the timescale of observation, and vice versa, are introduced first. This is followed by a physical interpretation of the glass transition, where viscoelastic behavior is most pronounced. The dependence of the glass transition temperature on time of measurement is pointed out, and its dependence on surrounding pressure examined in some detail. The hereditary constitutive equations are then generalized to include thermal strains. Two ways used in practice to represent the material property function that accounts for thermal strains are considered: independent of time, as for elastic solids, and time dependent, as other viscoelastic properties.

6.2   Time Temperature Superposition

Several constitutive and failure properties of viscoelastic substances—such as relaxation modulus, creep compliance, tensile strength, fracture resistance, and so on—are time and temperature dependent. In other words, the values of some properties depend on the time and temperature at which the measurements are taken. This is indicated conceptually in Fig. 6.1 for a generic property, P, such as the relaxation modulus or the tensile strength of a viscoelastic material [c.f. Chap. 1]. The same can be said of the dependence on frequency and temperature of some viscoelastic properties, as depicted in Fig. 6.2, for the real part of the complex modulus [1].

In principle, time–temperature characterization of a viscoelastic property would require that the pertinent tests—stress relaxation modulus, sustained load to failure, and so on—be carried out at several temperatures in the expected range of application, and the values be recorded for periods long enough to span that range. In similar fashion, the frequency–temperature characterization of a viscoelastic property function would require tests at several constant temperatures in which the forcing frequency spans the selected range of application.

Such direct approach to characterizing viscoelastic properties is not only time-consuming and expensive, but also impractical. Fortunately, however, viscoelastic materials of the type dealt within this text—amorphous polymers—can trade either

Fig. 6.1 Typical time and temperature dependence of a generic property, P, such as the relaxation modulus of an amorphous viscoelastic polymer (t = time; T = temperature; $T_1 < T_2 < T_3$)

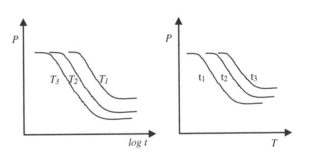

Fig. 6.2 Typical frequency
and temperature dependence
of the storage modulus, M', of
an amorphous viscoelastic
polymer (ω = frequency;
T = temperature;
$T_1 < T_2 < T_3$)

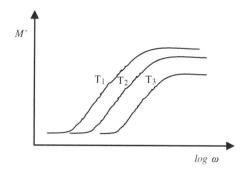

time or frequency for temperature and vice versa, which permits reducing or extending the actual timescale of observation through a corresponding change in test temperature.

Indeed, some functions describing properties of viscoelastic materials can be constructed by testing over smaller time intervals and at different temperatures than used in the direct approach and by shifting the curves describing those functions to one of the curves obtained at an arbitrarily selected temperature of interest. Experimental observation confirms that, within reason, the functions so constructed overlap the functions that would be obtained, were the tests carried out at the selected temperature and for the duration covered by the composite curve [2].

The process just described is shown schematically in Figs. 6.3 and 6.4. In the first of these figures are shown four curves of the same property function. Three of the curves are shorter than the fourth one because the time of observation used to measure them was shorter. The shorter curves correspond to tests carried out at temperatures T_1, T_2, and T_r, while the longer-duration curve was tested at temperature T_r.

In Fig. 6.4, the "short" curves at temperatures T_1 and T_2 are "shifted" to the right and left, respectively, to overlap the short curve that was tested at the same temperature as the long curve. As the figure indicates, by judiciously shifting test data collected at different temperatures, and for adequately short durations, one can construct the material property function at one of the test temperatures; which would thus be valid for a longer duration than that of the individual curves. For obvious reasons, the process described is referred to as time–temperature shifting or time–temperature superposition.

Fig. 6.3 Typical test data:
viscoelastic property
measured with short-duration
(laboratory timescale) tests at
different temperatures and
with a long-duration test at
one of the temperatures

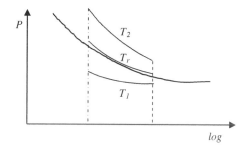

Fig. 6.4 Time-temperature shifting procedure. **a** Reference temperature, T_r, is preselected. **b** All *curves* are then shifted toward it

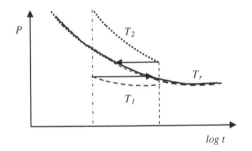

Clearly, the reference temperature, T_r, to which the partial material property functions are shifted, is arbitrary. Hence, by keeping track of the shift—the horizontal distance between a curve and that at the reference temperature—one can, equally well, construct the material property function corresponding to any other temperature among those used in the tests. Such curves succinctly represent the material property function for all temperatures and all times in the range of observation times covered by their extended composite forms; and are thus called *master curves* for the properties in question.

To get a feel for the usefulness of time–temperature shifting, consider the data in Fig. 6.5, which show the results of relaxation modulus testing of a natural rubber, carried out at three temperatures. As may be seen in the figure, each test lasted close to 1,000 s (about 20 min); yet, when the data were shifted—in this case to 22.8 °C—the "experimental time of observation" at that temperature was extended to about 1 million seconds (about 11 days), as indicated in Fig. 6.6. In practical terms, this indicates that, where it is not for the time–temperature superposition principle, one would have had to test the material at 22.8 °C, for slightly over 11 days to get the same information as obtained by means of just three 20 s duration tests, performed at the three temperatures listed in the figure.

Fig. 6.5 Relaxation modulus data for a natural rubber compound tested at three temperatures

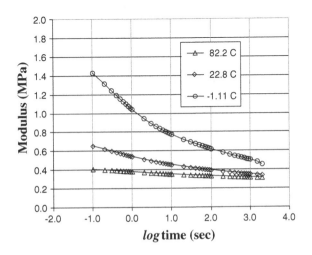

Fig. 6.6 Relaxation modulus data for a natural rubber compound tested at three temperatures and shifted to 22.8 °C

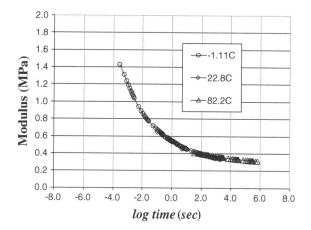

Fig. 6.7 Definition of time–temperature shift function, a_T, as a shift from time at temperature T, to time at temperature T_r

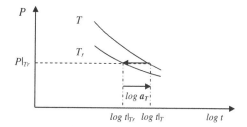

In the present case, the principle was actually corroborated by performing two 1,000-h (over 40 days) tests at the reference temperature of 22.8 °C.

It is not infrequent for practitioners in the field of viscoelasticity to be requested by their customers to demonstrate that the time–temperature superposition principle applies to the polymeric materials at hand. In some cases, such as for strategic and tactical missiles and even the Space Shuttle programs, corroboration tests, lasting several years, on polymeric materials have been carried out.

The time–temperature shift process can be formalized easily. This is done with reference to Fig. 6.7, which shows two partial curves of a material property function collected at temperatures, T and T_r, respectively. Without any loss of generality, T_r may be taken as the reference temperature and the other curve is shifted toward it.

In Fig. 6.7, the time–temperature shift allows the property, P, at time t and , T, (denoted $t|_T$ to emphasize the temperature T) to be established from test data collected at temperature T_r. In other words, the value sought can be read off the curve at temperature T_r and time $t|_{T_r}$. Based on the figure, the passage from $t|_T$ to $t|_{T_r}$ is provided by the shift $a_T (T, T_r)$:

$$\log a_T(T, T_r) \equiv \log t|_T - \log t|_{T_r} \qquad (6.1)$$

Fig. 6.8 Typical form of the
time–temperature shift
function, a_T, for amorphous
polymers

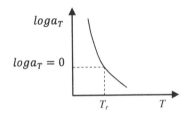

so that

$$\log t|_{T_r} \equiv \log t|_T - \left(\log t|_T - \log t|_{T_r} \right) \equiv \log t|_T - \log a_T(T, T_r) \qquad (6.2a)$$

Or, after rearranging

$$\log \left(t|_{T_r} \right) \equiv \log \left[\frac{t|_T}{a_T(T, T_r)} \right] \qquad (6.2b)$$

By its definition, $a_T (T = T_r, T_r) = 1$. In addition, it is required that its slope, that
is, its derivative with respect to temperature, be positive: $\frac{d}{dT} a_T > 0$ [1]. The above
expressions encompass the time–temperature superposition principle, that:

   t units of time at temperature T are equivalent to t/aT(T,Tr) units at temperature Tr

.The time–temperature shift function, a_T, for a viscoelastic polymer is depicted on
a semi-logarithmic timescale in Fig. 6.8.

   There are several analytical forms that capture the behavior shown in the figure.
Perhaps the most widely used expression to represent the shift function, a_T, is the
WLF equation, named after its developers [3]. Save for a different interpretation of
the parameters C_1 and C_2, entering the equation, which were once thought by the
proponents to be universal constants but are now taken more as curve-fit values,
the WLF equation is expressed as[1]

$$\log a_T(T, T_r) = - \frac{C_1 \cdot (T - T_r)}{C_2 + (T - T_r)} \qquad (6.3)$$

   From the preceding discussion, a master curve is expressed as a function of time
at an arbitrarily selected reference temperature. To obtain the value of the property
at an arbitrary time t and temperature T—in the range of the tests—the shift factor,
$a_T (T, T_r)$, is calculated first, and the value being sought of the property function is
read off the master curve at $t/a_T(T_r)$, that is,

---

[1] Note that, for the purposes of converting between different temperature scales, C_1 is non-
dimensional, but C_2 has the dimensions of temperature.

$$P(t, T) = P\left[\frac{t}{a_T(T, T_r)}, T_r\right] \tag{6.4}$$

In particular, application of the time–temperature superposition principle to the relaxation modulus, M, and creep compliance, C, produces that

$$M(t, T) = M\left[\frac{t}{a_T(T, T_r)}, T_r\right] \tag{6.5}$$

$$C(t, T) = C\left[\frac{t}{a_T(T, T_r)}, T_r\right] \tag{6.6}$$

Also, to simplify notation, or when it is clear from the context, the reference temperature, T_r, is usually omitted from the shift function so that $a_T(T)$ or even a_T is used instead of $a_T(T, T_r)$.

The time–temperature superposition principle applies to constitutive functions expressed in the frequency domain, as well. The analytical form of the principle in that case can be established using (6.4) in expression (4.3) for the complex modulus, M^* and (4.13), for the complex compliance, C^*. For the complex modulus, this would yield

$$M^*(j\omega, T) \equiv j\omega \int_{t=0}^{\infty} M(t/a_T, T_r)e^{-j\omega t}dt.$$

Introducing the change of variables $u = t/a_T$, and regrouping, produces:

$$M^*(j\omega, T) \equiv j\omega a_T \int_{t=0}^{\infty} M(u, T_r)e^{-j\omega a_T u}du \tag{6.7a}$$

In other words,

$$M^*(j\omega, T) = M^*(j\omega a_T, T_r) \tag{6.7b}$$

Entirely similar considerations, starting with (4.13), result in

$$C^*(j\omega, T) = C^*(j\omega a_T, T_r) \tag{6.8}$$

Materials for which the time–temperature superposition principle (6.4) holds are generically called thermorheologically simple [1]. The constitutive properties of these materials possess master functions, valid for any constant temperature state in the range of the test data used to establish the shift function. For this reason, if the relaxation modulus and creep compliance of a thermorheologically simple material are, respectively, $M(t)$ and $C(t)$, the time–temperature superposition principle, (6.4), can be used to write constitutive equations such as (2.1a) and (2.2a), using master material property functions which are valid for all constant temperature states, as

$$\sigma(t) = M\left(\frac{t}{a_T(T, T_r)}\right)\varepsilon(0) + \int_0^t M\left(\frac{t-s}{a_T(T, T_r)}\right)\frac{d}{ds}\varepsilon(s)ds \tag{6.9}$$

$$\varepsilon(t) = C\left(\frac{t}{a_T(T, T_r)}\right)\sigma(0) + \int_0^t C\left(\frac{t-s}{a_T(T, T_r)}\right)\frac{d}{ds}\sigma(s)ds \tag{6.10}$$

Although the time–temperature superposition principle was derived for constant temperature, it can be formally extended to non-constant temperature states by requiring the shift function to depend on time and position. This is done by first splitting the time difference argument $\frac{t-\tau}{a_T(T, T_r)}$ of the property function of interest in the equivalent form: $\frac{t-\tau}{a_T(T, T_r)} \equiv \frac{t}{a_T(T, T_r)} - \frac{\tau}{a_T(T, T_r)}$ and then introducing the dependence on time and position, as needed. On this basis, shifted time, ξ_t, is generalized as follows:

$$\xi_t = \int_0^t \frac{d\tau}{a_T[T(x, \tau), T_r]} \tag{6.11}$$

This notation allows writing constitutive equations for non-constant temperature states in succinct fashion. The constitutive equations for a thermorheologically simple material with master modulus M and master creep compliance C, equivalent to (6.9) and (6.10), which are applicable to varying temperatures, take the following forms:

$$\sigma(t) = M(\xi_t)\varepsilon(0) + \int_{0^+}^t M(\xi_t - \xi_s)\frac{d}{ds}\varepsilon(s)ds \tag{6.12}$$

$$\varepsilon(t) = C(\xi_t)\sigma(0) + \int_{0^+}^t C(\xi_t - \xi_s)\frac{d}{ds}\sigma(s)ds \tag{6.13}$$

Summarizing, the viscoelastic property functions of any thermorheologically simple material at a prescribed temperature and selected time may be obtained from the corresponding properties at any other temperature, used as a reference, dividing the time of interest by the value of the factor for shifting from the selected temperature to the reference temperature.

The reason why a viscoelastic material—such as a solid propellant or a rubber compound—can trade time for temperature, and vice versa, lies in its molecular structure. As pointed out in Chap. 1, a polymeric material is a network of molecular chains tied to one another by chemical bonds at discrete locations along their lengths—the cross-links. The chains are in constant, Brownian motion, which speeds up when the temperature goes up and slows down when the temperature is lowered. This type of behavior confers upon the polymer an internal clock with which it measures external events. At high temperatures—and fast molecular motions—the material's internal clock beats fast and interprets actual elapsed time as being longer than it really is. The reverse is true at low temperatures. The shift factor introduced above represents the number of clicks of the observer's clock per click of the material's internal clock at the selected temperature [4].

6.3   Phenomenology of the Glass Transition

The glass transition temperature is a very important concept in a variety of applications of viscoelastic materials. In Chap. 1, it was identified as a temperature, T_g, somewhere in the transition region, where properties change drastically [see Figs. 1.13 and 1.14]. There are several practical methods to establish T_g. Each method is based on the fact that the property used in the selected test (heat capacity, free volume, specific heat, viscosity, loss tangent) changes rather drastically near the glass transition. The method discussed here (calorimetry) is based on the volumetric thermal expansion of a polymer sample of known initial volume and mass, subjected to a uniform temperature change. The sample is quenched quickly to the target temperature and its volume and temperature recorded. The resulting change in volume per unit of original volume versus temperature is curve-fitted with two straight lines of different slopes, as shown in Fig. 6.9. The temperature at which the break in slope occurs is defined as the glass transition temperature [5]. T_g.

The total volume, v, of a sample of an amorphous polymer consists of the solid volume, v_o, occupied by the molecules—this is the occupied volume—and the interstitial volume, v_f, of the spaces between molecules—the free volume. That is, $v = v_o + v_f$. The fractional occupied volume, f_o, and fractional free volume, f_f, are, respectively, defined as $f_o \equiv v_o/v$ and $f_f \equiv v_f/v$.

Fig. 6.9 Glass transition temperature defined by thermal volume-change data

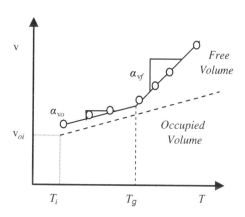

Fig. 6.10 Effect of time of measurement after quenching on the value of the glass transition. ($T_g|_{t1} < T_g|_{t2}$ whenever $t_1 \gg t_2$)

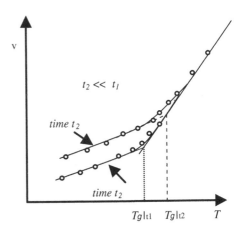

Figure 6.9 shows the division of the total volume of a sample into occupied volume and free volume. The volume occupied by the solid molecules is assumed to increase uniformly with temperature, so that using α_{vo} as the coefficient of thermal expansion for the occupied volume: $v_o = v_{oi} + \alpha_{vo} \cdot \Delta T$. By contrast, the free volume remains constant up to T_g, but increases linearly thereafter, due to an increase in the coefficient of free-volume thermal expansion, α_{vf}. That is,

$$v_f = \begin{cases} v_{fi}, & T < T_g \\ v_{fi} + \alpha_{vf}(T - T_g), & T \geq T_g \end{cases} \tag{6.14}$$

This discontinuity in the coefficient of volumetric thermal expansion is called a glass transition and the temperature associated with it, glass transition temperature. The glass transition marks the onset of molecular processes that control visco-elastic behavior. Also, in particular, as indicated by the previous expression, the glass transition is a state of constant free volume—a so-called iso-free-volume state.

The slope discontinuity is not as abrupt as suggested by Fig. 6.9, but gradual; as shown in Fig. 6.10, and extrapolation is usually needed to establish T_g. Also, as the figure suggests, the glass transition temperature depends on the specific time at which the measurements are taken, after quick quenching the sample to the target temperature. In other words, the glass transition temperature is rate sensitive: the shorter the time to take the measurement, the larger the value obtained for T_g [2].

The physical interpretation of the free-volume behavior of a viscoelastic polymer is that at low temperature, the sample shrinks and thus restricts the motion of the polymer chains. This manifests physically as an increase in modulus (the glassy behavior). By contrast, at high temperature, the sample expands, increasing free volume and molecular mobility. At the macroscopic level, this is seen as a reduction in modulus (the rubbery behavior).

6.4   Effect of Pressure on the Glass Transition Temperature

An important experimental observation about the glass transition temperature is that its value increases with surrounding pressure. This may be explained rather easily on purely physical grounds. Indeed, any external confining pressure on a polymer sample will tend to reduce its free volume through squeezing; but because the glass transition is an iso-free-volume state, the only way the volume can be reduced is by increasing the glass transition temperature—shifting T_g to the right in Fig. 6.9. The reduction in free volume due to a pressure change $\Delta P \equiv P - P_r$, where P_r is an arbitrary reference pressure is given by the product $C_{vf} \cdot \Delta P$; where C_{vf} denotes the isothermal compressibility of free volume of the polymer. Adding the reduction in free volume to (6.14) yields

$$v_f = v_{fi} + \alpha_{vf} \cdot \left(T - T_g\right) - C_{vf} \cdot \left(P - P_r\right) \tag{6.15}$$

The magnitude of the effect of a pressure change ΔP on T_g can now be established evaluating (6.15) at two pressures P_1 and P_2 at the corresponding glass transitions, T_{g1} and T_{g2}, where $v_{f1} = v_{f2}$. At that instant,

$$v_{f1} = v_{fi} + \alpha_{vf}\left(T_1 - T_{g1}\right) - C_{vf} \cdot \left(P_1 - P_r\right) \tag{6.15a}$$

Similarly,

$$v_{f2} = v_{fi} + \alpha_{vf}\left(T_2 - T_{g2}\right) - C_{vf} \cdot \left(P_2 - P_r\right) \tag{6.15b}$$

Subtracting (6.15a) from (6.15b), collecting terms, and rearranging produce the expression:

$$T_{g2} = T_{g1} + \frac{C_{vf}}{\alpha_{vf}} \cdot \left(P_2 - P_1\right) \tag{6.16a}$$

Identifying α_{vf} with the difference between the coefficients of thermal expansion above and below T_g : $\alpha_{vf} = \Delta\alpha$, and C_{vf} with the difference between the compressibility above and below T_g : $C_{vf} = \Delta C$ allows casting (6.16a) in the incremental form:

$$\Delta T_g = \frac{\Delta C}{\Delta\alpha} \cdot \Delta P \tag{6.16b}$$

Although very little data are available in the literature, amorphous polymers seem to obey this relation [2]. Also, because the WLF equation can be cast using T_g as reference,[2] it turns out that the shift function a_T depends on pressure as well.

---

[2] The WLF equation was originally proposed as: $\log a_T\left(T, T_g\right) = -\frac{C_1 \cdot (T - T_g)}{C_2 + (T - T_g)}$; implying different C_1 and C_2 from those corresponding to the case when $T_r = T_g$.

This fact becomes important in applications of polymers subjected to large hydrostatic pressures, such as in the case of elastomeric bearings. In general, a change in shift factor changes the timescale and hence the modulus and the compliance in accordance with the time–temperature superposition principle. This, in turn, leads to a change in the polymer's effective stiffness and consequently a change in the response of the component.

6.5 Hygrothermal Strains

Deformation of polymers is possible not only under the action of mechanical loads, but also upon change in temperature and upon absorption of moisture. The associated strains are referred to as mechanical, ε^M, thermal, ε^T, and moisture, ε^H, strains, respectively. The latter two are often generically termed hygrothermal strains [6]. When the strains are small, linear superposition is valid; the total strain, ε, is obtained as the sum of the mechanical and hygrothermal strains. Thus,

$$\varepsilon(t) = \varepsilon^M(t) + \varepsilon^T(t) + \varepsilon^H(t) \tag{6.17}$$

Hygrothermal strains do not produce stresses in bodies which are unconstrained and free to deform and accommodate changes in temperature or moisture. Mechanical strains, on the other hand, are always accompanied by internal stresses in the material and are thus called stress-producing strains. Using this and the appropriate mechanical constitutive equation from those discussed in Chaps. 2, 3, and 4, it is straightforward to generalize the stress–strain laws of viscoelasticity to account for hygrothermal strains. This is done by first rewriting Eq. (6.17) to express the mechanical strain in terms of the total and hygrothermal strains and by introducing the mechanical constitutive equation at hand. The mechanical strain is thus

$$\varepsilon^M(t) = \varepsilon(t) - \varepsilon^T(t) - \varepsilon^H(t) \tag{6.18}$$

For a viscoelastic material of relaxation–integral type, the constitutive equation is obtained inserting into the convolution integral form, $\varepsilon^M = C * d\sigma$, of the strain–stress Eq. (2.26): $\varepsilon^M(t) = \varepsilon(t) - \varepsilon^T(t) - \varepsilon^H(t) = C(t - \tau)^* d\sigma(\tau)$. Convolving the relaxation modulus M, with this expression, using that $M * dC = H(t)$, in accordance with (2.20) or (2.21) yields

$$\sigma(t) = M(t - \tau) * d[\varepsilon(\tau) - \varepsilon^T(\tau) - \varepsilon^H(\tau)] \tag{6.19}$$

For materials of differential operator type, for which, as presented in (3.9c), $P[\sigma] = Q[\varepsilon]$, the stress–strain equation generalized for hygrothermal strains is obtained by applying Q to expression (6.18), replacing $Q[\varepsilon^M]$ with $P[\sigma]$:

$$P[\sigma(t)] = Q[\varepsilon(t) - \varepsilon^T(t) - \varepsilon^H(t)] \qquad (6.20)$$

To complete the formulation, it is necessary to define the manner in which changes in temperature and moisture are related to hygrothermal strains. The needed relationships are, quite naturally, constitutive equations. The general thermomechanical constitutive equations of viscoelasticity can be derived from rigorous thermodynamic theory [1]. The approach followed here is based on the observations that

- Under a change in temperature from a reference value, a viscoelastic material deforms in a time-dependent creep-like fashion [7]. Therefore, the thermal strain may be defined in creep integral form, as

$$\varepsilon^T(t) = \int_{0^-}^{t} \alpha(t-\tau)\frac{d}{d\tau}\Delta T(\tau)d\tau; \quad \Delta T(t) \equiv T(t) - T_o \qquad (6.21a)$$

- Quite frequently, for simplicity, a non-hereditary, time-independent coefficient of thermal expansion is assumed, leading to a thermal strain–temperature constitutive equation of elastic type:

$$\varepsilon^T(t) = \alpha\Delta T(t) \quad \Delta T(t) \equiv T(t) - T_o \qquad (6.21b)$$

- Under a change in moisture concentration from a reference value, a viscoelastic material deforms in a time-independent fashion [6]. As a consequence, the moisture-induced strain may be defined by and expression of the following simple form:

$$\varepsilon^H(t) = \beta\Delta c(t) \quad \Delta c(t) \equiv c(t) - c_o \qquad (6.22)$$

In this relationship, β is the swelling coefficient, $c(t)$ is the moisture concentration at time t, and c_o is the reference moisture concentration.

6.6   Problems

1. The three coefficients of the WLF version of the shift function of a natural rubber are $T_r = 22.8$ °C, $C_1 \approx 5.499$, $C_2 = 74.96$. What would the coefficients be for this function in degrees Fahrenheit?
   Answer: $T_r = 73.0$ °F, $C_1 \approx 5.499$, $C_2 = 134.9$ °F
   Hint: By dimensional homogeneity, since a_T is dimensionless, the denominator is the sum of C_2 and a temperature difference $(T - T_r)$, and the numerator is the product of C_1 and the temperature difference $(T - T_r)$, it follows that C_2 has to have dimensions of temperature (difference) and C_1 be dimensionless. Hence, $C_{2(°F)} = 9/5 \cdot C_{2(°C)} - 134.9$ °F, and $T_{r(°F)} = 9/5 \cdot T_{r(°F)} + 32 - 73.0$ °F.

2. A flexible element is made of natural rubber whose master tensile stress endurance function is given by a power law of the form: $\sigma(t) = A\left(\frac{t}{a_T}\right)^{-p}$. To certify the design, a sustained load test has to be performed at a uniform temperature, T_1, and stress level σ_1. The design will be deemed satisfactory if the flexible element lasts at least t_1 units of time. Since a successful design is expected to last several years at the operational stress and temperature, the manufacturer wants to shorten the timescale of the test. Derive an expression that would allow changing the test duration in a meaningful way.

   Answer: $t_{test} = \frac{a_T(T_{test},\, T_r)}{a_T(T_{design},\, T_r)}\left(\frac{\sigma_{design}}{\sigma_{test}}\right)^{1/p} t_{design}$

   Hint: Solve for t from the master curve to get $t = a_T(T,\, T_r)\left(\frac{A}{\sigma}\right)^{1/p}$, evaluate this for the design and desired test conditions, divide one expression by the other and solve for the duration of the test. In practice, a suitable pair $(t_{test},\, T_{test})$ would be obtained by trial and error.

3. The coefficient of volumetric thermal expansion of a natural rubber compound is $\alpha = 560 \cdot 10^{-5}/\,°C$, and its bulk modulus established at 35 MPa is $K = 1{,}520$ MPa. Use these values as representative of the coefficient of free-volume thermal expansion α_f and the reciprocal of the isothermal compressibility of free volume, respectively, to estimate the change in glass transition temperature that could be expected at a pressure of 80 MPa.

   Answer: $\Delta T_g \approx 5.29\ °C$

   Hint: Use expression (6.16b) with the approximations $\Delta\alpha_f \approx \alpha = 560 \cdot 10^{-5}/°C$ and $\Delta c_{vf} \approx 1/K = 658 \cdot 10^{-6}$ /MPa to arrive at the result. Actual measured values of the ratio $\Delta c_{vf}/\Delta\alpha_f$ for natural rubber are of the order of 0.240 °C/MPa [2].

4. A slender bar made of a viscoelastic material of Kelvin type with elastic modulus E and viscous constant η is held at one end and subjected to a strain $\varepsilon(t) = \varepsilon_o \sin(\omega t)$ at its other end. Determine the stress in the bar if the bar is maintained at a constant thermal change, ΔT units above the stress-free temperature of the material, and its thermal response is elastic with coefficient of expansion α.

   Answer: $\sigma(t) = E\varepsilon_o\sin(\omega t) + \eta\omega\varepsilon_o\cos(\omega t) - E\alpha\Delta T$

   Hint: Combine (6.20) and (6.21b) with the constitutive equation of the Kelvin solid given in (3.16a) to write $\sigma(t) = [E + \eta\partial_t][\varepsilon_o\sin(\omega t) - E\alpha\Delta T]$. Perform the indicated operations and arrive at the stated result.

5. A slender bar of a viscoelastic material may be idealized as a standard linear solid. If the bar is held fixed at both ends and subjected to a constant temperature change, ΔT units above its stress-free temperature determine the stress that would be developed in the bar as a consequence of the restraint, assuming elastic thermal expansion.

   Answer: $\sigma(t) = -(q_o/p_1)\alpha\Delta T\left[1 - e^{-t/\tau}\right], \quad \tau \equiv p_1/p_0$

   Hint: Take the constitutive equation of the standard linear solid in the form given in (3.35b): $p_o\sigma + p_1\frac{d\sigma}{dt} = q_o\varepsilon^M + q_1\frac{d\varepsilon^M}{dt}$, where ε is the mechanical strain and combine it with (6.20) and (6.21b) to arrive at $p_o\sigma + p_1\frac{d\sigma}{dt} = q_o[\varepsilon - \alpha\Delta T] + q_1\frac{d}{dt}[\varepsilon - \alpha\Delta T]$, in which ε is now the total strain, which is identically zero, because the bar is restrained. Hence, obtain the differential equation: $p_o\sigma + p_1\frac{d\sigma}{dt} = -q_o\alpha\Delta T$. Cast this equation in the standard form: $\frac{d\sigma}{dt} + \frac{p_o}{p_1}\sigma = -\frac{q_o}{p_1}\alpha\Delta T$; using the integrating factor $u = e^{t/\tau}$, where $\tau \equiv p_1/p_0$, integrates the equation to arrive at the result.

References

1. R.M. Christensen, *Theory of Viscoelasticity*, 2nd edn. (Academic Press, Waltham, Massachusetts, 2003), pp. 90–94, 94–96, 77–87
2. J.J. Aklonis, W.J. MacKnight, *Introduction to Polymer Viscoelasticity*, (Wiley, Hoboken, 1983), pp. 44–47, 62–65
3. M.L. Williams, R.F. Landel, J.D. Ferry, Viscoelastic properties of polymers. J. Am. Chem. Soc. **77**, 3701 (1955)
4. A.C. Pipkin, *Lectures on Viscoelasticity Theory*, (Springer, Berlin, 1972), pp. 98–102
5. I.M. Ward, J. Sweeney, *An Introduction to the Mechanical Properties of Solid Polymers*, 2nd edn. (Wiley, Hoboken, 2012), pp. 108–113
6. S.W. Tsai, H.T. Hahn, *Introduction to Composite Materials*, (Technomic Pub. Co., Lancaster, 1980), pp. 329–344
W.G. Knauss, I. Emri, *Volume change and the nonlinearly thermo-viscoelastic constitution of polymers*, (Polymer Eng. Sci., 27, 1987), pp. 86–100.

Material Property Functions and Their Characterization

Abstract

This chapter examines four topics of practical importance. It begins with an introduction to material characterization testing, covering stress relaxation, creep, constant rate, and dynamic tests. The chapter then introduces two types of analytical forms, typically used to describe mechanical constitutive property functions. One type, usually referred to as a Dirichlet-Prony series, is expressed as a finite sum of exponentials; the other form is a power law in time. This treatment is followed by a discussion of methods of inversion of material property functions given in Prony series form; both exact and approximate methods of inversion are presented. The chapter is completed with a discussion of practical ways to establish the numerical coefficients entering the analytical forms used to represent the WLF shift relation, the relaxation modulus, and the creep compliance. The use of a computer application available with the book, which was specifically developed to obtain the exact convolution inverse of function in a Prony series form, is also presented and its use is illustrated by means of some examples.

Keywords

Characterization · Test · Constitutive · Kernel · Numerical · Inversion · Analytical · Parameter · Prony · Dirichlet · Power · Law · Hereditary · Logarithmic · Transition · Property function · Relaxation · Creep · Compliance · Complex · Laplace · Transform · Convolution · Retardation · Time · Volterra · Polynomial · Equation · Roots · Shift function · Least squares

7.1 Introduction

This chapter examines four topics of practical importance, beginning with an introduction to material characterization testing, covering stress relaxation, creep, constant rate, and dynamic tests. The chapter then introduces two types of analytical forms which are typically used to describe mechanical constitutive property functions, such as relaxation modulus, creep compliance and coefficient of thermal expansion. One form is expressed as a finite sum of exponentials, which is frequently referred to as a Dirichlet-Prony series. The other form is a power law function of time. We then present both approximate and exact methods of finding the convolution inverse of material property functions of Prony type. The chapter ends with a discussion of practical ways to establish the numerical coefficients entering the WLF function and the analytical forms used to represent the relaxation modulus and the creep compliance.

7.2 Experimental Characterization

The objective of experimental characterization is the construction of the kernel functions that enter the constitutive equations. What follows restricts attention to the tests used to establish the relaxation modulus and the creep compliance. Alternatively, either one of these functions may be derived from the other, by inversion, as discussed in Chap. 2 and in Sect. 7.4. Dynamic test methods used to establish property functions for steady-state conditions are only briefly mentioned, because the theory is treated extensively in Chap. 4.

7.2.1 Constant Strain Test

This test is generally known as stress relaxation. In it, a one-dimensional specimen is subjected to a specified target strain, ε_o, which is applied as fast as possible, and then held constant. This test is designed to yield the relaxation modulus, $M(t) = \sigma(t)/\varepsilon_o$, as the ratio of the stress response to the constant applied strain. The theoretical loading is $\varepsilon = \varepsilon_o H(t)$; the ensuing stress is calculated using Eq. (2.32a): $\sigma(t) = \int_{0^-}^{t} M(t - \tau)d\varepsilon(\tau)/d\tau$ and that $d\varepsilon/dt = \varepsilon_o\delta(t)$, which imply that $\sigma(t) = M(t)\varepsilon_o$. Consequently,

$$M(t) = \frac{\sigma(t)}{\varepsilon_o} \tag{7.1}$$

Quite simply, then, what is required in a stress relaxation test is to keep a record of the load as a function of time, and then divide that load by the cross-sectional area of the test specimen and by the enforced target strain. Since total stress relaxation typically occurs over a long time, and the difference between the short- and long-term

Fig. 7.1 Relaxation modulus data collected at five temperatures for a natural rubber compound

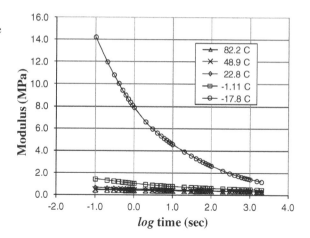

Fig. 7.2 Mismatch in response between theory and experiment in a stress relaxation test. The theoretical input is a step strain but the actual test is a ramp-and-hold strain

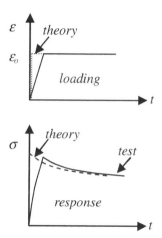

moduli may be several orders of magnitude, it is convenient to plot the test data on a semi-logarithmic or even, double-logarithmic scales. In practice, relaxation tests are performed at five or six different temperatures, each some 15 °C apart from its neighbors, so as to allow a sufficient expansion of the laboratory timescale of observation. To account for material variability, three to five replications are advisable. Table 7.1 presents relaxation modulus data for a natural rubber compound. The data were collected at five temperatures and averaged over three replications.

These data are displayed in a semi-logarithmic scale in Fig. 7.1 and will be used in later sections to demonstrate how to establish the coefficients of the WLF shift function, as well as of two analytical forms of the relaxation modulus.

In closing, it is pointed out that since it is not possible to apply an instantaneous strain, as the theory assumes, a discrepancy will always exist between theory and experimental data. This situation is illustrated in Fig. 7.2.

Table 7.1 Relaxation modulus data for a natural rubber compound

Temperature	−17.8 °C	−1.11 °C	22.8 °C	48.9 °C	82.2 °C
Test time, t (s)	Measured modulus, $M(t)$ (MPa)				
0.1	14.25	1.432	0.654	0.499	0.409
0.2	12.00	1.317	0.622	0.489	0.402
0.3	10.84	1.245	0.601	0.482	0.398
0.4	10.07	1.195	0.586	0.475	0.395
0.5	9.46	1.157	0.575	0.471	0.392
0.6	9.02	1.127	0.566	0.467	0.389
0.7	8.67	1.103	0.558	0.463	0.387
0.8	8.37	1.080	0.552	0.460	0.386
0.9	8.12	1.060	0.547	0.458	0.384
1.0	7.89	1.044	0.541	0.455	0.383
2.0	6.60	0.946	0.511	0.440	0.373
3.0	5.97	0.894	0.495	0.432	0.367
4.0	5.60	0.863	0.484	0.425	0.366
5.0	5.34	0.841	0.477	0.421	0.363
6.0	5.12	0.821	0.471	0.417	0.361
7.0	4.96	0.807	0.466	0.415	0.358
8.0	4.81	0.796	0.461	0.412	0.358
9.0	4.69	0.785	0.457	0.410	0.355
10.0	4.58	0.777	0.454	0.408	0.355
20.0	3.91	0.722	0.434	0.395	0.347
30.0	3.55	0.697	0.423	0.389	0.344
40.0	3.31	0.678	0.416	0.384	0.341
50.0	3.14	0.664	0.411	0.381	0.339
60.0	3.00	0.654	0.407	0.378	0.339
70.0	2.88	0.643	0.403	0.376	0.336
80.0	2.79	0.635	0.400	0.374	0.335
90.0	2.71	0.627	0.398	0.372	0.335
100.0	2.64	0.619	0.395	0.371	0.333
200.0	2.20	0.585	0.381	0.363	0.328
300.0	1.97	0.563	0.373	0.358	0.325
400.0	1.82	0.549	0.367	0.355	0.324
500.0	1.72	0.540	0.364	0.353	0.322
600.0	1.63	0.532	0.360	0.350	0.322

(continued)

Table 7.1 (continued)

Temperature	−17.8 °C	−1.11 °C	22.8 °C	48.9 °C	82.2 °C
Test time, t (s)	Measured modulus, $M(t)$ (MPa)				
700.0	1.56	0.526	0.358	0.350	0.320
800.0	1.51	0.515	0.356	0.348	0.319
900.0	1.46	0.511	0.354	0.346	0.319
1,000.0	1.42	0.508	0.352	0.346	0.318
1,500.0	1.27	0.485	0.346	0.342	0.315
2,000.0	1.19	0.458	0.342	0.339	0.315

The source data shown in this table, and elsewhere in the text, may be accessed at "extras.springer.com"

7.2.2 Constant Stress Test

This test is generally known as a creep compliance test. In a creep compliance test, a one-dimensional specimen is subjected to a specified target stress, σ_o, that is applied as fast as possible and then held constant. This test is designed to yield the creep compliance, $C(t) = \varepsilon(t)/\sigma_o$, as the ratio of the strain response to the constant applied stress. The loading is $\sigma = \sigma_o H(t)$; the strain is calculated per Eq. (2.2): $\varepsilon(t) = \int_0^t C(t - \tau)d\sigma(\tau)/d\tau$, noting that: $d\sigma/dt = \sigma_o\delta(t)$, and hence, that: $\varepsilon(t) = \int_0^t C(t - \tau)\delta(\tau)d\tau = C(t)\sigma_o$. Consequently,

$$C(t) = \frac{\varepsilon(t)}{\sigma_o} \tag{7.2}$$

Therefore, what is required in a creep compliance test is to keep a record of the strain response as a function of time and divide it by the applied target stress. Creep compliance test data are also plotted in double-logarithmic coordinates. A typical graph of the creep compliance of a viscoelastic material is shown in Fig 1.14. Three analytical forms to represent one or more aspects of creep compliance functions, which are the counter parts of those used to represent relaxation moduli, are presented later on.

7.2.3 Constant Rate Test

Although the forcing function in this type of test can be either strain or stress, strain is commonly used because it is easier to control under laboratory conditions. In a constant strain rate test, a uniaxial specimen is subjected to the loading is $\varepsilon(t) = R \cdot t$, in which the rate R is a selected constant. This test is designed to produce the relaxation modulus at the selected temperature, by calculating the response using Eq. (2.1) and that $d\varepsilon/dt = R$, one gets $\sigma(t) = R \int_0^t M(t - \tau)d\tau$.

Introducing the change in variable $u = t - \tau$ converts this expression into $\sigma(t) = R \int_0^t M(u) du$. Differentiating under the integral sign leads to $\frac{d\sigma}{dt} = RM(t)$. Now, given that $\varepsilon(t) = R \cdot t$, and thus, $\frac{d}{dt} = R \frac{d}{d\varepsilon}$, results in

$$M(t) = \frac{d\sigma}{d\varepsilon}\bigg|_{\varepsilon = Rt} \tag{7.3}$$

7.2.4 Dynamic Tests

Material property functions appropriate for steady-state conditions (M^*, M', M'', C^*, C', C'', $\tan\delta$) are obtained by means of dynamic tests. Dynamic tests are carried out by forcing a sinusoidal excitation of varying frequency on a polymer sample that is held at a given temperature. There are several methods of dynamic testing [1]. However, most of them target certain types of deformation, such as torsion or bending, or a specific frequency range, and even material stiffness.

In principle, any of the quantities associated with the either the complex modulus ($|M^*|, M^*, M', M'', \delta, \tan\delta$) or complex compliance ($|C^*|, C^*, C', C'', \delta, \tan\delta$) can be used in dynamic characterization testing. However, the absolute values, $|M^*|$ and $|C^*|$, and loss angle, δ, are not typically used.

7.3   Analytical Forms of Constitutive Functions

Two types of analytical forms are examined in what follows which comply with the fading memory hypothesis and closed cycle condition and, in addition, provide elastic response under fast and slow processes, as discussed in Chap. 2. One of these forms, referred to as a Dirichlet-Prony series, is expressed as a finite sum of decaying exponentials; the other is a power law in time.

7.3.1 Material Property Functions in Prony Series Form

A Prony series is a finite sum of decaying exponentials in time. The individual decaying exponentials allow modeling of relaxation—by adding the exponentials—as well as creep—by adding the complements to one, of each exponential term. The following Dirichlet–Prony series is frequently used to represent the relaxation modulus for viscoelastic solids:

$$M(t) = M_e + \sum_{i=1}^{N} M_i e^{-t/\tau_i} \tag{7.4}$$

The parameters τ_i have dimension of time and are usually called time parameters or time constants. In a phenomenological sense, these time constants may be thought of as characteristic relaxation times of the material, but in general have no more meaning than that of curve-fit parameters. The coefficients M_e and M_i are real, positive constants.

Expression (7.4) implies the following relationship between the extreme values of $M(t)$:

$$M(0) = M_e + \sum_{i=1}^{N} M_i \equiv M_g \tag{7.5}$$

Here, as discussed in Chap. 1:

$$M_e \equiv M_\infty = \begin{cases} \neq 0; & \text{for viscoelastic solids} \\ \equiv 0; & \text{for viscoelastic fluids} \end{cases} \tag{7.6}$$

Another Prony series form frequently used to describe relaxation functions is in terms of the glassy modulus, M_g, and may be derived from the previous one. Use (7.5) to express M_e in terms of M_g and insert the result into (7.4), to get:

$$M(t) = M_g - \sum_{i=1}^{N} M_i(1 - e^{-t/\tau_i}) \tag{7.7}$$

By contrast, the creep compliance function increases from its glassy value, C_g, as its argument increases or, conversely, decreases from its long-term value, C_e, as $t \to 0$. With this, the creep compliance counterparts of (7.4) and (7.7) for solids are, respectively,

$$C(t) = C_e - \sum_{i=1}^{N} C_i e^{-t/\lambda_i} \tag{7.8}$$

$$C(t) = C_g + \sum_{i=1}^{N} C_i(1 - e^{-t/\tau_i}) \tag{7.9}$$

For viscoelastic fluids, it is usual to add the term t/η to represent Newtonian viscous flow.

Just as for the relaxation function, the following expressions are derived from (7.8) and (7.9); and here, too, the coefficients C_e and C_i, are real and positive:

$$C_g \equiv C(0) = C_e - \sum_{i=1}^{N} C_i \tag{7.10}$$

$$C_e \equiv C(\infty) = C_g + \sum_{i=1}^{N} C_i \qquad (7.11)$$

Before leaving this subject, it is important to point out that since the hereditary form of thermal strains is expressed through a creep-like integral, the thermal expansion creep function may be modeled using Prony series of the same type as those in (7.8) or (7.9) [2].

7.3.2   Material Property Functions in Power-Law Form

A power law is an incomplete polynomial of fractional power. A popular power law form used to represent relaxation functions is as follows:

$$M(t) = M_e + \frac{M_g - M_e}{\left(1 + t/\theta\right)^p} \qquad (7.12)$$

In this expression, the time constant θ represents a value somewhere in the middle of the transition region; the exponent, p, is a positive real value, and the other symbols are as defined earlier. As can be seen, this form retrieves the glassy and equilibrium values of the relaxation modulus. The analogous form for the creep compliance is

$$C(t) = C_g + \frac{C_e - C_g}{\left(1 + \hat{\theta}/t\right)^{\hat{p}}} \qquad (7.13)$$

Here, the notations $\hat{\theta}$ and $\hat{p}$ for the corresponding time constant and fractional power are used to emphasize that they are different from those for the relaxation function.

For mathematical convenience, and in cases when only behavior in the transition region is of relevance, the following simplified forms of (7.12) and (7.13) are used:

$$M(t) = M_t t^{-p} \qquad (7.14)$$

$$C(t) = C_t t^{\hat{p}} \qquad (7.15)$$

7.4   Inversion of Material Property Functions

Two methods are presented which are especially useful to find the convolution inverse of relaxation modulus or creep compliance functions when they are expressed as finite sums exponentials. In both methods, the inverse function is sought by enforcing the equivalent inversion requirements expressed in (2.20) and (2.21) that

$$\int_{0^-}^{t} M(t-\tau)\frac{d}{d\tau}C(\tau)d\tau = M(t)C_g + \int_{0^+}^{t} M(t-\tau)\frac{d}{d\tau}C(\tau)d\tau = H(t) \qquad (7.16a)$$

$$\int_{0^-}^{t} C(t-\tau)\frac{d}{d\tau}M(\tau)d\tau = C(t)M_g + \int_{0^+}^{t} C(t-\tau)\frac{d}{d\tau}M(\tau)d\tau = H(t) \qquad (7.16b)$$

Here, $H(t)$ represents the unit step function [c.f. Appendix A]. The difference between the inversion methods lies in the manner in which the inversion requirement is met. The first method is approximate and enforces the inversion requirement in a least-squares sense in Laplace-transform space. The second method enforces the requirement in exact form in the time domain.

7.4.1 Approximate Inversion of Material Property Functions

Several approximate methods of Laplace transform inversion have been proposed in the literature. One of these, introduced by Schapery [3], is presented here, because it is especially attractive for finding approximate inverse Laplace transform of functions in general and not necessarily related by a convolution integral.

  In Schapery's approach, $\bar{f}(s)$ is taken as the known Laplace transform of the function $f(t)$ being sought. In addition, it is assumed that $f(t)$ can be approximated by a Prony series[1]:

$$f(t) \approx f_A(t) = \sum_{j=1}^{N} A_j e^{-t/\tau_j} \qquad (7.17)$$

In this expression, the τ_j's are prescribed positive constants,[2] and the coefficients A_j are to be determined in such a manner as to minimize the total squared error introduced by the approximation, $f_A(t)$, in the domain of $f(t)$:

$$e(f, f_A) \equiv \int_{0}^{\infty} [f(t) - f_A(t)]^2 dt \qquad (a)$$

---

[1] This condition is always met by relaxation or creep compliance functions, because a series of exponentials is complete in that it can represent any continuous function to any desired degree of accuracy, if enough terms are used in the representation.

[2] Typically, the time parameters are more or less arbitrarily taken at each of the several logarithmic cycles spanning the available data, such as at 10^{-5}, 10^{-4},...,10^3, 10^4 min, without worrying much about their relationship to any intrinsic response times (relaxation or creep) of the material in question.

Minimization with respect to the coefficients, A_i, demands that $\partial e/\partial A_i = 0$, for all i. Upon introducing (7.17) and simplifying, this requirement becomes

$$\int_0^\infty [f(t) - f_A(t)] e^{-t/\tau_i} dt = 0; \quad i = 1,\ldots,N \tag{b}$$

With the definition of the Laplace transform of an arbitrary function, f, of exponential order: $\bar{f}(s) \equiv \int_0^\infty e^{-st} f(t) dt$, (b) can be expressed as

$$\bar{f}_A(s)\big|_{s=1/\tau_i} = \bar{f}(s)\big|_{s=1/\tau_i} \tag{c}$$

Inserting the Laplace transform of the sum of exponentials in (7.17) leads to [c.f. Appendix A]:

$$\sum_{k=1}^n \frac{A_k}{s + 1/\tau_k}\bigg|_{s=1/\tau_i} = \bar{f}(s)\big|_{s=1/\tau_i}; \quad i = 1,\ldots,n \tag{7.18}$$

Quite clearly, when all the τ_i's are known, or assumed to be so, (7.18) represents a system of linear equations in the unknown coefficients A_k of the Dirichlet–Prony series. This system may be put in familiar matrix form as follows:

$$[F_{ik}]\{A_k\} = \{b_i\};$$

$$F_{ik} \equiv \frac{1}{\frac{1}{\tau_i} + \frac{1}{\tau_k}}; \quad b_i \equiv \bar{f}(s)\big|_{s=\frac{1}{\tau_i}}; \quad i = k = 1,\ldots,n \tag{7.19}$$

To obtain the convolution inverse of any viscoelastic material property function using this method, the target property function takes the place of the unknown function, $f_A(t)$, while $f(t)$ represents the known, source function. For the specific case of relaxation and creep compliance functions, assume for definiteness that the relaxation modulus, $M(t)$, is the source function, $f(t)$; the creep compliance, $C(t)$, is the target $f_A(t)$, being sought. Then, inverting (2.31): $\bar{M}(s)\bar{C}(s) = 1/s^2$, yields:

$$f_A(t) \equiv C(t) = \mathcal{L}^{-1}\left\{\frac{1}{s^2\bar{M}(s)}\right\} \equiv f(t) \tag{7.20}$$

In other words:

$$\bar{f}(s) = \bar{C}_A(s); \; and \; \bar{f}(s) = \frac{1}{s^2\,\bar{M}(s)} \tag{7.21}$$

As was evident in Chap. 3, the time parameters associated with relaxation functions differ from retardation or creep times. Thus, when dealing with relaxation moduli and creep compliance functions, the assumption made here, of utilizing the same time parameters for both the known function and the inverse function being sought, is not altogether correct. Another disadvantage of the method is that there is no guarantee that all the Prony coefficients of the inverse function will be positive, as they should be for real materials. The more general case, which does not require that the time parameters of the inverse function be known before hand, and which will yield positive coefficients for the Prony series of the inverse function, is discussed subsequently.

7.4.2   Exact Inversion of Material Property Functions

In practice, mutually inverse property functions, such as the relaxation modulus and creep compliance, are obtained from independent tests, from approximate inverse relationships pertinent to the processes at hand or by enforcing the inversion requirements (7.16a, b) in one form or another. The approach, here, is to enforce that requirement exactly.

The two expressions (7.16a, b) may be cast in the standard form of Volterra integral equations of the first kind in the unknown functions $M(t)$ and $C(t)$, respectively [c.f. Appendix A]. Given that such equations always admit a solution, it is possible, at least in principle, to obtain either function from the other.

Although Volterra integral equations of the first kind always admit a solution, it may not be possible to obtain it exactly. However, when the functions involved are represented by series of decaying exponentials, the target function is found by satisfying the inversion requirement exactly.

For ease of computation, the reciprocal relaxation times, $\alpha_i \equiv 1/\tau_i$, and the reciprocal retardation times, $\beta_i \equiv 1/\lambda_i$, will be used. This and the physically reasonable assumption that the Prony series of both the source and target functions have the same number of terms, N, transform the Prony series listed in (7.4) and (7.8) into the following forms:

$$M(t) = M_e + \sum_{i=1}^{N} M_i \cdot e^{-\alpha_i t} \tag{7.22}$$

$$C(t) = C_e + \sum_{r=1}^{N_J} (-C_r) e^{-\beta_i t_r} \tag{7.23}$$

Since, save for a sign, both functions have the same analytical representation, it does not matter which of the functions is known and which is not. To fix ideas, it is assumed that the relaxation modulus, $M(t)$, is the one available, and endeavor to obtain from it the creep compliance, $C(t)$.

For clarity, only a few intermediate steps of the lengthy but straightforward derivation are presented next. Taking (7.22) and (7.23) into (7.16a) leads to

$$(M_e + \sum_{i=1}^{N_G} M_i e^{-\alpha_i t}) \cdot C_g + \int_0^t [M_e + \sum_{i=1}^{N} M_i e^{-\alpha_i(t-\tau)}] \frac{\partial}{\partial \tau} C(\tau) d\tau = 1 \qquad (a)$$

Integrating directly the first portion of the integral and by parts the second and grouping like terms produce

$$C_g \sum_{i=1}^{N} M_i e^{-\alpha_i t} + M_e(C_g + \sum_{r=1}^{N} C_r) - M_e \sum_{r=1}^{N} C_r e^{-\beta_r t} + C(t) \sum_{i=1}^{N} M_i$$

$$- C_g \sum_{i=1}^{N} M_i e^{-\alpha_i t} - \sum_{i=1}^{N} M_i \alpha_i e^{-\alpha_i t} \int_0^t \{C_g + \sum_{r=1}^{N} C_r(1 - e^{-\beta_r \tau})\} e^{\alpha_i \tau} d\tau = 1 \qquad (b)$$

Using (7.10) and (7.11), the last integral evaluates to

$$\int_0^t \{C_g + \sum_{r=1}^{N} C_r(1 - e^{-\beta_r \tau})\} e^{\alpha_i \tau} d\tau = \frac{C_e}{\alpha_i}(e^{\alpha_i t} - 1) - \sum_{r=1}^{N} \frac{C_r}{\alpha_i - \beta_r}[e^{(\alpha_i - \beta_r)t} - 1]$$

$$(c)$$

Inserting this expression into the previous one, using relations (7.7) to (7.11), together with (2.22): $M_g = 1/C_g$ and (2.23): $M_e = 1/C_e$, as appropriate, and collecting like terms yield

$$\sum_{i=1}^{N} M_i e^{-\alpha_i t}\left\{C_e - \sum_{r=1}^{N} \frac{C_r \alpha_i}{\alpha_i - \beta_r}\right\} - \sum_{r=1}^{N} C_r e^{-\beta_r t}\left\{M_g - \sum_{i=1}^{N} \frac{M_i \alpha_i}{\alpha_i - \beta_r}\right\} \qquad (d)$$

Because the exponential functions $e^{-\alpha_i t}$ and $e^{-\beta_r t}$ are linearly independent, expression (d) will be satisfied for all choices of the original function only if the quantities in braces are independently equal to zero. This condition produces the equations [4]:

$$M_g - \sum_{i=1}^{N} \frac{M_i \cdot \beta_r}{\beta_r - \alpha_i} = 0; \quad r = 1, \ldots, N \qquad (7.24)$$

$$C_e - \sum_{r=1}^{N} \frac{C_r \cdot \beta_r}{\beta_r - \alpha_i} = 0; \quad i = 1, \ldots, N \qquad (7.25)$$

The first of these relations contains only the time constants, β_r, as unknowns; hence, it may be used to establish them. As a function of a single variable, β, expression (7.24) involves N different terms, each one having $(\beta - \alpha_i)$ in its denominator. Hence, (7.24) is a polynomial equation, $\Phi(\beta) = 0$, of degree N in β.

$$\Phi(\beta) \equiv M_g - \sum_{i=1}^{N} \frac{M_i \cdot \beta}{\beta - \alpha_i} = 0 \qquad (7.26)$$

The N roots of this equation are the characteristic times, β_r, of the target function, $C(t)$. Once the roots are known, Eq. (7.25) becomes a system of linear algebraic equations in the N unknown coefficients, C_r, of the Prony series of the target function. The solution of this linear system can be obtained with a linear solver, after it is noted that the equilibrium value of the inverse function is known: $C_e = 1/M_e$.

The N roots, β_r, of the polynomial equation may be located by examining the behavior of the polynomial $\Phi(\beta)$ as β approaches its extremes (0 and ∞), as well as when it approaches each characteristic time, α_i, of the source function. These considerations reveal that Φ tends to M_g and M_e, as β approaches 0 and ∞, respectively. In addition, Φ tends to $-\infty$ and $+\infty$, as β approaches α_i from above and from below, respectively. Since Φ is continuous in each subinterval (α_i, α_{i+1}), the graph of $\Phi = 0$ must be as depicted in Fig 7.3.

From this graph follows that the time constants α_i and β_i of the original and inverse functions are nested. Explicitly,

$$\alpha_i < \beta_i < \alpha_{i+1}; \quad for\ i = 1, N-1; \quad and: \ \beta_N > \alpha_N \qquad (7.27)$$

This result is the mathematical description of the fact that the creep times, $\lambda_i \equiv 1/\beta_i$, are shorter than the corresponding relaxation times, $\tau_i \equiv 1/\alpha_i$. In practical terms,

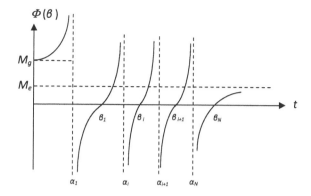

Fig. 7.3 Behavior of the polynomial equation in the retardation times. M is the source function and C, the target.

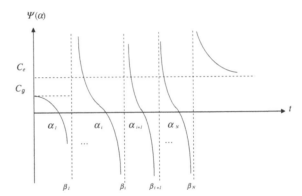

Fig. 7.4 Behavior of the polynomial equation in the relaxation times. C is the source function, and M, the target

nesting of the time constants allows efficient use of bisection to obtain the characteristic times as the roots of the polynomial equation $\Phi(\beta) = 0$.

As indicated before, the roles of functions M and C may be interchanged. If C were the source function and M the target, (7.25) would be a polynomial equation, $\Psi(\alpha) = 0$, in α, while (7.24) would represent the linear system in the unknown Prony coefficients, M_i. In this case, one has that $\Psi(\alpha = 0) = C_g$ and $\Psi(\alpha \to \infty) = C_e$. Also, Ψ tends to $+\infty$ and $-\infty$, as α approaches β_r from above and from below, respectively, and is continuous in each subinterval (β_i, β_{i+1}). Under these conditions, the graph of $\Psi = 0$ mirrors that of $\Phi = 0$, as indicated in Fig. 7.4.

Example 7.1　The relaxation modulus of a viscoelastic material is given by $M(t) = M_e + M_1 e^{-\alpha_1 t}$. Use the method presented here to find the exact creep compliance.
Solution:

Take the creep compliance in the form $C(t) = C_e - C_1 e^{-\beta_1 t}$. Although in this simple case it is straightforward to obtain the coefficients without resorting to the inversion method, we proceed as in the derivation of the method and insert the analytical form of the given creep compliance into (7.16a, b) to write

$$M(t)C_g + \int_{0+}^{t} \left[M_e + M_1 e^{-\alpha_1(t-\tau)} \right] \left[\beta_1 C_1 e^{-\beta_1 \tau} \right] d\tau = 1$$

We now replace $M(t)$ in the first term by its analytical form, perform the indicated integration, and collect terms to get

$$\left[M_1(C_e - C_1) - M_1 C_1 \frac{\beta_1}{\alpha_1 - \beta_1} \right] e^{-\alpha_1 t} - \left[M_1 C_1 \frac{\beta_1}{\alpha_1 - \beta_1} + M_e C_1 \right] e^{-\beta_1 t} = 0$$

Linear independence of the exponential functions implies the vanishing of each of the expressions in brackets. Setting the expression inside the second bracket to zero allows solving for the inverse retardation time, $\beta_1 = M_e \alpha_1 / (M_e + M_1) \equiv M_e \alpha_1 / M_g$. Now setting the expression inside the first bracket to zero, using the value of β_1 just established and simplifying, produces that $C_1 = M_1 C_g / M_e$; which upon using that $C_g = 1/M_g$, may be expressed as $C_1 = M_1/(M_e M_g)$. This is the same value one would obtain from (7.11) and the condition $C_g = 1/M_g$.

Accompanying this text is a computer application called "Prony-Inverse," which was specifically developed to invert Dirichlet–Prony series using the method presented in this section. Prony-Inverse is available at "extras.springer.com," and its use is explained in Sect. 7.6.

7.5   Numerical Characterization of Material Property Functions

This section describes how to use a standard spreadsheet program to curve-fit analytical models to material property function data. The models specifically examined include the WLF equation to fit shift function test data, and the Dirichlet–Prony series and power-law form to fit relaxation modulus and creep compliance data. In each case, the data provided in Table 7.1 will be used to exemplify the procedure. As it turns out, judicious treatment of the analytical forms in each case allows expressing them as multi-linear functions of the unknown coefficients, which can then be obtained by linear least-squares fitting. Therefore, in general, we will consider a multi-linear function, z, of the N_C variables x_k:

$$z = a_1 x_1 + a_2 x_2 + a_3 x_3 + \cdots a_{N_C} x_{N_C} \equiv \sum_{k=1}^{N_C} a_k x_k \qquad (7.28)$$

When this expression is evaluated at a data point, (x_i, z_i), an error, ε_i, results. Using (7.28), this error may be expressed as

$$\varepsilon_i = z_i - \sum_{k=1}^{N_C} a_k x_{ki} \qquad (7.29)$$

The method of least squares is based on minimizing the sum of the squared errors that result from the approximation. The mathematical condition for the summed squared error to attain a minimum is the vanishing of the first partial derivatives of the approximating function with respect to the unknown coefficients, a_j:

$$\frac{\partial}{\partial a_j} \sum_{i=1}^{N_p} \varepsilon_i^2 = 0, \quad j = 1, \ldots, N_C \qquad (7.30)$$

Using (7.29) with this expression and collecting terms lead to the following system of linear algebraic equations in the unknown coefficients:

$$\left[F_{jk}\right]\{a_k\} = \{b_j\}, \quad j,k = 1, \ldots, N_c \tag{7.31a}$$

$$F_{jk} \equiv \sum_{i=1}^{N_p} x_{ji} x_{ki}, \quad j,k = 1, \ldots, N_c \tag{7.31b}$$

$$b_j \equiv \sum_{i=1}^{N_p} x_{ji} z_i, \quad j,k = 1, \ldots, N_c \tag{7.31c}$$

7.5.1   Numerical Characterization of the Shift Function

The procedure described in Sect. 6.2 to shift viscoelastic property function test data may be readily implemented as follows:

1. Collect the available time-versus-property test data $[\log(t_i), P_i]$ at each of several temperatures, T_k. Use the data provided in Table 7.1 for this purpose.
2. Select the reference temperature, T_r. Without loss of generality, pick $T_r = 22.8$ °C.
3. For each test temperature, T_k:
   a. Provide an initial value of $\log(a_{T_k}) = 0$
   b. Add the data: $\log(t_i) - \log(a_{T_k})$.
   c. Plot the data: $[\log(t_i) - \log(a_{T_k}), P_i]$. At this time, $\log(a_{T_k}) = 0$, so that the plots correspond with the original data: $[\log(t_i), P_i]$.
   At this point, the shifted times corresponding to scheme should look as in Table 7.2 and the corresponding plots, as in Fig. 7.1, given earlier.
4. Starting with the temperature that is closest to T_r—on either side—guess a value of $\log(a_{T_k})$ and observe the shift of the curve relative to curve chosen as reference:

Table 7.2 Sample relaxation modulus test time data for a natural rubber before temperature shifting

i	T_k		Temperature			
		−17.8	−1.11	22.8	48.9	82.2
	$\log(a_{\mathrm{Tk}}) =$	0	0	0	0	0
	t (sec)		$\log(t/a_{\mathrm{Tk}})$			
1	0.1	−1.00000	−1.00000	−1.00000	−1.00000	−1.00000
2	0.2	−0.69897	−0.69897	−0.69897	−0.69897	−0.69897
3	0.3	−0.52288	−0.52288	−0.52288	−0.52288	−0.52288
...	0.4	−0.39794	−0.39794	−0.39794	−0.39794	−0.39794

a. If the guessed value shifts the curve away from the reference, change the sign of the guessed value.

b. Otherwise, fine-tune the guess until the shifted curve aligns satisfactorily with the reference curve—or its extension.

5. Repeat step 4 until the data for all available temperatures have been shifted. This completes the shifting process.

At this stage, the shifted times should look as indicated in Table 7.3, and the shifted data should look as shown in Fig. 7.5.

Although the pairs of temperature shift values $[T_k, log(a_{T_k})]$ fully characterize the shift function, it is sometimes possible and convenient to fit an analytical expression, such as the WLF function to the data. This can be done rather easily by rewriting (6.3) as

$$log a_T(T, T_r) \cdot (T - T_r) = -C_1 \cdot (T - T_r) - C_2 \cdot log a_T(T, T_r) \qquad (7.32)$$

Table 7.3 Sample relaxation modulus test time for a natural rubber shifted to construct master curve

i	T_k	Temperature				
		−17.8	−1.11	22.8	48.9	82.2
	$log(a_{Tk}) =$ t (sec)	6.5	2.6	0	−1.2	−2.5
		$log(t/a_{Tk})$				
1	0.1	−7.50000	−3.60000	−1.00000	0.20000	1.50000
2	0.2	−7.19897	−3.29897	−0.69897	0.50103	1.80103
3	0.3	−7.02288	−3.12288	−0.52288	0.67712	1.97712
	0.4	−6.89794	−2.99794	−0.39794	0.80206	2.10206

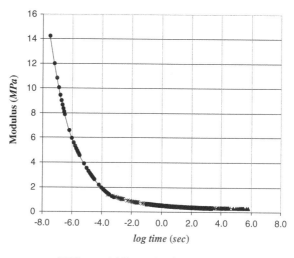

Fig. 7.5 Tensile relaxation modulus data for a natural rubber shifted to construct the master curve at 22.8 °C

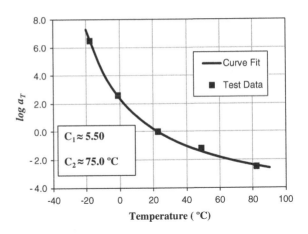

Fig. 7.6 Curve fit of the WLF shift function to tensile relaxation modulus data for a natural rubber compound

Introducing the change in variables

$$z \equiv \log a_T(T, T_r) \cdot (T - T_r); \quad x_1 \equiv -(T - T_r); \quad x_2 \equiv -\log a_T(T, T_r) \quad (7.33)$$

leads to the following expression in the standard form (7.28):

$$z = C_1 \cdot x_1 + C_2 \cdot x_2 \quad (7.34)$$

A linear least-square fit of this expression to the test data provides the coefficients C_1 and C_2 of the WLF equation. Using (7.31a, b, c) leads to the system:

$$\begin{bmatrix} \sum_i^{N_p} x_{1i}^2 & \sum_i^{N_p} x_{1i} x_{2i} \\ \sum_i^{N_p} x_{1i} x_{2i} & \sum_i^{N_p} x_{2i}^2 \end{bmatrix} \begin{Bmatrix} C_1 \\ C_2 \end{Bmatrix} = \begin{Bmatrix} \sum_i^{N_p} x_{1i} z_i \\ \sum_i^{N_p} x_{2i} z_i \end{Bmatrix} \quad (7.35)$$

The solution, $C_1 \approx 5.499$ and $C_2 = 74.96$, at $T_r = 22.8\ °C$, is depicted in Fig. 7.6.

7.5.2 Numerical Characterization of Modulus and Compliance

The analytical expressions used to represent the relaxation modulus, be they of Dirichlet-Prony series or power-law type, are of the same form as for the creep compliance. This can be seen by comparing expression (7.4) with (7.8), (7.12) with (7.13), and (7.14) with (7.15). Without loss of generality, we demonstrate how to numerically characterize either of these material property functions using expressions listed for the relaxation modulus only. We address both Prony series and power law forms.

(a) Functions in PronySeries Form

The Dirichlet-Prony series used to model relaxation and creep compliance, listed in (7.4) and (7.8), respectively, are of the same following general form:

$$P(t) = P_e + \sum_{k=1}^{N_c} P_k e^{t/\alpha_k} \tag{7.36}$$

A practical procedure to fit this expression to test data, when the time parameters are known, is as follows:

1. Shift the test data in accordance with the procedure described in Sect. 7.5.1 and plot them as in Fig. 7.5. In this case, use the data in Table 7.1, with the shift values listed in Table 7.2 in the row marked $log a_{T_k}$.
2. Use the plotted data to guess the equilibrium value, P_e. Even though this is not strictly necessary, it is physically meaningful and convenient to do so.
a. If you make this assumption, rewrite (7.36) as:

$$P(t) - P_e = \sum_{k=1}^{N_c} P_k e^{t/\alpha_k} \tag{7.37a}$$

and introduce the following change of variables:

$$z(t) \equiv P(t) - P_e; \quad a_k \equiv P_k; \quad x_k \equiv e^{t/\alpha_k}; \quad k = 1, \ldots N_C \tag{7.37b}$$

A good initial estimate of P_e for the present case is $P_e = 0.30$ MPa.
b. Otherwise, introduce the following change of variables:

$$z(t) \equiv P(t); \quad a_0 \equiv P_e; \quad a_k \equiv P_k; \quad x_k \equiv e^{t/\alpha_k}; \quad k = 1, \ldots N_C \tag{7.37c}$$

With these changes, expression (7.36) is put in the standard form given in (7.28).
3. Apply the least-squares procedure to (7.37a, b, c) and obtain the unknown coefficients as the solution of the resulting system, which is of the form in (7.31a, b, c).

Regarding the time parameters, α_k, one can either assume, more or less arbitrarily, the N_C values required to span the range of the data or make a more meaningful estimate of their values by trial and error. Although it is easy to do, arbitrarily picking the time parameters in the range of the test data will often lead to undesirable oscillations of the predicted property, which are absent from the data, but are due to the fact that each exponential dominates in the neighborhood of its time parameter [4]. An easy alternative to minimize this effect is to estimate the time parameters from a trial-and-error fit of the test data. This can be done per the following procedure:

1. Perform steps 1 and 2 of the previous procedure.
   a. Plot the analytical expression with initially arbitrarily selected values of the time parameters and coefficients, including P_e.
   b. Adjust P_e so that the prediction matches the long-term test data.
2. Sequentially guess a time parameter and a corresponding coefficient and adjust their values trying to bring the predicted curve close to the test data.

This trial-and-error process may be somewhat lengthy, the first couple of times it is tried. However, the end results are usually ready to use even if optimization of the coefficients through least squares is not carried out. In addition, unlike the unconstrained least-squares procedure presented earlier, which would yield the coefficients that best fit the data, irrespective of their sign, the user has control over the sign of the coefficients—which should always be positive, as is required by the fading memory hypothesis [c.f. Chap. 2]. In the present case, nine exponential terms and the equilibrium value were used to obtain a reasonably good fit to the test data, as indicated in Fig. 7.7. The figure also shows each of the exponential terms, used in the curve fit. The figure clearly shows the region of dominance of each exponential term.

The curve fit shown in the figure was obtained with the coefficients listed in Table 7.4 together with the originally suggested equilibrium value $P_e = 0.30$.

(b) Functions in PowerLaw Form.

As with the Prony series expressions, the power law forms used to model relaxation and creep compliance listed in (7.12) and (7.13) are of the following general form:

$$P(t) = P_e + P_t\left(1 + {}^t\!/_\theta\right)^p \tag{7.38}$$

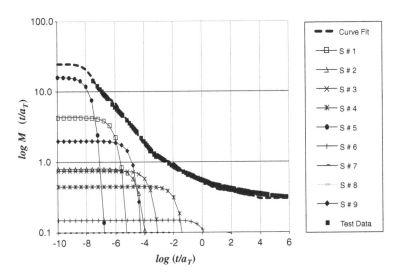

Fig. 7.7 Dirichlet–Prony series fit to relaxation modulus data of a natural rubber compound

Table 7.4 Trial-and-error curve fit of Prony series to relaxation modulus data of a natural rubber compound

P_k	16.0	4.3	0.8	2.0	0.75	0.45	0.09	0.15	0.10
a_k	$3.333.10^{-8}$	$1.538.10^{-6}$	$5.000.10^{-5}$	$2.000.10^{-5}$	$4.000.10^{-4}$	$2.857.10^{-2}$	$6.667.10^{-1}$	$3.333.10^{0}$	$2.500.10^{3}$

A practical procedure to fit this expression to test data is as follows:

1. Shift the test data in accordance with the procedure described in Sect. 7.5.1 and plot them as in Fig. 7.5. The data listed in Table 7.1 are used, after shifting, to illustrate the procedure.
2. Assume that the equilibrium value, P_e, the transient value, $P_t \equiv P_g - P_e$, and the time parameter θ are known and recast Eq. (7.38) in the form:

$$P(t) - P_e = P_t(1 + t/\theta)^p \tag{7.39}$$

3. Take the common logarithm of this expression and introduce the notation:

$$z(t) \equiv \log(P(t) - P_e);$$

$$x_1 \equiv 1, \quad x_2 \equiv \log(1 + t/\theta), \quad a_1 \equiv \log P_t, \quad a_2 \equiv p \tag{7.40a}$$

to transform the original Eq. (7.38) into the standard form (7.28), as

$$z(t) = a_1 x_1 + a_2 x_2 \tag{7.41a}$$

In practice, it is more advantageous to include the parameter P_t in the trial-and-error process and use the following notation, instead

$$z(t) \equiv \log(P(t) - P_e) - \log P_t, \quad x_1 \equiv \log(1 + t/\theta), \quad a_1 \equiv p \tag{7.40b}$$

to transform the original equation into the standard form as

$$z(t) = a_1 x_1 \tag{7.41b}$$

4. Make an initial estimate of P_e (and P_t, if needed) from the equilibrium and glassy values that you think the data would extrapolate to. Take the time parameter θ, as a value near where the data would show the steepest change. These two (alternatively, three) values will be adjusted during the fitting exercise.
5. Plot Eq. (7.38) with the initial guesses, and least-squares fit the data by means of (7.31a, b, c). This will yield the values of p and $\log(P_t)$ if (7.40a) is used or p, if (7.41b) is used instead.
6. Iteratively adjust the values of P_e, θ, and P_t, if needed, and perform the least-squares fit until the predictions match the data to your satisfaction.

Fig. 7.8 Trial-and-error least-squares fit of a power law form to relaxation modulus data of a natural rubber compound

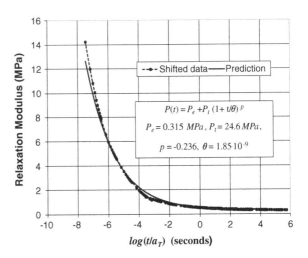

$$P(t) = P_e + P_t (1 + t/\theta)^P$$
$$P_e = 0.315 \, MPa, \, P_t = 24.6 \, MPa,$$
$$p = -0.236, \, \theta = 1.85 \cdot 10^{-9}$$

An acceptable solution, such as shown in Fig. 7.8, is obtained in just a couple of minutes. In the present case, the trial values used in the curve fit shown were $P_e = 0.315$ MPa, $P_t = 24.6$ MPa, and $\theta = 1.85 \cdot 10^{-9}$ s; the final least-square value of the time exponent was $p = -0.236$.

Standard linear regression may be used to fit power law functions of the forms listed in (7.14) and (7.15) to test data, using the following general form:

$$P(t) = P_t t^P \tag{7.42}$$

In this case, taking common logarithms and introducing the change in variables

$$z(t) \equiv \log P(t), \quad a_1 \equiv \log(P_t), \quad x_1 \equiv 1; \quad a_2 \equiv p, \quad x_2 = \log(t) \tag{7.43}$$

transform the original expression (7.38) into the standard form (7.28) as

$$z(t) = a_1 + a_2 x_2 \tag{7.44}$$

The system of equations presented earlier as (7.35), for the case of two variables, is directly applicable here and takes the following explicit form:

$$\begin{bmatrix} N_p & \sum_i^{N_p} x_{2i} \\ \sum_i^{N_p} x_{2i} & \sum_i^{N_p} x_{2i}^2 \end{bmatrix} \left\{ \begin{matrix} a_1 \\ a_2 \end{matrix} \right\} = \left\{ \begin{matrix} \sum_i^{N_p} z_i \\ \sum_i^{N_p} x_{2i} z_i \end{matrix} \right\} \tag{7.45}$$

7.6   Source Data Files and Computer Application

Accompanying this text are several files containing the source data used in some of the examples, as well as a computer application called Prony-Inverse, specifically developed to obtain the exact convolution inverse of a Dirichlet–Prony series. Both the files and the computer application can be accessed at "extras.springer.com." Here, we list the names of the data files, identify their contents, and provide a guide to using Prony-Inverse.

7.6.1   Source Data Files

The files available at "extras.springer.com" for this book are as follows:

(1) NR and PBAN data for engineering viscoelasticity.

This is an Excel file containing three worksheets:

   (a) Modulus data for NR. This worksheet contains the relaxation modulus data for a natural rubber compound, shown in Table 7.1 of the text.

   (b) C and M for NR. This worksheet contains the Prony coefficients of the compliance and relaxation functions—in that order—of the natural rubber compound of Problem P.7.5.

   (c) M and C for PBAN. This worksheet contains the Prony coefficients for the relaxation modulus and creep compliance of the solid propellant of Problem P.7.4.

(2) PBAN-Modulus.dat. This file contains the information in the form required by Prony-Inverse to invert the relaxation modulus of the solid propellant of Problem P.7.4.

(3) NR-Compliance.dat. This file contains the information in the form required by Prony-Inverse to invert the creep compliance of the natural rubber compound of Problem P.7.5.

(4) PBAN-Compliance.dat. This file contains the information in the form required by Prony-Inverse to invert the creep compliance of the solid propellant of Problem P.7.6.

(5) NR-Modulus.dat. This file contains the information in the form required by Prony-Inverse to invert the relaxation modulus of the natural rubber compound of Problem P.7.6.

7.6.2   Computer Application

Prony-Inverse is a computer application developed exclusively to obtain the convolution inverse of Dirichlet–Prony series, by the exact method of inversion discussed in Sect. 7.4. Prony-Inverse is an interactive code which explains, at run time, what information is necessary to use it. The application assumes the Prony series of the source function to be given as in Eq. (7.4):

$$S(t) = S_e + \sum_{i=1}^{N} S_i e^{-t/\tau_i} \qquad (7.46a)$$

in case it is a relaxation modulus and as in Eq. (7.8):

$$S(t) = S_e - \sum_{i=1}^{N} S_i e^{-t/\tau_i} \qquad (7.46b)$$

if the source is a creep compliance function.

As will be learned during execution of Prony-Inverse, the following information is asked for

1. Name of a data file on which the results will be stored
2. Input option:
   (a) 1, if input is manual (through screen)
   (b) 2, if input is through a data file
3. If input is manual:
   (a) Type of source function (1, if modulus; −1, if compliance)
   (b) Number of transient terms, N, and equilibrium value, Se
   (c) N pairs of Prony coefficients and time constants, (S_i, τ_i)
- The S_i and τ_i are always positive. The type of source function indicator (1 or − 1) is used to assign correct sign during execution.
- The τ_i's represent relaxation times if the source is a relaxation modulus and And retardation times if the source is a creep compliance. In either case, these time constants must be provided in ascending order of numerical value, to prevent erroneous solutions.
4. If input is through file:
   (a) Name of the file containing the source function data (the data file structure is as described before)

The accuracy of the inversion procedure hinges on that of the bisection method used to obtain the roots of the N-degree polynomial—either $\Phi(\beta) = 0$ or $\Psi(\alpha) = 0$. This may give rise to a slight discrepancy between a source function that is first used to obtain its inverse and the source function that is obtained by using as source the inverse that was initially calculated. Bearing this in mind, provide as accurate a set of values of the parameters of the source function as possible. This is illustrated in the problems at the end of the section.

7.7   Problems

P.7.1 Making the simple assumption that the relaxation and retardation times are the same, obtain the creep compliance of a viscoelastic material whose relaxation modulus in Prony series form is $M(t) = 1.00 + 0.50e^{-2t}$.

Answer : $C(t) = 1.00 - 1/3 \cdot e^{-2t}$

Hint: Take the creep compliance function in the form $C(t) = C_e - C_1 e^{-2t}$, as per Eq. (7.8), and calculate $M_g \equiv M(0) = 1.50$, and $M_e = 1.00$. Then, use (2.22): $M_g = 1/C_g$ and (2.23): $M_e = 1/C_e$, as well as (7.10): $C_g = C_e - C_1$ to obtain $C_e = 1/M_e = 1.00$, $C_g = 1/M_g = 1/1.50$, and $C_1 = 1/3$.

P.7.2 Solve Problem P.7.1 using the computer application "Prony-Inverse," and double-check your answer using the results of Example 7.1.

Answer: $C_e = 1.00$ kPa, $C_1 = 0.333$ kPa, $\lambda = 0.75$ s,   or
$C(t) = 1.00 - 1/3 \cdot e^{-4t/3}$

Hint: Double-click on the Prony-Inverse application and follow the directions on the screen, noting that the code expects relaxation times for modulus and retardation times for creep compliance. Hence, use $N = 1$, $S_e = 1.00$, $S_1 = 0.5$, $\tau_1 = 1/2 = 0.5$, when entering the data. Using that Prony-Inverse returns relaxation or retardation times, leads to the result presented.

P.7.3 Use the answer to Problem P.7.2 as the source input to Prony-Inverse to obtain the (initial) relaxation modulus.

Answer: $M_e = 1.00$ kPa, $M_1 = 0.499$ kPa, $\tau = 0.500$ s

Hint: Proceed as in Problem 7.2, this time using the creep compliance data. The discrepancy between the initial relaxation modulus data and the solution check is due to the fact that while the solution to P.7.2 was $C_1 = 1/3$, the approximation $C_1 = 0.333$ was used for this problem.

P.7.4 The equilibrium modulus of a solid propellant is 81.02, and the coefficients and time constants of the Dirichlet–Prony series form of its relaxation modulus are as follows:

M_k	2,057.4	1,014	458.4	212.8	98.00	44.88	21.50	7.860	36.78	3.820
τ_k	$1.00.10^{-5}$	$1.00.10^{-4}$	$1.00.10^{-3}$	$1.00.10^{-2}$	$1.00.10^{-1}$	$1.00.10^{0}$	$1.00.10^{1}$	$1.00.10^{2}$	$1.00.10^{3}$	$1.00.10^{4}$

Use Prony-Inverse to obtain the coefficients and time constants of the creep compliance.

Answer: $C_e = 1.2343\mathrm{E}{-}02$; with Prony coefficients and time constants given by

C_k	$2.018.10^{-4}$	$4.562.10^{-4}$	$8.630.10^{-4}$	$1.400\mathrm{E}.10^{-3}$	$1.765.10^{-3}$	$1.607.10^{-3}$	$1.152\mathrm{E}.10^{-3}$	$4.844.10^{-4}$	$3.555\mathrm{E}.10^{-3}$	$6.115\mathrm{E}.10^{-4}$
λ_k	$1.923.10^{-5}$	$2.034.10^{-4}$	$1.929.10^{-3}$	$1.757.10^{-2}$	$1.528\mathrm{E}.10^{-1}$	$1.311.10^{0}$	$1.171.10^{1}$	$1.064.10^{2}$	$1.434.10^{3}$	$1.050.10^{4}$

Hint: To avoid typographical errors, use a text editor to prepare a data file with the information given. This file should have the following structure:

1 Source function type (1 = modulus, −1 = compliance)

10, 81.02 Number of transient terms and equilibrium value (S_e). Other lines: (S_i, α_i)

2057.4	1.0E−5
1014.0	1.0E−4
458.4	1.0E−3
212.8	1.0E−2
98.0	1.0E−1
44.88	1.0E+0
21.50	1.0E+1
7.86	1.0E+2
36.78	1.0E+3
3.820	1.0E+4

It is important to note that annotations are only permissible on each data line, but only following the expected data for the line.

Run Prony-Inverse and select file input mode when prompted. The result should be as stated.

P.7.5 The equilibrium compliance of a natural rubber is 3.333, and the coefficients and time constants of the Dirichlet–Prony series form of its creep compliance are as follows:

C_k	$6.741.10^{-2}$	$9.126.10^{-2}$	$6.979.10^{-2}$	$2.193.10^{-1}$	$4.200.10^{-1}$	$6.357E.10^{-1}$	$2.295E.10^{-1}$	$7.263.10^{-1}$	$8.341E.10^{-1}$
λ_k	$9.116.10^{-8}$	$2.902.10^{-6}$	$3.072.10^{-5}$	$8.107.10^{-5}$	$7.085.10^{-4}$	$4.862.10^{-2}$	$7.706.10^{-1}$	$4.641.10^{0}$	$3.334.10^{3}$

Use Prony-Inverse to obtain the coefficients and time constants of the relaxation modulus.

Answer: $M_e = 0.3000$; with Prony coefficients and time constants given by

M_k	$6.741.10^{-2}$	$9.126.10^{-2}$	$6.979.10^{-2}$	$2.193.10^{-1}$	$4.200.10^{-1}$	$6.357.10^{-1}$	$2.295.10^{-1}$	$7.263.10^{-1}$	$8.341.10^{-1}$
τ_k	$9.116.10^{-8}$	$2.902.10^{-6}$	$3.072.10^{-5}$	$8.107.10^{-5}$	$7.085.10^{-4}$	$4.862E{-}02$	$7.706.10^{-1}$	$4.641.10^{0}$	$3.334.10^{3}$

Hint: To avoid typographical errors, use a text editor to prepare a data file with the information given. This file should have the following structure:

−1 Source function type (1 = modulus, 0 = compliance)

9, 3.3333 Number of transient terms and equilibrium value (S_e). Other lines: (S_i, τ_i)

6.7406E−02	9.1159E−08
9.1256E−02	2.9020E−06
6.9792E−02	3.0715E−05
2.1931E−01	8.1073E−05
4.1998E−01	7.0845E−04
6.3568E−01	4.8618E−02
2.2947E−01	7.7063E−01
7.2629E−01	4.6410E+00
8.3405E−01	3.3338E+03

P.7.6 Proceed as in Problem P.7.3 and use "Prony-Inverse" to double-check the results obtained in problems P.7.4 and P.7.5.

Hint: For each problem in turn, prepare a Prony-Inverse input file and run the application; then, compare the results obtained for the inverse to the initial input given in each case.

References

1. B.E. Read, G.D. Dean, The determination of dynamic properties of polymer composites (Wiley, London, 1978) pp. 1–12
2. W.G. Knauss, I. Emri, Volume change and the nonlinearly thermo-viscoelastic constitution of polymers. Polymer Eng. Sci. **27**, 86–100 (1987)
3. R.A. Schapery, Approximate methods of transform inversion for viscoelastic stress analysis, in Proceedings of 4th U.S. national congress of Appl. Mech., 1075 (1962)
4. D. Gutierrez-Lemini, Exact inversion of viscoelastic property functions of exponential type, JANNAF, JSF, San Diego, CA (2005)

Three-Dimensional Constitutive Equations

8

Abstract

This chapter generalizes to three dimensions the one-dimensional viscoelastic constitutive equations derived in earlier chapters. The concepts of homogeneity, isotropy, and anisotropy are introduced and the principle of superposition is used to construct three-dimensional constitutive equations for general anisotropic, orthotropic, and isotropic viscoelastic materials. So-called Poisson's ratios are introduced, and it is shown that uniaxial tensile and shear relaxation and creep tests suffice to characterize orthotropic viscoelastic solids. A rigorous treatment extends applicability of the Laplace and Fourier transforms to three-dimensional conditions, and constitutive equations in both hereditary integral form and differential form for compressible and incompressible isotropic solids are developed and discussed in detailed.

Keywords

Anisotropic · Orthotropic · Isotropic · Symmetry · Tensor · Matrix · Summation · Indicial · Transform · Spherical · Deviatoric · Incompressible

8.1 Introduction

The one-dimensional viscoelastic constitutive equations derived up to this point in the book are generalized here to three dimensions. For ease of reference, Sect. 8.2 gives an overview of the notation used, regarding the representation of stresses, strains, and constitutive properties. The subject matter is then developed in Sects. 8.3 to 8.8.

Section 8.3 introduces the concepts of material homogeneity, as well as isotropy and anisotropy and uses the principle of superposition to construct three-dimensional constitutive equations for general anisotropic viscoelastic solids.

D. Gutierrez-Lemini, *Engineering Viscoelasticity*, DOI: 10.1007/978-1-4614-8139-3_8,   193
© Springer Science+Business Media New York 2014

Section 8.4 introduces material symmetry planes, on which the direct or normal stresses decouple from shear strains and vice versa, leading to orthotropic materials and their constitutive equations. Contraction or Poisson's ratios are introduced also, and it is shown that uniaxial tensile and shear relaxation and creep tests suffice to characterize orthotropic viscoelastic solids. Using the argument that the constitutive tensors defining the three-dimensional stress–strain equations are ordered arrays of scalar relaxation or creep functions, Sect. 8.5 extends the applicability of the Laplace and Fourier transforms to three-dimensional conditions.

Sections 8.6, 8.7 and 8.8 are devoted to constitutive equations for isotropic viscoelastic solids. Sections 8.6 and 8.7 develop constitutive equations in hereditary integral form, applicable to compressible and incompressible solids, respectively. Section 8.8, on the other hand, treats compressible and incompressible isotropic viscoelastic solids of differential type.

8.2   Background and Notation

Three types of notation are used in this book: symbolic notation, indicial tensor notation, and matrix notation. In symbolic notation, the same symbols used in the one-dimensional case are used to represent the corresponding tensors. Therefore, the one-, two-, and three-dimensional equations—constitutive and otherwise—look exactly alike, and this applies to the Laplace and Fourier transforms too.

In indicial tensor notation, tensors, as ordered arrays of quantities, are denoted by subscripted symbols, such as A_{ijk} or B_r. The number of non-repeated indices, also called free indices, denotes the order of the tensor, which is related to the number of elements it has. In three-dimensional Euclidean space, a tensor with p free indices has 3^p elements and is said to be a tensor of order p. For instance, M_{ijkl} would represent a tensor of fourth order, having $3^4 = 81$ components, while D_{ij} would represent a tensor of second order, with $3^2 = 9$ components; A_{ii}, having no free indices, would represent a tensor of order zero, also called a scalar (c.f. Appendix B).

Just like matrices, tensors of the same order—with the same number of free indices—may be added together, term by term, to yield a tensor of the same order as the addends. Thus, for instance, $D_{ij} + \sigma_{ij} = A_{ij}$, etc. Tensors can also feature repeated indices. However, no index can be repeated more than once in a tensor. Thus, A_{ijj}, M_{ijkk}, C_{ii}, and M_{iikk} are valid tensors, but A_{jjj} and M_{ikkk} are not.

Also like matrices, tensors of the same order can be multiplied together. This is done according to the summation convention: "when an index appears twice in a tensor, a summation is implied over all the terms that are obtained by letting that index assume all its possible values, unless explicitly stated otherwise" [1]. This convention is suspended if an index appears three or more times in a term or a tensor. According to the summation convention then, the explicit meaning of the tensor expressions $A_i B_i$ and $A_{ij} x_j$, for instance, is

$$A_i B_i = A_1 B_1 + A_2 B_2 + A_3 B_3; \quad A_{ij} x_j = A_{i1} x_1 + A_{i2} x_2 + A_{i3} x_3; \quad j = 1,2,3 \quad \text{(a)}$$

As a consequence of the summation convention, the presence of a repeated index reduces the order of a tensor by two. For instance, setting $k = l$ in the fourth-order tensor M_{ijkl} leads to $M_{ijkk} = M_{ij11} + M_{ij22} + M_{ij33}$, which is a sum of three second-order tensors.

The symbol δ_{ij} is reserved for a special second-order tensor, with the properties of the identity matrix: $\delta_{ij} = 1$, if $i = j$; and $\delta_{ij} = 0$, if $i \neq j$. This tensor is also usually called the Kronecker delta and, on occasion, the substitution operator. The latter name is due to its property that, for example, $A_{ij}\delta_{jk} = A_{i1}\delta_{1k} + A_{i2}\delta_{2k} + A_{i3}\delta_{3k}$, which, based on the definition of the Kronecker delta, evaluates to either A_{i1}, A_{i2}, or A_{i3}, depending on the value of k. Thus, $A_{ij}\delta_{jk} = A_{ij}$, where dummy index j has been "substituted" with free index k.

All quantities entering the one-dimensional constitutive equations, that is the stress, the strain, and the material property functions relating them—relaxation modulus and creep compliance—are considered scalar quantities, each fully described by a single entity. In three dimensions, the state of stress at a point requires an ordered set of nine functions to be fully specified and similarly for the state of strain at a point. The ordered collection of nine functions forms the stress or the strain tensor. The nine components of the stress and strain tensors are arranged in 3×3 arrays, or matrices with entries $\sigma_{ij}(t)$ and $\varepsilon_{ij}(t)$, respectively, as follows:

$$\sigma_{ij}(t) = \begin{bmatrix} \sigma_{11}(t) & \sigma_{12}(t) & \sigma_{13}(t) \\ \sigma_{21}(t) & \sigma_{22}(t) & \sigma_{23}(t) \\ \sigma_{31}(t) & \sigma_{32}(t) & \sigma_{33}(t) \end{bmatrix} \quad \varepsilon_{ij}(t) = \begin{bmatrix} \varepsilon_{11}(t) & \varepsilon_{12}(t) & \varepsilon_{13}(t) \\ \varepsilon_{21}(t) & \varepsilon_{22}(t) & \varepsilon_{23}(t) \\ \varepsilon_{31}(t) & \varepsilon_{32}(t) & \varepsilon_{33}(t) \end{bmatrix} \quad \text{(b)}$$

In this notation, the components of the stress tensor that are listed on anyone of its three rows denote the components of a traction vector acting on a surface whose outward unit normally points in the direction of the axis represented by the index of the row—that is, by the first index of the tensor component. The second index indicates the axis or direction in which that stress component acts. The sign convention for the components of the stress tensors is presented in Fig. 8.1. According to it, a stress component may act on either a positive plane or a negative plane at a material point. A stress tensor component on a positive plane is positive if it acts in the positive direction of a coordinate axis. A stress component on a negative plane is positive if it acts in a negative coordinate direction.

In somewhat similar fashion, the components of the strain tensor, which are listed on the main diagonal of the matrix of the strain tensor, measure the change in length of a material element lying along the direction of the axis represented by the first index, per unit length of the material element along the axis represented by the second index which, being on the main diagonal, is the same as the first. The off-diagonal elements in the strain tensor measure the decrease in angle of a material plane with edges parallel to the axis represented by the two indices of the strain tensor component [2].

Fig. 8.1 Sign convention for
the components of the stress
tensor

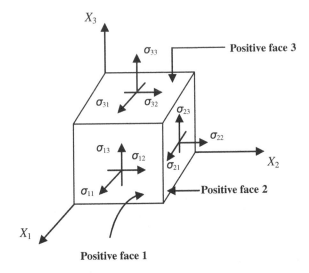

Positive face 1

Explicit matrix notation is very similar to indicial tensor notation, including the use of the summation convention. The difference is that the stress and strain tensors are represented as extended (9-by-1) vectors, and the elements of the constitutive tensors are arranged in 9-by-9 square matrices.

The reader is assumed familiar with the usual rules of operation with matrices: addition, multiplication, transposition, and inversion. Regarding differentiation and integration, which enter viscoelastic expressions, suffice it to say that they are defined in the same fashion for matrices and tensors, so that "the integral of a matrix is the matrix of the integrals of its elements" and "the derivative of a matrix is the matrix of the derivatives of its elements".

8.3   Constitutive Equations for Anisotropic Materials

Material property functions, such as relaxation modulus or creep coefficient of thermal expansion, may depend on the orientation the test specimen had before it was extracted from the parent material and even on the position it occupied in the bulk. If the value of a material property function is the same, within acceptable material variability, when the test specimen is extracted in the same orientation, but from different positions, such as (a_1) to (a_3) in Fig. 8.2, the material is said to be homogeneous with respect to the property measured. If that is not the case, the material is said to be inhomogeneous. Additionally, if the value of a material property function does not depend on the orientation that the test article had in the bulk—such as (b) or (c) in Fig. 8.2, the material is said to be isotropic with respect to the property function being measured. Otherwise, the material is said to be anisotropic. Only homogeneous materials are considered here.

Fig. 8.2 Dependence of
material properties on
position and orientation

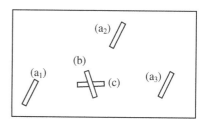

The approach followed to derive the three-dimensional constitutive equations
for general anisotropic materials is essentially identical to the one-dimensional
case, except that for anisotropic materials, attention has to be paid to the orientation of the stresses and strains relative to the parent material itself.

The principle of superposition, which applies to any linear material, requires
that each component of stress must be linearly related to each of the strain com-
ponents and vice versa. Letting $\sigma_{ij}(t)\big|_{\varepsilon_{kl}}$ represent the stress σ_{ij} due to strain ε_{kl}
acting alone; similarly for $\varepsilon_{ij}(t)\big|_{\sigma_{kl}}$, the principle of superposition may be expressed
in either of the following forms:

$$\sigma_{ij}(t) = \sum_{k=1}^{3}\sum_{l=1}^{3}\sigma_{ij}(t)\big|_{\varepsilon_{kl}}; \quad i, j = 1, 3 \tag{8.1}$$

$$\varepsilon_{ij}(t) = \sum_{k=1}^{3}\sum_{l=1}^{3}\varepsilon_{ij}(t)\big|_{\sigma_{kl}}; \quad i, j = 1, 3 \tag{8.2}$$

Each term in these expressions relates one single stress or strain component to
one single strain or stress component. In other words, each term in each linear sum
in (8.1) and (8.2) is a one-dimensional stress–strain (strain–stress) viscoelastic
relationship. Denoting by $M_{ijkl}(t)$, the scalar modulus-type function giving the
stress response σ_{ij} to a strain input ε_{kl} and $C_{ijkl}(t)$ to represent the scalar compli-
ance-like function for strain response ε_{ij} to a stress input σ_{kl} allows each term in
(8.1) and (8.2) to be represented by mutually independent one-dimensional
expressions such as $\sigma_{11}\big|_{\varepsilon_{12}} = M_{1112} * d\varepsilon_{12}$ or $\varepsilon_{33}\big|_{\sigma_{11}} = C_{3311} * d\sigma_{11}$, .

Using indicial tensor notation, the sums (8.1) and (8.2), which represent the
three-dimensional constitutive equations of a general anisotropic viscoelastic
material, may be cast as

$$\sigma_{ij}(t) = M_{ijkl}(t - \tau) * d\varepsilon_{kl}(\tau) = \int_{0^-}^{t} M_{ijkl}(t - \tau) * \frac{d}{d\tau}\varepsilon_{kl}(\tau)d\tau \tag{8.3}$$

$$\varepsilon_{ij}(t) = C_{ijkl}(t - \tau) * d\sigma_{kl}(\tau) = \int_{0^-}^{t} C_{ijkl}(t - \tau) * \frac{d}{d\tau}\sigma_{kl}(\tau)d\tau \tag{8.4}$$

Before proceeding, it must be pointed out that although each material property function in the sets M_{ijkl} and C_{ijkl} is a scalar function relating a single component of stress (strain) to a single component of strain (stress), the individual functions in M and C are generally not the same as the moduli or compliances of the one-dimensional theory. The relationships between the two sets of functions will be made clear later on.

Since each of the four indices in these expressions ranges from 1 to 3, $M_{ijkl}(t)$ and $C_{ijkl}(t)$ each contain $3 \cdot 3 \cdot 3 \cdot 3 = 3^4 = 81$ scalar relaxation and compliance functions, respectively. These two ordered sets of 81 scalar property functions each are the fourth-order material property tensors for so-called polar anisotropic viscoelastic materials, for which the stress and strain tensors are non-symmetric [5]. For non-polar materials—which are the typical engineering materials considered in this text—the stress and strain tensors are symmetric and each contains only 6 independent components. Hence, the constitutive tensors M and C for general non-polar anisotropic materials are symmetric in their first and second pairs of indices, respectively, which indicate that the number of material property functions of a general anisotropic material is, at most, 36.[1]

Switching to matrix notation for simplicity, the relaxation moduli and the creep compliances for anisotropic viscoelastic materials become the 6-by-6 matrices $[M_{ij}]$ and $[C_{ij}]$, or $[M]$ and $[C]$, for short. Expressions (8.3) and (8.4) then take the equivalent forms [3]:

$$\sigma_i = M_{ij}(t - \tau) * d\varepsilon_j(\tau); \quad i = j = 1, 6 \tag{8.5}$$

$$\varepsilon_i = C_{ij}(t - \tau) * d\sigma_j(\tau); \quad i = j = 1, 6 \tag{8.6}$$

In these expressions, the 6 independent components of the stress and strain tensors are arranged in the 6-by-1 extended column vectors[2]: $\{\sigma_i\}^T \equiv \{\sigma\}^T \equiv \{\sigma_{11}, \sigma_{22}, \sigma_{33}, \sigma_{12}, \sigma_{13}, \sigma_{23}\}$ and $\{\varepsilon_i\}^T \equiv \{\varepsilon\}^T \equiv \{\varepsilon_{11}, \varepsilon_{22}, \varepsilon_{33}, \varepsilon_{12}, \varepsilon_{13}\}$. And clearly, as in the one-dimensional case, combining (8.5) and (8.6) proves that the moduli and compliance matrices are convolution inverses of each other[3]:

$$M_{ik}(t - \tau) * dC_{kj}(\tau) = C_{ik}(t - \tau) * dM_{kj}(\tau) = H(t)\delta_{ij} \tag{8.7}$$

Now, the work per unit volume is $\sigma_i * d\varepsilon_i \equiv M_{ij} * d\varepsilon_j * d\varepsilon_i = d\varepsilon_i * M_{ij} * d\varepsilon_j$. After interchanging the dummy indices, $i \leftrightarrow j$, on the right-hand side and reordering,

---

[1] For each pair of indices, ij and kl, ranging from 1 to 3, the number of independent components is $3 \cdot 4/2 = 6$.

[2] In matrix notation, the superscript T is used to denote the "transpose" of the matrix it is appended to.

[3] The equivalent tensor expression, derived combining (8.3) and (8.4), is $M_{ijkl} * dC_{klpq} = H(t)\delta_{ip}\delta_{jq}$

leads to $M_{ij} * d\varepsilon_j * d\varepsilon_i = M_{ji} * d\varepsilon_i * d\varepsilon_j = M_{ji} * d\varepsilon_j * d\varepsilon_i$, which implies $M_{ij} = M_{ji}$. In other words, the 6-by-6 material property matrix, M, is symmetric, requiring only $6 \times 7/2 = 21$, independent coefficients. The same can be said of the matrix of compliances, C_{ij}. From all these arguments follow that the number of independent material property functions of a general non-polar anisotropic viscoelastic material is 21.

8.4   Constitutive Equations for Orthotropic Materials

By definition, the mechanical response of anisotropic materials is always coupled in that their response to any action will generally involve stretching, contraction, and shearing. Accordingly, 21 independent tests would be required to characterize linear anisotropic materials, viscoelastic or otherwise.

A material symmetry plane is one defined by two perpendicular directions along which the direct stresses decouple from shearing in the plane and vice versa. The directions along which decoupling occurs are termed material principal directions.[4] By definition, then, a direct pull in a principal material direction produces a stretch in that direction and, quite generally, a contraction in the material principal directions that are perpendicular to the direction of pull. This deformation, however, occurs without shearing in the material plane of symmetry.

An orthotropic material is defined as one with three material symmetry planes, meaning that in these planes, the three direct stresses (strains) decouple from the three shear strains (stresses). This reduces the number of independent constitutive coefficients by 6. Also, since shear stresses (strains) applied on a material principal plane decouple from other shear strains (stresses), the constitutive equations for shear involve only shear terms of the same type and decouple completely from the rest. This reduces the number of independent constitutive functions by an additional 3 per shear component, to 9 overall. In explicit matrix notation, omitting the time arguments, for conciseness, the constitutive equations for general orthotropic viscoelastic materials take the form:

$$
\begin{Bmatrix} \sigma_1 \\ \sigma_2 \\ \sigma_3 \end{Bmatrix} = \begin{bmatrix} M_{11} & M_{12} & M_{13} \\ M_{12} & M_{22} & M_{23} \\ M_{13} & M_{23} & M_{33} \end{bmatrix} * d \begin{Bmatrix} \varepsilon_1 \\ \varepsilon_2 \\ \varepsilon_3 \end{Bmatrix};
$$

$$
\begin{Bmatrix} \sigma_4 \\ \sigma_5 \\ \sigma_6 \end{Bmatrix} = \begin{bmatrix} M_{44} & 0 & 0 \\ 0 & M_{55} & 0 \\ 0 & 0 & M_{66} \end{bmatrix} * d \begin{Bmatrix} \varepsilon_4 \\ \varepsilon_5 \\ \varepsilon_6 \end{Bmatrix} \qquad (8.8)
$$

---

[4] Material principal directions are not to be confused with the principal directions of stress or the principal directions of strain, the latter so-named because along them the stress and, respectively, the strain attain their extreme numerical values.

$$\begin{Bmatrix} \varepsilon_1 \\ \varepsilon_2 \\ \varepsilon_3 \end{Bmatrix} = \begin{bmatrix} C_{11} & C_{12} & C_{13} \\ C_{12} & C_{22} & C_{23} \\ C_{13} & C_{23} & C_{33} \end{bmatrix} * d \begin{Bmatrix} \sigma_1 \\ \sigma_2 \\ \sigma_3 \end{Bmatrix};$$

$$\begin{Bmatrix} \varepsilon_4 \\ \varepsilon_5 \\ \varepsilon_6 \end{Bmatrix} = \begin{bmatrix} C_{44} & 0 & 0 \\ 0 & C_{55} & 0 \\ 0 & 0 & C_{66} \end{bmatrix} * d \begin{Bmatrix} \sigma_4 \\ \sigma_5 \\ \sigma_6 \end{Bmatrix} \tag{8.9}$$

As pointed out previously, the relaxation functions M_{ij} are not in general the one-dimensional relaxation moduli in the indicated directions, except for shear response. In this case, for instance, $\sigma_{12} \equiv \sigma_4 = M_{44}*d\varepsilon_4 \equiv 2G_{12}*d\varepsilon_{12}$ and so $M_{44} = 2G_{12}$. In similar fashion, $M_{55} = 2G_{13}$ and $M_{66} = 2G_{23}$. The diagonal form of the constitutive equations for shear, together with (8.7), implies that the shear compliances C_{44} to C_{66} are convolution inverses of M_{44} to M_{66}, respectively. Clearly, standard shear relaxation or shear creep tests performed in material principal planes would suffice to establish the three material property functions in shear.

The uniaxial relaxation modulus in each material principal direction may be related to the components of the creep compliance matrix for direct strain–stress response. This is done using (8.9) to evaluate the strain response to each of three separate uniaxial states of stress: $\{\sigma_1 \neq 0, \ \sigma_2 = \sigma_3 = 0\}$, $\{\sigma_2 \neq 0, \ \sigma_1 = \sigma_3 = 0\}$, and $\{\sigma_3 \neq 0, \ \sigma_1 = \sigma_2 = 0\}$ and noting the analogy with the uniaxial conditions.

Using the first state of stress, (8.9) produces $\varepsilon_1 = C_{11}*d\sigma_1$, $\varepsilon_2 = C_{12}*d\sigma_1$, $\varepsilon_3 = C_{13}*d\sigma_1$. The first of these relations gives $\sigma_1 = C_{11}^{-1} * d\varepsilon_1 \equiv E_{11} * d\varepsilon_1$, where E_{11} is the uniaxial (tensile) relaxation modulus. In other words, $E_{11} = C_{11}^{-1}$. Inserting $\sigma_1 = E_{11}*d\varepsilon_1$ in the second and third relationships produces $\varepsilon_2 = C_{12}*dE_{11}* d\varepsilon_1$ and $\varepsilon_3 = C_{13}*dE_{11}* d\varepsilon_1$. The negative of the functions $C_{12}*dE_{11}$ and $C_{13}*dE_{11}$ are given the special symbols v_{12} and v_{13}, respectively. Then, ε_2 and ε_3 are simply written as $\varepsilon_2(t) = C_{12}*dE_{11}*d\varepsilon_1 \equiv -v_{12}*d\varepsilon_1$ and $\varepsilon_3(t) = C_{13}*dE_{11}*d\varepsilon_1 \equiv -v_{13}*d\varepsilon_1$. From these expressions follow that $C_{12} = -E_{11}^{-1} * dv_{12}$ and $C_{13} = -E_{11}^{-1} * dv_{13}$. These newly introduced functions are called contraction ratios or, more generally, Poisson's ratios.

Exactly the same arguments can be followed with the other two states of direct stress listed before. Hence, collecting the findings for the pure shear and pure direct states of stress, the following relationships are obtained:

$$C_{11}(t) = E_{11}^{-1}(t); C_{22}(t) = E_{22}^{-1}(t); C_{33}(t) = E_{33}^{-1}(t) \tag{8.10a}$$

$$C_{12} = -E_{11}^{-1} * dv_{12}; C_{13} = -E_{11}^{-1} * dv_{13}; C_{23} = -E_{22}^{-1} * dv_{23} \tag{8.10b}$$

$$C_{44}(t) = \frac{1}{2}G_{12}^{-1}(t); C_{55}(t) = \frac{1}{2}G_{13}^{-1}(t); C_{66}(t) = \frac{1}{2}G_{23}^{-1}(t) \tag{8.10c}$$

Fig. 8.3 Expected time dependence of Poisson's ratio for orthotropic viscoelastic solids

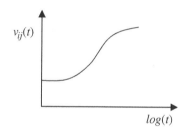

$$v_{ij}(t)$$

$$log(t)$$

Also, had the functions C_{21}, C_{31}, and C_{32} been used in the derivations, instead of C_{12}, C_{13}, and C_{23}, the following would have resulted:

$$C_{21} = -E_{22}^{-1} * dv_{21}; C_{31} = -E_{33}^{-1} * dv_{31}; C_{32} = -E_{33}^{-1} * dv_{32} \qquad (8.11)$$

Because of the symmetry of the compliance matrix: $C_{ij} = C_{ji}$, (8.11) and (8.10b) imply that $E_{11}^{-1} * dv_{12} = E_{22}^{-1} * dv_{21}$ or, equivalently, due to the commutative property of the Stieltjes convolution (see Appendix A) $v_{12} * dE_{22} = v_{21} * dE_{11}$. Since the same is true for the other Poisson's ratios, we see that just as for orthotropic elastic materials, suspending the summation convention,[5] these relations can be expressed as [4]:

$$v_{ij} * dE_{\underline{jj}} = v_{ji} * dE_{\underline{ii}}; \text{ no sum on } i \text{ or } j; \quad i, j = 1, 3; \ i \neq j \qquad (8.12)$$

Using (8.10a) to (8.12) with (8.9) leads to the following matrix of compliances for orthotropic viscoelastic solids in terms of tensile and shear relaxation moduli and Poisson's ratios:

$$\left\{ \begin{matrix} \varepsilon_{11} \\ \varepsilon_{22} \\ \varepsilon_{33} \end{matrix} \right\} = \begin{bmatrix} E_{11}^{-1} & -E_{11}^{-1} * dv_{12} & -E_{11}^{-1} * dv_{13} \\ -E_{11}^{-1} * dv_{12} & E_{22}^{-1} & -E_{22}^{-1} * dv_{23} \\ -E_{11}^{-1} * dv_{13} & -E_{22}^{-1} * dv_{23} & E_{33}^{-1} \end{bmatrix} * d \left\{ \begin{matrix} \sigma_{11} \\ \sigma_{22} \\ \sigma_{33} \end{matrix} \right\} \qquad (8.13a)$$

$$\left\{ \begin{matrix} \varepsilon_{12} \\ \varepsilon_{13} \\ \varepsilon_{23} \end{matrix} \right\} = \left\{ \begin{matrix} \frac{1}{2} G_{12}^{-1} * d\sigma_{12} \\ \frac{1}{2} G_{13}^{-1} * d\sigma_{13} \\ \frac{1}{2} G_{23}^{-1} * d\sigma_{23} \end{matrix} \right\} \qquad (8.13b)$$

From their definitions, each Poisson's ratio functions, v_{12}, v_{13} and v_{23}, relates a strain response to a strain caused by a single stress component. Their functional behavior should therefore be like that of a compliance function, as indicated in Fig. 8.3.

---

[5] In indicial tensor notation, the summation convention is typically suspended by adding an underscore to the indices excluded from the summation.

In practical applications, the Poisson's functions, $v_{ij}(t)$ of an orthotropic vis-coelastic material are established by subjecting a slender test specimen to a step strain $\varepsilon_i(t) = \varepsilon_{oi}H(t)$ and by monitoring the strain response, $\varepsilon_j(t)$, in material principal direction j; according to its definition

$$\varepsilon_j(t) \equiv -v_{ij} * d\varepsilon_i = -\int_{o^-}^{t} v_{ij}(t-\tau)\frac{d}{d\tau}\varepsilon_i(\tau)d\tau \qquad (8.14)$$

Inserting the step strain history input and rearranging yield the practical expression:

$$v_{ij}(t) = -\frac{\varepsilon_j(t)}{\varepsilon_{oi}} \qquad (8.15)$$

8.5   Constitutive Equations in Integral Transform Space

Because the constitutive tensors and associated matrices are ordered collections of scalar functions, any linear transformation of them will yield tensors or matrices of transforms. This allows writing the Laplace and Fourier transforms of the con-stitutive equations of anisotropic viscoelastic materials directly, as the tensors or matrices of the corresponding transforms of the individual material property functions. Thus, for instance, expressions (8.3) and (8.4) in Laplace-transformed space become

$$\bar{\sigma}_{ij}(s) = s\bar{M}_{ijkl}(s)\bar{\varepsilon}_{kl}(s) \qquad (8.16)$$

$$\bar{\varepsilon}_{ij}(s) = s\bar{C}_{ijkl}(s)\bar{\sigma}_{kl}(s) \qquad (8.17)$$

In like fashion, the Fourier transform of the same expressions read

$$\bar{\sigma}_{ij}(j\omega) = (j\omega) \cdot \bar{M}_{ijkl}(j\omega)\bar{\varepsilon}_{kl}(j\omega) \qquad (8.18a)$$

$$\bar{\varepsilon}_{ij}(j\omega) = (j\omega) \cdot \bar{C}_{ijkl}(j\omega)\bar{\sigma}_{kl}(j\omega) \qquad (8.19a)$$

Equivalently, in terms of the complex stress, strain, modulus, and compliance, as is more customary in dealing with steady-state conditions (see Chap. 4)

$$\sigma_{ij}^*(j\omega) = M_{ijkl}^*(j\omega)\varepsilon_{kl}^*(j\omega) \qquad (8.18b)$$

$$\varepsilon_{ij}^*(j\omega) = C_{ijkl}^*(j\omega)\sigma_{kl}^*(j\omega) \qquad (8.19b)$$

According to the derivations in this section, the results in (8.16) to (8.19b) apply to each term of a modulus or compliance matrix.

Example 8.1 Using the quantity $\varepsilon_{11}(t)|_{\sigma_{22}} = -E_{11}^{-1} * dv_{12} * d\sigma_2$, appearing as one of the addends in the strain–stress constitutive equation of an orthotropic visco-elastic solid, according to (8.13a)– demonstrate by direct evaluation of the con-volution integrals involved, that for steady-state loading, $\varepsilon_{11}^*(j\omega t)|_{\sigma_{22}^*}$

$= -[E_{11}^*(j\omega)]^{-1} \cdot v_{12}^*(j\omega) \cdot \sigma_2^*(j\omega t)$; in accordance with the developments of this section.

Solution:

The steady-state strain response $\varepsilon_{11}(t)|_{\sigma_{22}} = -E_{11}^{-1} * dv_{12} * d\sigma_2$ to the cyclic stress $\sigma_2(t) = \sigma_{o2}\sin(\omega t)$ is obtained as the imaginary component of the complex stress $\sigma_2^*(j\omega t) = \sigma_{o2}e^{j\omega t}$, as explained in Chap. 4. Hence, integrating the given strain component of the constitutive relation from $-\infty$ to t, using the complex input, leads to

$$\varepsilon_{11}^*(j\omega t)|_{\sigma_{22}^*} \equiv - \int_{\tau=-\infty}^{t} E_{11}^{-1}(t-\tau)\frac{d}{d\tau} \int_{s=-\infty}^{\tau} v_{12}(\tau-s)\frac{d}{ds}\sigma_{o2}e^{j\omega s}ds$$

$$= -(j\omega)\sigma_{o2} \int_{\tau=-\infty}^{t} E_{11}^{-1}(t-\tau)\frac{d}{d\tau} \int_{s=-\infty}^{\tau} v_{12}(\tau-s)e^{j\omega s}ds$$

Introducing the change in variable $u = \tau - s$, using $ds = -du$, to invert the resulting limits of integration and taking the derivative of the inner integral with respect to τ:

$$\varepsilon_{11}^*(j\omega t)|_{\sigma_{22}^*} = -(j\omega)\sigma_{o2} \int_{\tau=-\infty}^{t} E_{11}^{-1}(t-\tau)\frac{d}{d\tau}e^{-j\omega\tau} \int_{u=0}^{\infty} v_{12}(u)e^{-j\omega u}du$$

$$= -(j\omega)\sigma_{o2} \int_{\tau=-\infty}^{t} E_{11}^{-1}(t-\tau)e^{-j\omega\tau}(j\omega) \int_{u=0}^{\infty} v_{12}(u)e^{-j\omega u}du$$

Defining $v_{12}^*(j\omega) \equiv (j\omega) \int_{s=0}^{\infty} v_{12}(u)e^{-j\omega u}du$, changing variables of integration on the outer integral, to $z = t - \tau$, and inverting the limits of integration, as before, yield

$$-E_{11}^{-1} * dv_{12} * d\sigma_2 = \left[(j\omega) \int_{z=0}^{\infty} E_{11}^{-1}(z)e^{-j\omega z}dz\right]v_{12}^*(j\omega)\sigma_{o2}e^{-j\omega t}$$

Recognizing that the quantity in brackets is the complex compliance $[E_{11}^{-1}(j\omega)]^*$ and $\sigma_{22}^*(j\omega) = \sigma_{o2}e^{-j\omega t}$ leads to $\varepsilon_{11}^*(j\omega t)|_{\sigma_{22}^*} = -[E_{11}^*(j\omega)]^{-1} \cdot v_{12}^*(j\omega) \cdot \sigma_{22}^*(j\omega t)$. By the arbitrariness of the term selected, the same is true of any other component of the compliance matrix and similarly for the matrix of moduli.

8.6   Integral Constitutive Equations for Isotropic Materials

There is a very important and large class of materials, which are called isotropic, whose material property functions are the same irrespective of the orientation in which the test specimens used to establish them are extracted from the bulk. The mechanical response of isotropic materials is fully described by means of two independent material property functions. This is so for the following reasons [5]:
1. The most general isotropic representation of a fourth-order tensor, such as M or C, can be expressed in terms of only three independent scalar property functions

$$
\begin{aligned}
M_{ijkl}(t) = \lambda(t) \cdot (\delta_{ij}\delta_{kl} + \delta_{ik}\delta_{jl}) + \mu(t) \cdot (\delta_{ik}\delta_{jl} + \delta_{il}\delta_{jk}) \\
+ v(t) \cdot (\delta_{il}\delta_{jk} - \delta_{ik}\delta_{jl})
\end{aligned}
\tag{8.20}
$$

2. The stress and strain tensors are symmetric: $\sigma_{ij} = \sigma_{ji}$ and $\varepsilon_{ij} = \varepsilon_{ji}$. Using this and that the repeated indices k and l are dummy and thus interchangeable—omitting the time dependence for conciseness:

$$
\sigma_{ij} = M_{ijkl} * d\varepsilon_{kl} = M_{ijlk} * d\varepsilon_{lk} = \sigma_{ji} = M_{jilk} * d\varepsilon_{lk}
\tag{a}
$$

This implies that

$$
M_{ijkl} = M_{ijlk} = M_{jilk}
\tag{b}
$$

The rate of work, $W = \sigma_{ij} * d\varepsilon_{ij}$, is a scalar quantity, so that using the associative property of convolution integrals, interchanging dummy indices, and invoking the commutative property of convolutions, in that order:

$$
W = M_{ijkl} * d\varepsilon_{kl} * d\varepsilon_{ij} = M_{klij} * d\varepsilon_{kl} * d\varepsilon_{ij}
\tag{c}
$$

From these two expressions follow the symmetry relationships satisfied by the relaxation and compliance material property tensors, M and C:

$$
M_{ijkl}(t) = M_{ijlk}(t) = M_{jikl}(t) = M_{klij}(t)
\tag{8.21}
$$

$$
C_{ijkl}(t) = C_{ijlk}(t) = C_{jikl}(t) = C_{klij}(t)
\tag{8.22}
$$

Using (8.21) in (8.20), together with the properties of the identity tensor,[6] adding, in turn, the subscripts M and C to the root symbols λ and μ to distinguish between modulus and compliance material property functions, there result

$$
M_{ijkl}(t) = \lambda_M(t)(\delta_{ij}\delta_{kl} + \delta_{ik}\delta_{jl}) + \mu_M(t)(\delta_{ik}\delta_{jl} + \delta_{il}\delta_{jk})
\tag{8.23}
$$

---

[6] $\delta_{ij} = 1$ if $i = j$, $\delta_{ij} = 0$ if $i \neq j$; $\delta_{ii} = 3$; $A_{ik}\cdot\delta_{kj} = A_{i1}\delta_{1j} + A_{i2}\delta_{2j} + A_{i3}\delta_{3j} = A_{ij}$.

$$C_{ijkl}(t) = \lambda_C(t)(\delta_{ij}\delta_{kl} + \delta_{ik}\delta_{jl}) + \mu_C(t)(\delta_{ik}\delta_{jl} + \delta_{il}\delta_{jk}) \tag{8.24}$$

This leads to the following form of the constitutive equation for isotropic viscoelastic materials:

$$\sigma_{ij}(t) = \int_{0^-}^{t} \lambda_M(t-\tau)\frac{\partial}{\partial\tau}\varepsilon_{kk}(\tau)\delta_{ij}d\tau + 2\int_{0^-}^{t} \mu_M(t-\tau)\frac{\partial}{\partial\tau}\varepsilon_{ij}(t-\tau)d\tau \tag{8.25}$$

$$\varepsilon_{ij}(t) = \int_{0^-}^{t} \lambda_C(t-\tau)\frac{\partial}{\partial\tau}\sigma_{kk}(\tau)\delta_{ij}d\tau + 2\int_{0^-}^{t} \mu_C(t-\tau)\frac{\partial}{\partial\tau}\sigma_{ij}(t-\tau)d\tau \tag{8.26}$$

More simply, using convolution notation, and omitting the time dependence, for brevity:

$$\sigma_{ij}(t) = \lambda_M * d\varepsilon_{kk}\delta_{ij} + 2\mu_M * d\varepsilon_{ij} \tag{8.27}$$

$$\varepsilon_{ij}(t) = \lambda_C * d\sigma_{kk}\delta_{ij} + 2\mu_C * d\sigma_{ij} \tag{8.28}$$

Likewise, introducing the vector $\{e\}^T \equiv \{1,1,1,0,0,0\}$ and combining convolution and matrix notations:

$$\{\sigma_i(t)\} = \{e_i\} \cdot \lambda_M * d\varepsilon_{kk} + 2\mu_M * \{d\varepsilon_i\} \tag{8.29}$$

$$\{\varepsilon_i(t)\} = \{e_i\} \cdot \lambda_C * d\sigma_{kk} + 2\mu_C * \{d\sigma_i\} \tag{8.30}$$

The manner in which the λ's and μ's are related to the uniaxial direct and shear property functions may be established from the observation of the possible forms of the entries in the strain–stress matrix equations (8.28) and from the use of the correspondence between the Carson transforms of elastic properties and the transforms of the homologous viscoelastic material property functions.

Since the response of an isotropic solid is independent of orientation, it follows that the uniaxial moduli should all be equal: $E_{11} = E_{22} = E_{33} \equiv E$. By the same token, the three Poisson's ratios must also be equal to each other: $v_{12} = v_{13} = v_{23} \equiv v$; the same has to be true of the three shear moduli: $G_{12} = G_{13} = G_{23} \equiv G$. Since only two of the material property functions are supposed to be independent, E, G, and v must be related. The relationships among these properties are established by means of the elastic relationship $E = 2(1 + v)G$ and the elastic–viscoelastic correspondence. This leads first to the relation $s\bar{E} = 2[1 + s\bar{v}]s\bar{G}$; cancelling out the common factor, s, and taking the inverse Laplace transform yield the result:

$$E(t) = 2[1 + v] * dG(\tau) = 2G(t) + 2\int_{0^-}^{t} v(t-\tau)\frac{d}{d\tau}G(\tau)d\tau \tag{8.31}$$

In similar fashion, the relationship between λ_M and the uniaxial property functions E and v is established using the elastic relation $\lambda = \frac{vE}{(1+v)(1-2v)}$ and the correspondence between elastic and viscoelastic constitutive relations. Using this correspondence, the previous expression may be cast as $(1 + s\bar{v})(1 - 2s\bar{v})$ $s\bar{\lambda} = s\bar{v}s\bar{E}$. Cancelling out the common term s, taking the inverse Laplace transform, and rearranging yield the result:

$$v * dE = \lambda(t) - v * d\lambda - 2v * dv * d\lambda \tag{8.32}$$

As will become apparent later shortly, the constitutive equations for isotropic viscoelastic materials can be expressed by using material property functions other than the λ's and μ's, or E, v, and G. The relations among the corresponding elastic properties listed in Appendix B may be used together with the elastic–viscoelastic correspondence to establish the viscoelastic expressions.

For isotropic materials, it is sometimes convenient to separate the stress and strain tensors into their spherical (σ_{Sij}, ε_{Sij}) and deviatoric (σ_{Dij}, ε_{Dij}) parts and to split the constitutive equations accordingly. This is done as follows.

The spherical part, A_S, of the 3×3 square matrix of a second-order tensor, A, is defined as the diagonal matrix with non-zero entries equal to the average of the diagonal elements of the given matrix. Thus, $A_{Sij} \equiv (A_{kk}/3)\delta_{ij} = 1/3(A_{11} + A_{22} + A_{33})\delta_{ij}$. The deviatoric part, A_D, is what is left over, that is, $A_{Dij} \equiv A_{ij} - A_{Sij}$. In explicit matrix notation, $[A] \equiv [A_S] + [A_D]$ or more simply

$$A \equiv A_S + A_D \tag{d}$$

$$[A_S] \equiv \frac{A_{kk}}{3}\delta_{ij} = \frac{(A_{11} + A_{22} + A_{33})}{3}\begin{bmatrix} 1 & 0 & 0 \\ 0 & 1 & 0 \\ 0 & 0 & 1 \end{bmatrix} \tag{e}$$

$$[A_D] \equiv A_{ij} - \frac{A_{kk}}{3}\delta_{ij} = \begin{bmatrix} (A_{11} - A_s) & A_{12} & A_{13} \\ A_{21} & (A_{22} - A_s) & A_{23} \\ A_{31} & A_{32} & (A_{33} - A_s) \end{bmatrix} \tag{f}$$

By definition then, the spherical part of a second-order tensor is obtained by making its two free indices the same—a tensor operation called contraction—and dividing the result by 3. Applying this operation to (8.27) produces

$$\sigma_{kk}(t) = (3\lambda_M + 2\mu_M) * d\varepsilon_{kk} = 3\left(\lambda_M + \frac{2}{3}\mu_M\right) * d\varepsilon_{kk} \tag{g}$$

and

$$\sigma_{Dij}(t) = \lambda_M * d\varepsilon_{kk}\delta_{ij} + 2\mu_M * d\varepsilon_{ij} - \left(\lambda_M + \frac{2}{3}\mu_M\right) * d\varepsilon_{kk} \tag{h}$$

Grouping terms and using the definition of spherical and deviatoric strains, (8.27) and (8.28), may be cast as

$$\sigma_S(t) = 3(\lambda_M + 2/3\mu_M) * d\varepsilon_S \tag{8.33a}$$

$$\sigma_{Dij}(t) = 2\mu_M * d\varepsilon_{Dij} \tag{8.33b}$$

$$\varepsilon_S(t) = 3(\lambda_C + 2/3\mu_c) * d\sigma_S \tag{8.34a}$$

$$\varepsilon_{Dij}(t) = 2\mu_C * d\sigma_{Dij} \tag{8.34b}$$

By its definition, the spherical stress is the average of the three direct stresses that act at a point in a body in any three mutually perpendicular directions such as (x, y, z) or (r, θ, z). For this reason, the spherical stress is also called the pressure stress. When the strains are infinitesimal, the spherical strain measures the volume strain, that is, the change in volume at a point after deformation, relative to the volume before application of the loads (c.f. Appendix B). Based on these observations, the function $(\lambda_M + 2/3 \mu_M)$ entering (8.33a) relates pressure and volume strain; hence, it represents a volumetric or bulk relaxation modulus. Since (8.33a) accounts for all changes in volume, (8.33b) must account for all changes in shape. Hence, the material property function μ_M represents the uniaxial shear relaxation modulus, G, so that $\mu_M(t) = G(t)$ and $\mu_C(t-\tau)*dG(\tau) = H(t)$.

Expressions (8.33a) and (8.34a) for bulk response are also often written in terms of the bulk relaxation modulus, K_M, and the bulk creep compliance, K_C, as

$$\sigma_{kk}(t) = 3(\lambda_M + 2/3\mu_M) * d\varepsilon_{kk} \equiv 3K_M * d\varepsilon_{kk} \tag{8.35a}$$

$$\sigma_{Dij}(t) = 2\mu_M * d\varepsilon_{Dij} \tag{8.35b}$$

$$\varepsilon_{kk}(t) = 3(\lambda_C + 2/3\mu_c) * d\sigma_S \equiv 3K_C * d\sigma_{kk} \tag{8.36a}$$

$$\sigma_{Dij}(t) = 2\mu_M * d\varepsilon_{Dij} \tag{8.36b}$$

in which the bulk modulus and bulk compliance have been defined as K_M and K_C, through

$$K_M(t) \equiv 3(\lambda_M + 2/3\mu_M) \tag{8.37a}$$

$$K_C(t) \equiv 3(\lambda_C + 2/3\mu_c) \tag{8.37b}$$

Example 8.1 Write the explicit integral form of the stress–strain equations for an isotropic viscoelastic solid with shear relaxation modulus, G, and bulk relaxation modulus K.

Solution:

Insert (8.37a) into (8.35a) and use the definition of convolution to get

$$\sigma_{kk}(t) = 3 \int_{0^-}^{t} K(t - \tau) \frac{\partial}{\partial \tau} \varepsilon_{kk}(\tau) d\tau$$

Likewise take the convolution of (8.33b) to write

$$\sigma_{Dij}(t) = 2 \int_{0^-}^{t} G(t - \tau) \frac{\partial}{\partial \tau} \varepsilon_{Dij}(t - \tau) d\tau.$$

There are also many viscoelastic materials of practical interest whose spherical response is elastic or approximately so. For such materials, (8.35a, b) take the form:

$$\sigma_{kk}(t) = 3K\varepsilon_{kk}(t); \sigma_{Dij}(t) = 2 \int_{0^-}^{t} \mu(t - \tau) \frac{\partial}{\partial \tau} \varepsilon_{Dij}(t - \tau) d\tau \qquad (8.38a)$$

Or, equivalently, but more simply

$$\sigma_{kk}(t) = 3K\varepsilon_{kk}(t); \sigma_{Dij}(t) = 2\mu * d\varepsilon_{Dij} \qquad (8.38b)$$

Example 8.2 Write the matrix form of the plane-stress constitutive equations of a viscoelastic material of hereditary integral type, whose volumetric response is elastic with bulk modulus K and shear relaxation modulus G.

Solution: In plane stress, all stresses associated with one reference direction, the 3-axis, say, are zero. Specifically, $\sigma_{31} = \sigma_{32} = \sigma_{33} = 0$. Therefore, from the definitions listed in (d) and (e): $\sigma_{kk} = \sigma_{11} + \sigma_{22}$, $\sigma_{D11} = 2/3\sigma_{11} - 1/3\sigma_{22}$, $\sigma_{D22} = -1/3\sigma_{11} + 2/3\sigma_{22}$ and hence $\sigma_{kk} = (\sigma_{11} + \sigma_{22})/3$ and $\varepsilon_{D12} = \varepsilon_{12}$. Using these with (8.37a, b) and simplifying give one form of the result, as

$$\frac{1}{9K} \begin{bmatrix} 6K & -3K & 0 \\ -3K & 6K & 0 \\ 0 & 0 & 0 \end{bmatrix} \begin{Bmatrix} \sigma_{11} \\ \sigma_{22} \\ \sigma_{12} \end{Bmatrix} + \begin{Bmatrix} \frac{2}{9K}G * d\sigma_{11} \\ \frac{2}{9K}G * d\sigma_{22} \\ \sigma_{12} \end{Bmatrix} = 2 \begin{Bmatrix} G * d\varepsilon_{11} \\ G * d\varepsilon_{22} \\ G * d\varepsilon_{12} \end{Bmatrix}$$

The constitutive property functions as well as the components of the stress and strain are assumed to be piecewise smooth, bounded, and with, at most, a finite number of finite amplitude discontinuities. Under these conditions, their Laplace and Fourier transforms are well defined (c.f. Appendix A). Hence, either transform may be applied to the constitutive equations listed in this section. For instance, the Laplace- and Fourier-transformed stress–strain relations corresponding to (8.27) are

$$\bar{\sigma}_{ij}(s) = s\bar{\lambda}_M(s) \cdot \bar{\varepsilon}_{kk}(s) \cdot \delta_{ij} + 2s\bar{\mu}_M \cdot \bar{\varepsilon}_{ij}(s) \tag{8.39}$$

$$\sigma_{ij}^*(j\omega) = \lambda_M^*(j\omega) \cdot \varepsilon_{kk}^*(j\omega) \cdot \delta_{ij} + 2\mu_M^*(j\omega) \cdot \varepsilon_{ij}^*(j\omega) \tag{8.40}$$

Example 8.3 Compute the steady-state response of a slender bar subjected to the harmonic loading $\{\varepsilon\}^T = \{\varepsilon_{11}^o, \varepsilon_{22}^o, \varepsilon_{33}^o, 0, \varepsilon_{13}^o, 0\}\sin(\omega_o t)$. Assume the material's volumetric response to occur at a constant bulk modulus, K, and its deviatoric response to be that of Kelvin solid with modulus, E, and viscosity η.

Solution:

To apply the constitutive equations, we need the complex form of the applied strain, which is $\{\varepsilon^*(j\omega_o t)\}^T = \{\varepsilon_{11}^o, \varepsilon_{22}^o, \varepsilon_{33}^o, 0, \varepsilon_{13}^o, 0\}e^{j\omega_o t}$, but expressed in terms of its spherical and deviatoric components, using (e) and (f); thus, $\varepsilon_S = (1/3)(\varepsilon_{11}^o + \varepsilon_{22}^o + \varepsilon_{33}^o)e^{j\omega_o t}$

$$[\varepsilon_D] = \frac{1}{3}\begin{bmatrix} (2\varepsilon_{11}^o - \varepsilon_{22}^o - \varepsilon_{33}^o) & 0 & \varepsilon_{13}^o \\ 0 & (2\varepsilon_{22}^o - \varepsilon_{33}^o - \varepsilon_{11}^o) & 0 \\ \varepsilon_{13}^o & 0 & (2\varepsilon_{33}^o - \varepsilon_{11}^o - \varepsilon_{22}^o) \end{bmatrix}e^{j\omega_o t}$$

From the statement of the problem, $K^* = K$, while G^* must be derived for a Kelvin solid. To do this, use the stress-strain equations for a Kelvin solid (3.16a) and the applied strain $\varepsilon^*(j\omega t) = \varepsilon_o \, e^{j\omega t}$ to write $\sigma^*(j\omega t) \equiv \sigma_o e^{j(\omega t+\delta)} = (E + j\omega\eta)\varepsilon_o e^{j\omega t}$. After regrouping, using Eq. (4.2) and dividing by ε_o, get: $G^*(j\omega) = (E + j\omega\eta)$. Inserting K^* and G^*—evaluated at the forcing frequency—together with the applied strain, the following is obtained:

$$\sigma_S^*(j\omega_o t) = K\{\varepsilon_{11}^o + \varepsilon_{22}^o + \varepsilon_{33}^o\}e^{j\omega_o t}$$

$$[\sigma_D^*(j\omega_o t)] = 2(E + j\omega_o\eta)\frac{1}{3}\begin{bmatrix} (2\varepsilon_{11}^o - \varepsilon_{22}^o - \varepsilon_{33}^o) & 0 & \varepsilon_{13}^o \\ 0 & (2\varepsilon_{22}^o - \varepsilon_{33}^o - \varepsilon_{11}^o) & 0 \\ \varepsilon_{13}^o & 0 & (2\varepsilon_{33}^o - \varepsilon_{11}^o - \varepsilon_{22}^o) \end{bmatrix}e^{j\omega_o t}$$

Since the applied strain is the imaginary part of the complex strain used, the shear stress response will correspond to the imaginary part of the complex stress and the volumetric response, being elastic, will be proportional to the applied strain. That is,

$$\sigma_S(t) = K(\varepsilon_{11}^o + \varepsilon_{22}^o + \varepsilon_{33}^o)\sin(\omega_o t)$$

$$[\sigma_D(t)] = \frac{2}{3}(E\sin(\omega_o t) + \omega_o\eta\cos(\omega_o t))\begin{bmatrix} (2\varepsilon_{11}^o - \varepsilon_{22}^o - \varepsilon_{33}^o) & 0 & \varepsilon_{13}^o \\ 0 & (2\varepsilon_{22}^o - \varepsilon_{33}^o - \varepsilon_{11}^o) & 0 \\ \varepsilon_{13}^o & 0 & (2\varepsilon_{33}^o - \varepsilon_{11}^o - \varepsilon_{22}^o) \end{bmatrix}$$

8.7   Constitutive Equations for Isotropic Incompressible Materials

An important class of viscoelastic materials also exists—such as most rubbe compounds—which are very nearly incompressible, in that they respond to loading by changing their shape but preserving their volume. Incompressible materials preserve volume throughout deformation. In geometrically linear problems, volume preservation requires that the volume strain, ε_{kk}, be identically zero throughout the loading process. Incompressibility acts as a constraint on the equations of motion. This means that one can add a spherical stress—uniform in all directions—of any magnitude to the acting stress field, without altering the strains. That is, for incompressible materials, the stress tensor is determined from the strains, only up to a spherical stress. The spherical stress is usually called the pressure stress, p. The stress–strain constitutive equations for isotropic incompressible materials take the form:

$$\sigma_{ij}(t) = -p(t)\delta_{ij} + \int_0^t 2\mu(t-\tau)\frac{\partial}{\partial\tau}\varepsilon_{ij}(t-\tau)d\tau;\, \varepsilon_{kk}(t) = 0 \qquad (8.41a)$$

Or, in convolution notation, omitting the time arguments of the convolution integrals:

$$\sigma_{ij}(t) = -p(t)\delta_{ij} + 2\mu * d\varepsilon_{ij};\, \varepsilon_{kk}(t) = 0 \qquad (8.41b)$$

Example 8.4 Strip biaxial test. A short, thin, and very wide sheet of rubber is stretched along its height direction, x_2, as depicted in Fig. 8.4. Because of its large aspect ratio, the sheet does not contract appreciably in the wide direction. As a consequence, $\varepsilon_{11} = 0$ in its central region. Using this fact, determine the stress in the direction of pull, in the center of the specimen, assuming that the shear relaxation modulus of the material is $\mu(t)$.
Solution
   In this case, $\varepsilon_{11} = 0$ and $\sigma_{33} = 0$. The incompressibility condition, $\varepsilon_{kk} = 0$, requires that $\varepsilon_{33} = -\varepsilon_{22}$, while, according to (8.41b), $\sigma_{33} = 0 = -p(t) + 2\mu * d\varepsilon_{33} \equiv -p(t) - 2\mu * d\varepsilon_{22}$. Using this to evaluate σ_{22} with (8.41b) again and simplifying lead to $\sigma_{22} = 4\mu * d\varepsilon_{22}$.

Fig. 8.4 Example 8.4

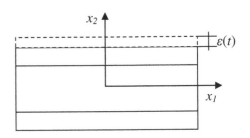

All material property functions as well as the components of the stress and strain tensors are assumed well behaved for their Laplace and Fourier transform to exist. Under these conditions, the Laplace and Fourier transforms of the stress–strain relations listed in (8.41b) are:

$$\bar{\sigma}_{ij}(s) = -\bar{p}(s)\delta_{ij} + 2s\bar{\mu}(s)\bar{\varepsilon}_{ij}(s); \bar{\varepsilon}_{kk}(s) = 0 \qquad (8.42)$$

$$\bar{\sigma}_{ij}(j\omega) = -\bar{p}(j\omega)\delta_{ij} + 2j\omega\bar{\mu}(j\omega)\bar{\varepsilon}_{ij}(j\omega); \bar{\varepsilon}_{kk}(j\omega) = 0 \qquad (8.43a)$$

Using the notation $j\omega \cdot \bar{\mu} \equiv \mu^*$, $\bar{p} \equiv p^*$; $\bar{\sigma}_{ij} \equiv \sigma_{ij}^*$, $\bar{\varepsilon}_{ij} \equiv \varepsilon_{ij}^*$, as is customary, (8.43a) may be expressed alternatively as

$$\sigma^*(j\omega) = -p^*(j\omega)\delta_{ij} + 2\mu^*(j\omega)\varepsilon^*(j\omega); \varepsilon_{kk}^*(j\omega) = 0 \qquad (8.43b)$$

8.8   Differential Constitutive Equations for Isotropic Materials

As with integral models, generalization of the constitutive equations to three-dimensional conditions is straightforward. Using indicial notation and the summation convention,[7] for instance, the most general constitutive equation in differential operator form becomes

$$P\sigma_{ij}(t) = Q\varepsilon_{ij}(t) \qquad (8.44)$$

For homogeneous isotropic materials, which require only two property functions, the viscoelastic constitutive equations in differential operator form—in analogy with the hereditary integral forms in (8.35a) and (8.36a)—read

$$P_D\sigma_{Dij} = 2Q_D\varepsilon_{Dij}; P_S\sigma_{kk} = 3Q_S\varepsilon_{kk} \qquad (8.45)$$

In general, then, four differential operators would be required to define the constitutive equations of a viscoelastic material in differential form. On the other hand, for materials whose volumetric response may be idealized as elastic, the stress strain Eq. (8.38a) takes the form:

$$P_D\sigma_{Dij} = 2Q_D\varepsilon_{Dij}; \sigma_{kk} = 3K\varepsilon_{kk} \qquad (8.46)$$

Following the same reasoning, the constitutive equations for isotropic incompressible viscoelastic materials of differential type are written as

$$P_D\sigma_{Dij} = 2Q_D\varepsilon_{Dij}; -p(t) \equiv \frac{1}{3}\sigma_{kk}; \varepsilon_{kk} = 0 \qquad (8.47)$$

---

[7] According to the summation convention, terms in an expression are summed over the range of their repeated indices, so that, for instance, $A_{ikk} = A_{i11} + A_{i22} + A_{i33}$, for $i = 1, 2, 3$, etc.

Example 8.5 A slender bar made of a Kelvin material having constant bulk modulus K and spring and dashpot parameters G and η is subjected to a uniaxial stress $\sigma_{11}(t) = \sigma_o H(t)$. Determine the strain $\varepsilon_{11}(t)$ and its long-term value.

Solution:

Use (8.46) to get $P_D \sigma_{D11} \equiv 2 Q_D \varepsilon_{D11}$ and $\sigma_{kk} = 3K \varepsilon_{kk}$. For a Kelvin material, $P = 1$ and $Q = G + \eta \partial_t$ so that $\sigma_{D11} \equiv 2[G + \eta \partial_t] \varepsilon_{D11}$; $\varepsilon_{kk} = \frac{1}{3K} \sigma_{kk} H(t)$. Or, in strain–stress form, introducing the retardation time $\lambda = \eta/G$ of the differential model $\frac{d}{dt} \varepsilon_{D11}(t) + \frac{1}{\lambda} \varepsilon_{D11}(t) \equiv \frac{1}{2\eta} \sigma_{D11}(t)$; $\varepsilon_{kk} = \frac{1}{3K} \sigma_{kk} H(t)$.

In this case, the only non-zero stress is σ_{11}. Hence, $\sigma_{kk} = \sigma_o H(t)$; $\sigma_{D11}(t) = \sigma_{11}(t) - \sigma_{kk}/3$, that is, $\sigma_{D11}(t) = (2/3)\sigma_o H(t)$ and $\varepsilon_{D11}(t) = \varepsilon_{11}(t) - \varepsilon_{kk}/3 \equiv \varepsilon_{11}(t) - \sigma_o H(t)/(9 K)$. Taking these results into the previous expression, cancelling the common factor, and reordering, the following is obtained:

$$\frac{d}{dt} \varepsilon_{11} + \frac{1}{\lambda} \varepsilon_{11} = \sigma_o H(t) \left[\frac{1}{3\eta} + \frac{G}{9K\eta} \right] + \frac{\eta}{9K} \sigma_o \delta(t); \ \lambda \equiv \frac{\eta}{G}; \ \varepsilon_{kk} = \frac{1}{3K} \sigma_o H(t)$$

Integrating between 0^- and t, using the integrating factor $e^{t/\lambda}$ with the initial condition $\varepsilon_{11}(0^-) = 0$, and rearranging produce $\varepsilon_{11}(t) = \frac{3K+G}{9KG} \left[1 - e^{-t/\lambda} \right] \sigma_o + \frac{\eta}{9K} e^{-t/\lambda} \sigma_o$. The long-term response is obtained as the limit $t \to \infty$: $\varepsilon_{11}(\infty) = \frac{3K+G}{9KG} \sigma_o \equiv \frac{\sigma_o}{E}$. Where the expression $9 KG/(3 K + G) = E$, for the elastic, Young's modulus was borrowed from Appendix B.

Physically realistic differential constitutive equations admit Laplace and Fourier integral transformations. As with one-dimensional models, the integral transform of a differential operator of order n turns out to be a polynomial of degree n in the transform variable (s or $j\omega$), as seen in (3.13). Using this, and letting v stand for either s or $j\omega$, the Laplace and Fourier transforms of (8.45) to (8.47) can be succinctly written as

$$\bar{P}_D(v) \bar{\sigma}_{Dij}(v) = 2 \bar{Q}_D(v) \bar{\varepsilon}_{Dij}(v); \bar{P}_S(v) \bar{\sigma}_{kk}(v) = 3 \bar{Q}_S(v) \bar{\varepsilon}_{kk}(v) \tag{8.48}$$

$$\bar{P}_D(v) \bar{\sigma}_{Dij}(v) = 2 \bar{Q}_D(v) \bar{\varepsilon}_{Dij}; \bar{\sigma}_{kk}(v) = 3 K \bar{\varepsilon}_{kk}(v) \tag{8.49}$$

$$\bar{P}_D(v) \sigma_{Dij}(v) = 2 \bar{Q}_D(v) \bar{\varepsilon}_{Dij}(v); \tag{8.50a}$$

$$-\bar{p}(v) \equiv \frac{1}{3} \bar{\sigma}_{kk}(v); \ \bar{\varepsilon}_{kk}(v) = 0 \tag{8.50b}$$

The relationships between the various property functions may be easily established by means of the Laplace transformation. Taking the Laplace transform of the corresponding relationships (8.35a) and (8.45) to do this, for instance, and equating the results, the following is obtained:

$$\bar{\mu}_M(s) = \frac{\bar{Q}_D(s)}{s\bar{P}_D(s)}; \bar{K}_M(s) = \frac{\bar{Q}_S(s)}{s\bar{P}_S(s)} \tag{8.51}$$

8.9   Problems

P.8.1 Write the convolution integral form of the two-dimensional constitutive equation for an isotropic incompressible viscoelastic solid of shear relaxation modulus G.

$$\text{Answer}: \begin{Bmatrix} \sigma_{xx}(t) \\ \sigma_{yy}(t) \\ \sigma_{xy}(t) \end{Bmatrix} = -p(t) \begin{Bmatrix} 1 \\ 1 \\ 0 \end{Bmatrix} + 2G(t-\tau)*d \begin{Bmatrix} \varepsilon_{xx}(\tau) \\ \varepsilon_{yy}(\tau) \\ \varepsilon_{xy}(\tau) \end{Bmatrix}$$

$$\equiv \begin{Bmatrix} -p(t) + 2G*d\varepsilon_{xx} \\ -p(t) + 2G*d\varepsilon_{yy} \\ 2G*d\varepsilon_{xy} \end{Bmatrix}$$

Hint: Use the extended vector form $A_{ij} \to \{A_{ij}\}^T \equiv \{A_{xx}, A_{yy}, A_{xy}\}$ of the stress and strain matrices to rewrite (8.41b): $\sigma_{ij}(t) = -p(t)\delta_{ij} + 2G*d\varepsilon_{ij}; \varepsilon_{kk}(t) = 0$.

P.8.2 Use matrix notation to extend to three dimensions the concept of effective elastic modulus for "constant rate" loading.

$$\text{Answer}: [E_{eff}(t)] \equiv \frac{1}{t} \int_0^t [M(u)] du$$

Hint. Note that, in this case, even though the prescribed boundary conditions may be changing at a constant rate, the induced strains may vary not only from point to point, but also from coordinate direction to coordinate direction at every point in the body. Using this, take $\varepsilon_i(t) = R_i \cdot t$ for each coordinate direction, use matrix algebra, and proceed as in Example 2.1 to arrive at the matrix form.

P.8.3 Use the Laplace transform to establish the tensile relaxation modulus of an elastomer for which a uniaxial Maxwell model is available with spring and dashpot parameters E and η, respectively.

Hint: Write the uniaxial constitutive equation of the model in Laplace-transformed space for both the differential form $P\sigma = Q\varepsilon$ and the convolution form $\sigma = M*d\varepsilon$. Then, apply the Laplace transform to both expressions and equate them to get $\bar{M} = \frac{\bar{Q}}{s\bar{P}}$. Now, insert the transformed operators, $\bar{P} = E + s\eta$ and $\bar{Q} = E\eta s$, and arrive at $\bar{M} = \frac{E\eta s}{s(E+s\eta)} \equiv \frac{E}{1/\tau+s}; \tau \equiv \eta/E$. The inverse transform of this expression [see Appendix] yields the desired result $M(t) = Ee^{-t/\tau}$.

P.8.4 Write the matrix form of the plane-stress constitutive equations of a viscoelastic material of differential type, whose volumetric response is elastic, with bulk modulus K.

$$\text{Answer}: \frac{1}{9K} \begin{bmatrix} (6KP_D + 2Q_D) & (2Q_D - 3KP_D) & 0 \\ (2Q_D - 3KP_D) & (6KP_D + 2Q_D) & 0 \\ 0 & 0 & 9KP_D \end{bmatrix} \begin{Bmatrix} \sigma_{11} \\ \sigma_{22} \\ \sigma_{12} \end{Bmatrix}$$

$$= 2Q_D \begin{Bmatrix} \varepsilon_{11} \\ \varepsilon_{22} \\ \varepsilon_{12} \end{Bmatrix}$$

Hint:

In plane stress, all stresses associated with one reference direction, the 3-axis, say, are zero. Specifically, $\sigma_{31} = \sigma_{32} = \sigma_{33} = 0$. Therefore, from the definitions given in (e) and (f) of Sect. 8.6: $\sigma_{kk} = \sigma_{11} + \sigma_{22}$, $\sigma_{D11} = 2/3\sigma_{11} - 1/3\sigma_{22}$, $\sigma_{D22} = -1/3\sigma_{11} + 2/3\sigma_{22}$, and also $\varepsilon_{kk} = (\sigma_{11} + \sigma_{22})/3$. Use these facts with (8.46) and perform the indicated operations to arrive at the result.

P.8.5 Solve Example 8.5 assuming that the Kelvin material is incompressible.

Hint:

The constitutive applicable constitutive equation in this case is (8.47). Because of incompressibility, $\varepsilon_{kk}(t) = 0$, so that $\varepsilon_{D11}(t) = \varepsilon_{11}(t)$. In addition, $-p(t) = \sigma_o H(t)/3$. Hence, σ_{D11} becomes $\sigma_{D11}(t) = \sigma_{11}(t) + p(t) \equiv (2/3)\sigma_o H(t)$. Taking this into (8.47), recalling that $P = 1$ and $Q = G + \eta\partial_t$ yields $\frac{d}{dt}\varepsilon_{11}(t) + \frac{1}{\lambda}\varepsilon_{11}(t) \equiv \frac{1}{\eta}\sigma_o H(t)$. Integrating this equation between 0^- and t, using the integrating factor $e^{t/\lambda}$ and the initial condition $\varepsilon_{11}(0^-) = 0$, and rearranging produce $\varepsilon_{11}(t) = \frac{1}{G}\left[1 - e^{-t/\lambda}\right]\sigma_o$. Evaluating this expression in the limit as $t \to \infty$ gives the long-term solution as $\varepsilon_{11}(\infty) = \frac{\sigma_o}{G}$.

P. 8.6 Determine the differential operator expression for the Poisson's ratio of a linear isotropic viscoelastic material having constant bulk modulus.

$$\text{Answer}: \nu = \frac{3KP - 2Q}{6KP + 2Q}$$

Hint: Here, the target is an operator expression for the ratio $\nu = -\varepsilon_{22}/\varepsilon_{11}$, resulting from a stress σ_{11}, when all other stresses are zero. Therefore, select a one-dimensional tension test, for which the only non-zero stress would be σ_{11}. Because of symmetry, $\varepsilon_{22} = \varepsilon_{33}$. Hence, $\sigma_{kk} = (1/3)\sigma_{11}$, $\varepsilon_{kk} = (\varepsilon_{11} + 2\ \varepsilon_{22})/3$, together with $\sigma_{D11} = (2/3)\sigma_{11}$ and $\varepsilon_{D11} = (2/3)(\varepsilon_{11} - \varepsilon_{22})$. Use expression (8.46) to obtain $P\sigma_{11} = 2Q\varepsilon_{11} - 2Q\varepsilon_{22}$ and $\sigma_{11} = 3\ K(\varepsilon_{11} + 2\varepsilon_{22})$. Combine these relationships and perform the necessary algebra to obtain the differential operator form of the Poisson's ratio.

P.8.7 Under two-dimensional conditions, the direction of the maximum principal stress is given by the relation: $\tan(2\theta^\sigma_{max}) = 2\sigma_{xy}/(\sigma_{xx} - \sigma_{yy})$; similarly, the direction of the maximum principal strain is given by $\tan(2\theta^\varepsilon_{max}) = 2\varepsilon_{xy}/(\varepsilon_{xx} - \varepsilon_{yy})$. Use these facts to demonstrate that, in general, for linear isotropic viscoelastic materials, the directions of maximum principal stress and strain do not coincide.

Answer : $\theta^{\sigma}_{max} = \dfrac{2\mu * d\varepsilon_{xy}}{\mu * d(\varepsilon_{xx} - \varepsilon_{yy})} \neq \dfrac{2\varepsilon_{xy}}{(\varepsilon_{xx} - \varepsilon_{yy})} = \theta^{\varepsilon}_{max}$

Hint: Evaluate θ^{σ}_{max} by inserting any one of the general constitutive equations for linear isotropic viscoelastic solids, such as (8.27) or (8.41a), cancel the common terms, and arrive at the proof. This problem shows that in viscoelastic materials, the directions of maximum principal stress and maximum principal strain do not coincide in general. As may be seen from the answer to this problem, the directions of principal stress and strain in a viscoelastic solid would coincide if the stress response were elastic like, that is, simply proportional to the strain, to allow cancelling the material property function. Conditions under which this happens are discussed in Chap. 9.

P. 8.8 A rigid cylindrical container is filled with a Maxwell material having elastic bulk modulus, K, and spring and dashpot parameters G and η, respectively. Determine the axial strain if a rigid plunger is used to apply a pressure $p_oH(t)$ at its open end, as indicated in Fig. 8.5.

Answer : $\varepsilon_{zz}(t) = -p_o\left[\dfrac{1}{K} + \left(\dfrac{1}{K} - \dfrac{1}{2G\eta + K} - \dfrac{3}{4G + 3K}\right)e^{-t/\tau}\right]$; τ

$\equiv \eta\left(\dfrac{4G + 3K}{3GK}\right)$

Hint: Write the three-dimensional constitutive equations for a Maxwell material with elastic bulk response: $Po_{Dij} = 2Q\varepsilon_{Dij}$ and $\sigma_{kk} = 3\,K\varepsilon_{kk}$, in cylindrical coordinates (r,θ,z). In this case, $\varepsilon_{rr} = \varepsilon_{\theta\theta} = 0$, $\varepsilon_{kk} = \varepsilon_{zz}$, $\sigma_{kk} = 3\,K\varepsilon_{zz}$, $\varepsilon_{Dzz} = (2/3)\varepsilon_{zz}$, and $\sigma_{Dzz} = -p_oH(t) - K\varepsilon_{zz}$ lead to $[G + \eta\partial_t][-p_oH(t) - K\varepsilon_{zz}] = 2[G\eta\partial_t]\frac{2}{3}\varepsilon_{zz}$. Collect terms and arrive at the equation $\frac{d\varepsilon_{zz}}{dt} + \frac{1}{\tau}\varepsilon_{zz} = -\frac{3p_o}{\eta(4G+3K)}[GH(t) + \eta\delta(t)]$. Solve this equation using the initial condition $\varepsilon_{zz}(0) = \frac{-p_o}{2G\eta+K}$, which can be obtained from $Po_{Dij} = 2Q\varepsilon_{Dij}$, following the procedure discussed in Chap. 3 (see Sect. 3.3.4).

P.8.9 Prove that the Poisson's ratio of an incompressible viscoelastic material is 1/2.

Hint: Consider a uniaxial tensile test on a general incompressible material. Invoke its incompressibility $\varepsilon_{kk} = 0$ and symmetry of strain response ($\varepsilon_{22} = \varepsilon_{33}$) to establish that $\varepsilon_{22} = -(1/2)\varepsilon_{11}$. Use this result and the definition of Poisson's ratio that $\nu = -\varepsilon_{22}/\varepsilon_{11}$ to complete the proof.

Fig. 8.5 Problem 8.7

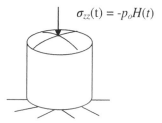

$\sigma_{zz}(t) = -p_oH(t)$

P.8.10 Write the stress–strain constitutive equations applicable to plane-stress steady-state conditions for an orthotropic viscoelastic solid.

$$\text{Answer}: \left\{ \begin{matrix} \sigma_{11}^* \\ \sigma_{22}^* \end{matrix} \right\} = \frac{1}{1 - v_{12}^* v_{21}^*} \begin{bmatrix} E_{11}^* & v_{12}^* E_{22}^* \\ v_{21}^* E_{11}^* & E_{22}^* \end{bmatrix} \left\{ \begin{matrix} \varepsilon_{11}^* \\ \varepsilon_{22}^* \end{matrix} \right\}, \sigma_{12}^* = 2G_{12}^* \varepsilon_{12}^*$$

Hint: Write the complex form of the strain constitutive Eq. (8.13a, b), as discussed in Sect. 8.5: $\left\{ \begin{matrix} \varepsilon_{11}^* \\ \varepsilon_{22}^* \end{matrix} \right\} = \begin{bmatrix} 1/E_{11}^* & -v_{21}^*/E_{22}^* \\ -v_{12}^*/E_{11}^* & 1/E_{22}^* \end{bmatrix} \left\{ \begin{matrix} \sigma_{11}^* \\ \sigma_{22}^* \end{matrix} \right\}, \varepsilon_{12}^* = \sigma_{12}^*/(2G_{12}^*).$ Invert the matrix and the shear equation and use the complex form of the symmetry relations (8.12) to arrive at the result.

P.8.11 A rectangular plate made of a linear isotropic viscoelastic solid is supported along its long sides while subjected to a strain $\varepsilon_1(t) = 0.05 \cdot \cos(0.25\,t)$ on its short edges, as shown in Fig. 8.6. Determine the reaction stress at the support, for a material with relaxation modulus $E(t) = 100 + 1{,}000e^{-0.5t}$ MPa and Poisson's ratio $v(t) = 0.40 – 0.15e^{-0.25t}$.

$$\text{Answer}: \ \sigma_2(t) \approx 1.540\,\text{MPa}$$

Hint: Note that for an isotropic solid, $v_{12} = v_{21} = v$ and $E_{11} = E_{22} = E$ and use the result of P. 8.10, to write $\left\{ \begin{matrix} \sigma_{11}^* \\ \sigma_{22}^* \end{matrix} \right\} = \frac{E^*}{1-(v^*)^2} \begin{bmatrix} 1 & v^* \\ v^* & 1 \end{bmatrix} \left\{ \begin{matrix} \varepsilon_{11}^* \\ \varepsilon_{22}^* \end{matrix} \right\}, \sigma_{12}^* = 2G^* \varepsilon_{12}^*.$ Since the plate is restrained in the 2-direction but free to extend in the 1-direction, no shearing would develop in it, the strain field would be $\varepsilon_{11}^* = \varepsilon_{o11}\cos(0.25t)$ and $\varepsilon_{22}^* = 0 = \varepsilon_{12}^*$, and as a consequence, the stress in the 2-direction becomes $\sigma_{22}^* = \frac{v^* E^*}{1-(v^*)^2} \varepsilon_{11}^*$. Next, evaluate the complex modulus and the complex Poisson's ratio using the results of Example 4.1. For E^*, use $E_e = 100$, $E_1 = 1{,}000$, and $\tau = 2$; and for v, $v_e = 0.40$, $v_1 = -0.15$, and $\tau = 4$. So that $E^*(0.25j) = 100 + \frac{1{,}000(0.25*2)^2}{1+(0.25*2)^2} + j\frac{1{,}000(.25*2)}{1+(0.25*2)^2} = 300 + 200j$ and $v^*(0.25j) = 0.4 - \frac{0.15(0.25*4)^2}{1+(0.25*4)^2} - j\frac{-0.15(.25*4)}{1+(0.25*4)^2} = 0.325 - 0.075j.$ Take these values and the applied strain field into the expression for σ_{22}^* and perform the complex algebra and since the actual

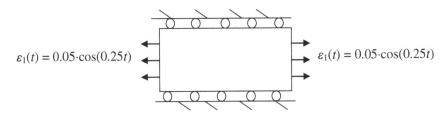

$\varepsilon_1(t) = 0.05 \cdot \cos(0.25t)$ $\varepsilon_1(t) = 0.05 \cdot \cos(0.25t)$

Fig. 8.6 Problem 8.10

applied strain is the real part of the complex strain ε_{22}^*, take the real part of the result to arrive at the answer sought.

References

1. A.E. Green, W. Zerna, *Theoretical Elasticity*, 2nd edn. (Dover, NY, 1968), pp. 1–3
2. L.E. Malvern, *Introduction to the Mechanics of a Continuous Medium* (Prentice-Hall, Englewood Cliffs, 1963), pp. 64–80
3. B.E. Read, G.D. Dean, *The Determination of Dynamic Properties of Polymer Composites* (Wiley, New York, 1978), pp. 12–17
4. R.M. Christensen, *Mechanics of Composite Materials* (Wiley, New York, 1979), pp. 150–160
5. W.R. Little, *Elasticity* (Prentice Hall, Englewood Cliffs, 1973), pp. 62–63. 16–17

Isothermal Boundary-Value Problems

9

Abstract

This chapter contains a comprehensive discussion of the types of boundary-value problems encountered in linear viscoelasticity. The chapter presents detailed solution methods for compressible and incompressible solids, including materials with synchronous moduli, whose property functions are assumed to have the same time dependence. The method of separation of variables in the time domain and frequency domains is also described in full, as is the use of the Laplace and Fourier transformations. The elastic–viscoelastic correspondence principle, which allows viscoelastic solutions to be constructed from equivalent elastic ones and as a consequence of the applicability of integral transforms, is also developed and examined in detail.

Keywords

Boundary-value · Balance · Conservation · Momentum · Energy · Equilibrium · Motion · Quasi-static · Compatibility · Synchronous · Separation of variables · Correspondence principle

9.1 Introduction

This chapter introduces several methods of solving boundary-value problems of isotropic linear viscoelastic continua for which the temperature is assumed constant in space and time. Thus, all required material property functions are assumed known at the arbitrary but constant temperature of the discussions.

The chapter begins by listing the differential equations of motion, which embody the principles of balance of linear and angular momenta and are postulated to be obeyed by all materials in bulk. The boundary-value problems of viscoelasticity are presented next, restricting attention to quasi-static conditions,

D. Gutierrez-Lemini, *Engineering Viscoelasticity*, DOI: 10.1007/978-1-4614-8139-3_9,   219
© Springer Science+Business Media New York 2014

for which the acceleration, or inertia terms are negligible. Specific methods of solution of boundary-value problems are discussed after that. The methods presented include the separation of variables and integral transformations, as well as the special case of materials with synchronous moduli, for which the viscoelastic property functions in shear and bulk are assumed to have the same dependence on time.

9.2   Differential Equations of Motion

There are certain laws of physics which are obeyed by all substances in the bulk, be they elastic, viscous or viscoelastic, irrespective of whether their response to external stimuli is linear or non-linear [1]. These laws proclaim the conservation of mass, linear momentum, angular momentum, and energy. All the mentioned laws are discussed in Appendix B, but for continuity with the topic of this chapter, the balance laws of linear and angular momenta are presented here as well.

9.2.1   Balance of Linear Momentum

Newton's second law of motion requires a balance between the external resultant load acting on a system and the rate of change of its linear momentum. The integral version of this law is due to Cauchy and gives rise to the equations of motion, which are valid for all materials in the bulk. The derivation of the motion equations is straightforward; and, as shown in Appendix B, for instance, in unabridged notation, the x-component of the equations of motion for a body with density ρ and body force ρb_x, reads:

$$\frac{\partial}{\partial x}\sigma_{xx} + \frac{\partial}{\partial y}\sigma_{xy} + \frac{\partial}{\partial z}\sigma_{xz} + \rho b_x = \rho\frac{\partial^2}{\partial t^2}v_x \qquad (a)$$

The y- and z-component equations are written with appropriate permutations of the coordinates. Thus, using indicial tensor notation, where terms with repeated indices are summed over their full range, the equations of balance of linear momentum take the form:

$$\frac{\partial}{\partial x_j}\sigma_{ij} + \rho b_i = \rho\frac{\partial^2}{\partial t^2}u_i; \quad i, j = 1, 3 \qquad (9.1)$$

When inertia terms are zero, as in static problems, or can be neglected, which leads to so-called quasi-static problems, the acceleration term on the right-hand side of the equations is dropped. This leads to the equations of static equilibrium:

$$\frac{\partial}{\partial x_j}\sigma_{ij} + \rho b_i = 0_i \qquad (9.2)$$

Where differentiation is understood with respect to coordinates in the undistorted state. The comma notation: $\sigma_{ij,k} \equiv \partial\sigma_{ij}/\partial x_k$, is also frequently used to denote differentiation with respect to the coordinate(s) following the comma.

9.2.2   Balance of Angular Momentum

Non-polar materials are defined as those without intrinsic body couples or spin, and for which the resultant internal moment on the surface of any infinitesimal material element is zero. For such materials, the principle of conservation of angular momentum—that the resultant external moment on a body is equal to the time rate of change of its angular momentum—yields the simple requirement that the stress tensor, σ_{ij}, be symmetric:

$$\sigma_{ij} = \sigma_{ji} \tag{9.3a}$$

Using unabridged notation and the standard x, y, z Cartesian coordinates, these expressions take the form:

$$\sigma_{xy} = \sigma_{yx}; \quad \sigma_{xz} = \sigma_{zx}; \quad \sigma_{yz} = \sigma_{zy} \tag{9.3b}$$

The implication of the balance of angular momentum for non-polar materials is that the stress tensor has only 6 independent components. This is so because a symmetric N-by-N matrix has $N \cdot (N+1)/2$ independent components (6, when $N = 3$).

9.3   General Boundary-Value Problem

The isothermal boundary-value problem of linear isotropic viscoelasticity consists of the three equations of motion (9.1) together with their initial and boundary conditions. Taking the initial configuration of the body to be free of stress and at rest prior to the application of the loading (all field variables are identically zero for $t < 0$), the equations of motion and the associated boundary conditions are as follows[1]:

$$\frac{\partial}{\partial x_j} \sigma_{ij}(t) + \rho b_i(t) = \rho \frac{\partial^2}{\partial t^2} u_i(t); \qquad \text{in} \quad V$$
$$u_i(t) = u_i^o(t); \qquad\qquad \text{on } S_u; \quad t \geq 0 \tag{9.4}$$
$$n_j \sigma_{ji}(t) = T_i^o(t); \qquad\qquad \text{on } S_T; \quad t \geq 0$$

---

[1] The components of the displacement vector, the stress and strain tensors, the material density, the boundary values, and the normal to the surface, will in general depend on position. For clarity of exposition, however, dependence on position is omitted most of the time, but shall be understood.

In these expressions, n_i represents the unit outward normal to the boundary, S, formed of S_u and S_T, where displacements and tractions, respectively, are prescribed. Here, n_i is not a function of time. In addition, in a well-posed problem, S_u and S_T do not intersect (that is: $S_u \cap S_T = \Phi$). This last requirement means that one cannot prescribe different types of boundary conditions at the same point and in the same direction.

The system of equations in (9.4) contains 9 unknown quantities—3 displacement components and 6 independent stress components. For a solution of this system to exist, the system has to be complemented by other relations. The displacements u_i, are related to the strains, ε_{ij} through 6 relationships (c.f. Appendix B):

$$\varepsilon_{ij}(t) = \frac{1}{2}\left[\frac{\partial}{\partial x_j} u_i(t) + \frac{\partial}{\partial x_i} u_j(t) \right]$$ (9.5)

Finally, the 6 constitutive equations relate 6 additional strains to the 6 components of the stress tensor. The constitutive equations may be taken in hereditary integral form, in differential form, or, for steady-state problems, in terms of complex quantities. As indicated in Chap. 8, the constitutive equations of isotropic materials are completely defined by any two of five interrelated relaxation functions: the tensile, shear, and bulk moduli, $E(t)$, $G(t)$, $K(t)$, respectively; the Poisson's or contraction ratio, $v(t)$; and Lame's function $\lambda(t)$. As discussed in Chap. 8, the relationships among these functions may be derived using their elastic counterparts (as listed in [2]) and the elastic–viscoelastic correspondence. For the sake of presentation, the constitutive equations developed as (8.27) and (8.41) for compressible and incompressible materials, respectively, and listed here:

$$\sigma_{ij}(t) = \lambda_M(t - \tau) * d\varepsilon_{kk}(\tau)\delta_{ij} + 2\mu_M(t - \tau) * d\varepsilon_{ij}(\tau)$$ (9.6)

$$\sigma_{ij}(t) = -p(t)\delta_{ij} + 2\mu_M(t - \tau) * d\varepsilon_{ij}(\tau)$$ (9.7)

Equations (9.4), (9.5), and any of the sets of 6 constitutive equations derived in Chap. 8, such as either (9.6) or (9.7), represent 15 equations in 15 unknowns: 3 equilibrium equations, 6 strain–displacement relations, and 6 stress–strain equations; and 3 unknown displacements, 6 unknown stresses and 6 unknown strains. These are the field equations of isothermal viscoelasticity, which will possess a unique solution if the initial values of the material relaxation functions are non-negative [3].

The strain–displacement equations listed in (9.5) relate the three components of the displacement field to the six components of strain. These expressions result in a unique set of strains for any prescribed set of displacements, but in general, do not suffice to produce a unique displacement field from an arbitrarily prescribed set of strains. The system of equations in the latter case is over-determined, as it has six equations in three unknowns. This prevents the six components of strain to be prescribed arbitrarily. The additional conditions that the strain tensor must satisfy to allow a unique displacement field upon integration of the strain–displacement relations are known as the integrability or compatibility conditions.

The equations of compatibility are obtained by differentiating the strain–displacement relations twice and permuting indices (c.f. Appendix B). This process yields the following 81 equations, of which only six are independent [4]:

$$\varepsilon_{ij,kl} + \varepsilon_{kl,ij} = \varepsilon_{ik,jl} + \varepsilon_{jl,ik} \tag{9.8}$$

Using the standard Cartesian coordinates x, y, z, for subscripts 1,2,3, respectively, the six independent compatibility equations in unabridged notation are of the following form:

$$\frac{\partial^2}{\partial x^2} \varepsilon_{yy} + \frac{\partial^2}{\partial y^2} \varepsilon_{xx} = 2 \frac{\partial^2}{\partial x \partial y} \varepsilon_{xy}; \quad \frac{\partial^2}{\partial y \partial z} \varepsilon_{xx} = \frac{\partial}{\partial x} \left[-\frac{\partial \varepsilon_{yz}}{\partial x} + \frac{\partial \varepsilon_{zx}}{\partial y} + \frac{\partial \varepsilon_{xy}}{\partial z} \right] \tag{b}$$

Similar permutations of x, y, and z produce the other four independent relations. In the case of two-dimensional problems in the x–y plane, the only non-trivially satisfied relation is the first one listed above.

On occasion, the compatibility conditions are required in terms of stresses. Since stresses and strains are connected through the constitutive equations, the integrability conditions in terms of stresses depend on material properties. These equations may be obtained by direct substitution of the strain–stress constitutive equations into (9.8). For a viscoelastic material of hereditary type, the result is the convolution form of the so-called Beltrami-Michell relations of the theory of elasticity [5].

9.4   Quasi-Static Approximation

In what follows, inertia effects are ignored but it is recognized that all field variables, u_i, $\sigma_{ij,}$ and ε_{ij} will generally change with time. This is the quasi-static approximation, in which problems remain time-dependent, because the stress response depends, through the strains, on the complete history of displacements, and not only on the current state, even when the boundary conditions remain constant. In the latter case, the step functions that define the boundary displacements and tractions produce disturbances that propagate into the body at high speed. Stress relaxation and heat conduction in viscoelastic materials, however, tend to damp out this wave motion rather quickly. In using the quasi-static approximation, it is assumed that the wave motion generated by any sudden changes in the boundary conditions is damped out more quickly than the boundary data change afterward. The quasi-static approximation, then, assumes that at any instant the body is in equilibrium with the concurrent boundary data [6].

9.5   Classification of Boundary-Value Problems

The differential equations of motion (9.4) and associated boundary conditions suggest that the boundary conditions may be of one of the three types. The first class corresponds to problems where only tractions are prescribed. A second class involves pure displacement boundary values, while a third class allows displacements and tractions to be prescribed on the boundary. The three types of boundary-value problems are examined next; and for simplicity, only the quasi-static approximation is considered, so that the inertia terms are set to zero.

9.5.1   Traction Boundary-Value Problem

In this case, (9.4) takes the quasi-static form:

$$\frac{\partial}{\partial x_j} \sigma_{ij}(t) + \rho b_i(t) = 0; \quad \text{in } V$$
$$n_j \sigma_{ji}(t) = T_i^o(t); \qquad \text{on } S \tag{9.9}$$

It is important to note that in this case the differential equations of equilibrium and the boundary data involve only stresses. Since this system does not involve material properties, its primary solution, that is the stress field, has to be independent of material constitution. In other words:

> The stresses in a linear viscoelastic body—irrespective of whether it is isotropic or not— subjected only to tractions on its boundary, are exactly the same as those which the same body would experience if its material were replaced with any other linear material.

The strain and displacement solutions, however, will in general depend on the material in question, and the equations of compatibility need to be employed to ensure that the strains, calculated from the stress field through the constitutive equations, will yield single-value displacements in simply connected domains— when the body has no internal holes.

9.5.2   Displacement Boundary-Value Problem

In this case, only displacements are prescribed on the surface of the body, and the governing equations and boundary conditions become:

$$\frac{\partial}{\partial x_j} \sigma_{ij}(t) + \rho b_i(t) = 0; \quad \text{in } V \tag{9.10}$$

$$u_i(t) = u_i^o(t); \quad \text{on } S$$

Here, since strains are determined uniquely by the displacement field, strain compatibility is automatically satisfied. Also, because stresses and displacements are connected by the strain–displacement relations and constitutive equations, the equations of equilibrium may be cast in terms of displacements only.

Unlike the traction-only problem, no savings can be realized in the present case for viscoelastic materials in general. However, as with the traction-only problem, the viscoelastic solutions for incompressible isotropic materials, and for isotropic materials of constant Poisson's ratio can be constructed from those of elastic solids, as shown in Sect. 9.6.

9.5.3 Mixed Boundary-Value Problem

This is the most general type of boundary-value problem in that both displacements and tractions are specified on the bounding surface. Just as the traction-only problem, the constitutive equations and compatibility conditions are required. In addition, however, the strain–displacement relations are needed for the problem to be mathematically well posed. The field equations for this case take the form:

$$
\begin{aligned}
\frac{\partial}{\partial x_j}\sigma_{ij}(t) + \rho b_i(t) = 0; &\quad \text{in } V \\
n_j \sigma_{ji}(t) = T_i^o(t); &\quad \text{on } S_T \\
u_i(t) = u_i^o(t); &\quad \text{on } S_u
\end{aligned}
\tag{9.11}
$$

No savings in solution effort are available in general for the mixed boundary-value case. Also, as with the other two types of problem, viscoelastic solutions can be constructed from elastic solutions for isotropic solids of constant Poisson's ratio, which include as a special case, incompressible materials.

Example 9.1 A disk of diameter D, made of a viscoelastic material is subjected to a pair of diametrically opposing concentrated loads, $P(t)$, as indicated in Fig. 9.1. Find the stresses along the load line, if the corresponding values for a linear elastic solid are [2]:

$$
\sigma_x^e(x = 0, y) = \frac{2P(t)}{\pi D};
$$

$$
\sigma_y^e(x = 0, y) = -\frac{2P}{\pi}\left[\frac{2}{D - 2y} + \frac{2}{D + 2y} - \frac{1}{D}\right]; \quad \sigma_{xy}^e(x = 0, y) = 0;
$$

Solution:

This is a traction-only boundary-value problem. Consequently, the stress picture is independent of material properties, and the viscoelastic solution must be equal to the elastic one given.

Fig. 9.1 Example 9.1

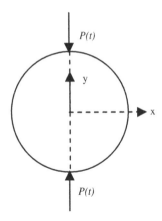

9.6   Incompressible Materials

The mechanical response of isotropic incompressible materials is fully described by one single material property function, as indicated by (9.7). The solution of a displacement boundary-value problem for this material type can be constructed from the solution of the same problem posed for an isotropic incompressible elastic material of arbitrary shear modulus. The method followed here to accomplish this, normalizes the shear relaxation modulus, μ_M, as:

$$\mu_M(t) \equiv \mu_R.m_N(t) \tag{9.12}$$

This normalization introduces an elastic material of shear modulus μ_R that is meaningful to the problem at hand. At the same time, through (2.20) or (2.21), it defines the creep compliance, C_μ and its normalized counterpart c_N, and establishes the relationship between the normalized functions m_N and c_N, as:

$$C_\mu(t) = C_R \cdot c_N(t) = \frac{1}{\mu_R} c_N(t) \tag{9.13}$$

$$m_N(t - \tau) * dc_N(\tau) = c_N(t - \tau) * dm_N(\tau) \tag{9.14}$$

The method also introduces auxiliary displacement, u_{Vi}, in terms of the actual displacements, u_i, by means of the definition:

$$u_{Vi}(t) \equiv m_N(t - \tau) * du_i(\tau) = \int_{0^-}^{t} m_N(t - \tau) \frac{\partial}{\partial \tau} u_i(\tau) d\tau \tag{9.15}$$

According to (9.5), the strains, ε_{Vij}, corresponding to this auxiliary field would be given by:

$$\varepsilon_{Vij}(t) = \frac{1}{2}\left(\frac{\partial}{\partial x_j}u_{Vi} + \frac{\partial}{\partial x_i}u_{Vj}\right) \equiv \frac{1}{2}\left(u_{Vi,j} + u_{Vj,i}\right) \tag{9.16}$$

and, in particular:

$$\mu_M * d\varepsilon_{ij} = \mu_R m_N * d\left(u_{i,j} + u_{j,i}\right) \equiv \mu_R\left(m_N * du_{i,j} + m_N * du_{j,i}\right) = \mu_R\varepsilon_{Vij}(t),$$

Using this result in (9.7), yields

$$\sigma_{ij}(x_k, t) = -p(x_k, t)\delta_{ij} + 2\mu_R\varepsilon_{Vij}(x_k, t) = \sigma_{Vij}(x_k, t) \tag{9.17}$$

That $\sigma_{ij} = \sigma_{Vij}$, follows from the fact that, considered as a function of the strains, $\sigma(\varepsilon) = -p + 2\mu\varepsilon$, implies that $\sigma(\varepsilon_V) = -p + 2\mu\varepsilon_V \equiv \sigma_V$. From this, follows that the equations of equilibrium in (9.10) are identically satisfied by the auxiliary elastic stress (9.17). To satisfy the displacement boundary conditions also, it is necessary to convolve them with the normalized shear modulus m_N, and use (9.15) to convert them into the auxiliary elastic form: $m * du_i \equiv u_{Vi} = m * du_i^o$. This transforms the displacement boundary-value problem (9.10) for an incompressible viscoelastic material to the auxiliary elastic form:

$$\frac{\partial}{\partial x_j}\sigma_{Vij} + \rho b_i = 0, \, in \, V; \quad u_{Vi} = m * du_i^o, \quad on \, S \tag{9.18}$$

Once the auxiliary elastic displacements are obtained, the actual displacements are determined inverting (9.15)—convolving c_N with u_{Vi}. The strains follow from the strain–displacement relationships.

9.7   Materials with Synchronous Moduli

The mechanical response of viscoelastic materials for which the relaxation functions in bulk and shear have the same time dependence are said to have synchronous moduli. The proportionality between the shear and bulk relaxation moduli implies that the stress–strain behavior for this type of material is fully described by a single relaxation or compliance function. Synchronous moduli are possible if the Poisson's ratio is constant. In this case, several options are available regarding the time-dependent material properties used: λ_M and $\mu_M \equiv G$; G and v; the uniaxial tensile modulus, E, and Poisson's ratio, v; and so on.

Since materials with synchronous moduli have only one viscoelastic property function, the same approach can be followed as for incompressible materials to transform a displacement boundary-value problem of a viscoelastic material to that of an auxiliary elastic one. The difference lies in the specific form of the constitutive equation and the choice of the material property functions used.

The approach used introduces the normalized relaxation—or creep—functions entering the problem, as follows:

$$\lambda_M(t) \equiv \lambda_R \cdot m_N(t);\ \mu_M(t) \equiv \mu_R \cdot m_N(t);\ E(t) \equiv E_R \cdot m_N(t) \tag{9.19}$$

$$C_\lambda(t) \equiv \frac{1}{\lambda_R} \cdot c_N(t);\ C_\mu(t) \equiv \frac{1}{\mu_R} \cdot c_N(t);\ C_E(t) \equiv \frac{1}{E_R} \cdot m_N(t) \tag{9.20}$$

This normalization, for which (9.14): $m_N(t - \tau) * dc_N(\tau) = c_N(t - \tau) * dm_N(\tau)$, still holds, and the constancy of the Poisson's ratio imply that the normalizing moduli λ_R, G_R, and E_R, are related elastically. That is (c.f. Appendix B):

$$\lambda_R = \frac{v}{(1+v)(1-2v)};\ E_R = \frac{2v}{(1-2v)} G_R \tag{9.21}$$

From this point on, the method is the same as for incompressible materials. It uses the auxiliary elastic displacements, u_{Vi}, and the corresponding strains, ε_{Vij}, listed in (9.15) and (9.16), introduces (9.19) into the selected constitutive equations and arrives at the same result as for incompressible materials, that the displacement boundary-value problem (9.10) of viscoelasticity is converted to that of an auxiliary elastic solid, as in (9.18).

Example 9.2 Write the explicit form of the auxiliary elastic stress field for use in a displacement boundary-value problem of viscoelasticity, for a viscoelastic solid with synchronous moduli whose constitutive equation is given in the form: $\sigma_{ij}(t) = \lambda(t - \tau) * d\varepsilon_{kk}(\tau)\delta_{ij} + 2\mu(t - \tau) * d\varepsilon_{ij}(\tau)$.
Solution:
Use the normalization suggested (9.19) with the selected constitutive equation to write that: $\sigma_{ij}(t) = \lambda_R m_N * d\varepsilon_{kk}\delta_{ij} + 2\mu_R m_N * d\varepsilon_{ij}$. Then, interchange the order of differentiation and integration and invoke the definitions of the auxiliary elastic displacement and strain fields to arrive at the form:

$$\sigma_{ij}(x_k, t) = \lambda_R \varepsilon_{Vkk}(x_k, t)\delta_{ij} + 2\mu_R \varepsilon_{Vij}(x_k, t) = \sigma_{Vij}(x_k, t).$$

9.8   Separation of Variables in the Time Domain

The method of separation of variables is applied to construct solutions to viscoelastic boundary-value problems for isotropic materials of constant Poisson's ratio. The basis of the method is the assumption that the solution variables have the same dependence on time as their boundary data counterparts. That is, the stresses and body forces have the same time variation as the prescribed surface tractions and the displacements and strains, the same time variation as the boundary displacements. As will be seen, the method of separation of variables—valid only for materials with constant Poisson's ratio—applies to any of the three types of boundary-value problems. Specifically, it is assumed that to boundary data[2]:

$$T_i^o(x_k, t)\big|_{S_T} \equiv T_{oi}^o(x_k)f_T(t); \quad u_i^o(x_k, t)\big|_{S_u} \equiv u_{oi}^o(x_k)f_u(t) \tag{9.22}$$

There correspond solution variables in the separable forms:

$$\sigma_{ij}(x_k, t) = \sigma_{oi}(x_k)f_T(t); \quad b_i(x_k, t) \equiv b_{oi}(x_k)f_T(t) \tag{9.23}$$

$$u_i(x_k, t) = u_{oi}(x_k)f_u(t); \quad \varepsilon_{ij}(x_k, t) = \varepsilon_{oij}(x_k)f_u(t) \tag{9.24}$$

Where the spatial part of the strain, ε_{oij}, is computed from the spatial part of the displacement field through the strain–displacement relations (9.5), as:

$$\varepsilon_{oij} = \frac{1}{2}\left(\frac{\partial}{\partial x_j}u_{oi} + \frac{\partial}{\partial x_i}u_{oj}\right) \tag{9.25}$$

Using (9.22) to (9.24) with the equations of equilibrium under mixed boundary data (9.11), canceling the common factors $f_T(t)$ and $f_u(t)$, results in a mixed boundary-value problem for the spatial parts σ_{oij}, u_{oi}, and ε_{oij} of the viscoelastic solution fields. Explicitly:

$$\frac{\partial}{\partial x_j}\sigma_{oij}(x_k) + b_{oi}(x_k) = 0; \quad \text{in } V$$
$$n_j(x_l)\sigma_{oji}(x_k) = T_{oi}^o(x_k); \quad \text{on } S_T \tag{9.26}$$
$$u_{oi}(x_k) = u_{oi}^o(x_k); \quad \text{on } S_u$$

The viscoelastic solution is retrieved from the elastic one through (9.23) and (9.24). Since the stress and strain fields are related through the constitutive equations, the functions f_T and f_u cannot be prescribed independently. The relation between them is established by means of the constitutive equations. Taking the strain field (9.24) into the constitutive equation, such as (9.6), using (9.23) and (9.25), and separating the spatial and temporal parts, there results:

$$\sigma_{ij}(t) = \sigma_{oij}f_T(t) = \left[\lambda_R\varepsilon_{okk}\delta_{ij} + 2\mu_R\varepsilon_{oij}\right]m_N(t-\tau) * df_u(\tau) \tag{9.27}$$

Since the quantity inside the braces is the spatial part, σ_{oij}, of the stress tensor, this expression leads to the relation sought between the time functions f_u and f_T:

$$f_T(t) = m_N(t-\tau) * df_u(\tau) = \int_{0^-}^{t} m_N(t-\tau)\frac{d}{d\tau}f_u(\tau)d\tau \tag{9.28}$$

---

[2] The reader is reminded that all boundary data as well as the displacement, stress and strain fields are dependent on position but that, sometimes, such dependence has been omitted for clarity.

The same approach may be applied when the boundary conditions are either pure traction or pure displacement. The difference is that in problems with boundary data of a single type only one of the time functions is given, and the other has to be determined from (9.28). The approach described applies equally well to isotropic incompressible viscoelastic materials, because their constitutive equations involve a single material property function.

Example 9.3 A small hole of radius, r_i, in a very large block of a viscoelastic material of constant Poisson's ratio, v, is subjected to an internal pressure $P \cdot f(t)$, as indicated in Fig. 9.2. Determine the stresses and displacements in the block, if the solution for the cavity in an infinite elastic medium under a pressure of magnitude, P_i, is given by the relations [2]:

$$\sigma_r^e(r,\theta) = -\left(\frac{r_i}{r}\right)^2 P; \quad \sigma_\theta^e(r,\theta) = \left(\frac{r_i}{r}\right)^2 P; \quad \text{and} \quad u_r^e(r,\theta) = \frac{(1+v)}{E}\frac{r_i^2}{r}P$$

Solution:
If the cavity in question is small compared with the dimensions of the visco-elastic solid, the elastic boundary-value problem approximates the viscoelastic one, and its solution may be used to construct the one to the problem posed. Although only tractions are prescribed on the boundary of the hole, the remote displacement field is zero. That is, this problem is a mixed boundary-value problem. Hence, according to (9.23) the elastic and viscoelastic stresses at each point and time are the same; that is:

$$\sigma_r(r,\theta,t) = -\left(\frac{r_i}{r}\right)^2 Pf(t); \quad \sigma_\theta(r,\theta,t) = \left(\frac{r_i}{r}\right)^2 Pf(t)$$

The actual displacements are obtained from (9.24): $u_{io}(t) = u_i f_u(t)$, and the convolution inverse of (9.28): $f_u(t) = m_N^{-1} * df_T(\tau)$, using the normalization $E(t) = E_R \cdot m_N(t)$. This gives the solution $u_r(r,\theta,t) = \frac{r_i^2}{r}P\frac{(1+v)}{E_R}\int\limits_{0^-}^{t} m_N^{-1}(t-\tau)\,df(\tau)d\tau$.

Fig. 9.2 Example 9.3

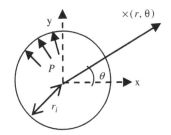

9.9   Integral-Transform Correspondence Principles

Elastic–viscoelastic correspondence principles are a convenient means of constructing solutions to quasi-static viscoelastic boundary-value problems from elastic solutions. The steady-state case falls under this general definition.

Inasmuch as the Laplace and Fourier transformations are mathematically related, the elastic–viscoelastic correspondence principle presented here may be interpreted as derived from either integral transform. For definiteness, we let s denote the transform parameter and understand it to mean s in case of the Laplace transform, or $j\omega$ if the Fourier transform is meant. In the latter case, j stands for the imaginary unit: $j \equiv \sqrt{-1}$. Also, the same symbol is used to represent a quantity and its transform, distinguishing the latter with an over-bar.

Viscoelastic boundary-value problems in transformed space can be derived by applying the corresponding transform—Laplace or Fourier—to the field, compatibility, and constitutive equations of linear viscoelasticity. Proceeding thus, interchanging the order of integration and differentiation, implied, respectively, by the integral transforms of the field equations, there result:

(a) For the equilibrium equations and boundary conditions (9.4):

$$\frac{\partial}{\partial x_j}\bar{\sigma}_{ij}(s) + \bar{B}_i(s) = 0; \quad n_j\bar{\sigma}_{ji}(s)\big|_{S_T} = \bar{T}_i^o(s); \quad \bar{u}_i(s)\big|_{S_u} = \bar{u}_i^o(s) \qquad (9.29)$$

(b) For the general stress–strain constitutive equations, such as (9.6) or (9.7):

$$\bar{\sigma}_{ij}(s) = s\bar{\lambda}_M(s)\bar{\varepsilon}_{kk}(\tau)\delta_{ij} + 2s\bar{\mu}_M(s)\bar{\varepsilon}_{ij}(s) \qquad (9.30)$$

$$\bar{\sigma}_{ij}(s) = -\bar{p}(s)\delta_{ij} + 2s\bar{\mu}_M(s)\bar{\varepsilon}_{ij}(s) \qquad (9.31)$$

(c) For the strain-displacement relationships (9.5):

$$\bar{\varepsilon}_{ij}(s) = \frac{1}{2}\left[\frac{\partial}{\partial x_j}\bar{u}_i(s) + \frac{\partial}{\partial x_i}\bar{u}_j(s)\right] \qquad (9.32)$$

(d) For the strain-compatibility conditions (9.8):

$$\bar{\varepsilon}_{ij,kl}(s) + \bar{\varepsilon}_{kl,ij}(s) = \bar{\varepsilon}_{ik,jl}(s) + \bar{\varepsilon}_{jl,ik}(s) \qquad (9.33)$$

As is apparent from these expressions, the transformed viscoelastic field equations are identical to those of linear elasticity, if the transforms of all viscoelastic variables are associated with their elastic counterparts, and if the products of the transforms of each material property function and the transform variable— s or $j\omega$—is associated with the corresponding elastic constant. From this analogy follows that the transformed solution to a viscoelastic boundary-value problem may be obtained directly from the solution of the same problem posed for an elastic material, by replacing the boundary data with their transforms, and each elastic constant with the transform-parameter multiplied transform of the corresponding viscoelastic property function. The solution to the original problem is then obtained by inverting the transformed solution back to the time domain.

In closing, it is remarked that the present approach is applicable to displacement, traction and mixed boundary-value problems, and that, unlike the methods presented for incompressible materials and for materials with synchronous moduli, or the separation of variables method, the correspondence principle does not require the Poisson's ratio to be a constant. One limitation of elastic–viscoelastic correspondence principles that are based on integral transforms, however, is that the prescribed conditions must not change type during the deformation process.[3] This excludes from application cases such as contact, where the data prescribed on some parts of the boundary change from displacement to traction and vice versa, during deformation.

Example 9.4 A cantilever beam of length l and constant cross-sectional area, made of a viscoelastic material, is subjected to an oscillating load $P(t) = P_o sin(\omega t)$ applied at its free end, as indicated in Fig. 9.3. Use the correspondence principle for steady-state conditions to obtain an expression for the steady-state displacement of the tip of the beam, if the material's relaxation modulus is $E(t)$.

Fig. 9.3 Example 9.4

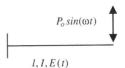

$P_o sin(\omega t)$

$l, I, E(t)$

Solution:
Following the correspondence principle, first write the elastic solution: $\Delta_o = \frac{P_o l^3}{3E_o I}$. Next, replace the amplitudes of the displacement, load, and modulus by their complex forms, using that $E^* C^* = 1$. This produces: $\Delta^* = \frac{P^* l^3}{3E^* I} = C^* P^* \frac{l^3}{3I}$. The viscoelastic solution is obtained by noting that the load actually applied

---

[3] This limitation is implicit also in the methods presented in Sects. 9.6 and 9.7, for displacement-only boundary conditions. .

corresponds to the imaginary part of the complex load $P^* = P_o\{cos(\omega t) + j\cdot sin(\omega t)\}$. The solution sought is, therefore, $\Delta(t) = \text{Im}\{\Delta^*\} = \text{Im}\left\{C^* P^* \frac{l^3}{3I}\right\} = \frac{P_o l^3}{3I}\,\text{Im}\{[C' - jC''][cos(\omega t) + j\sin(\omega t)]\}$. Or, carrying out the indicated algebra: $\Delta(t) = \frac{P_o l^3}{3I}[C'(\omega)\sin(\omega t) - C''(\omega)\cos(\omega t)]$.

Example 9.5 The stress and displacement in the direction of pull in a very wide, centrally cracked plate of an elastic material, in the vicinity of each crack tip are given by the following relations [7]: (Fig. 9.4) $\sigma_{22}^e(t) = \sigma_o(t)\sqrt{\frac{a}{2r}}\cos\left(\frac{\theta}{2}\right)\left[1 + \sin\left(\frac{\theta}{2}\right)\sin\left(3\frac{\theta}{2}\right)\right]$ and

$$u_2(t) = \frac{\sigma_o(t)}{G^e}\sqrt{\frac{ar}{2}}\sin\left(\frac{\theta}{2}\right)\left[2 - 2v^e - \cos^2\left(\frac{\theta}{2}\right)\right].$$

Fig. 9.4 Example 9.5

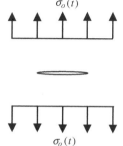

In these expressions, r and θ represent the radial and angular coordinates measured from a polar system centered at the crack tip, and G^e stands for the material's shear modulus. Using the correspondence principle obtain the solution for a plate of a linear viscoelastic solid with shear relaxation modulus $G(t)$ and constant Poisson's ratio v.

Solution:

According to the correspondence principle, the transformed solution takes the form:

$$\bar{\sigma}_{22}(s) = \bar{\sigma}_o(s)\sqrt{\frac{a}{2r}}\cos\left(\frac{\theta}{2}\right)\left[1 + \sin\left(\frac{\theta}{2}\right)\sin\left(3\frac{\theta}{2}\right)\right]$$

$$\bar{u}_2(s) = \frac{\bar{\sigma}_o(s)}{s\bar{G}}\sqrt{\frac{ar}{2}}\sin\left(\frac{\theta}{2}\right)\left[2 - 2s\bar{v}(s) - \cos^2\left(\frac{\theta}{2}\right)\right]$$

$$\equiv s\bar{C}_G(s)\bar{\sigma}_o(s)\sqrt{\frac{ar}{2}}\sin\left(\frac{\theta}{2}\right)\left[2 - 2s\bar{v}(s) - \cos^2\left(\frac{\theta}{2}\right)\right]$$

Where $sG = \frac{1}{sC_G}$ has been used to recast the expression for the transformed displacement.

Since Poisson's ratio is a constant, noting that the s-multiplied value of the transform of a constant is the constant itself, the viscoelastic solution becomes:

$$\sigma_{22}(t) = \sigma_o(t)\sqrt{\frac{a}{2r}}\cos\left(\frac{\theta}{2}\right)\left[1 + \sin\left(\frac{\theta}{2}\right)\sin\left(3\frac{\theta}{2}\right)\right]$$

$$u_2(t) = \sqrt{\frac{ar}{2}}\sin\left(\frac{\theta}{2}\right)\left[2 - 2v - \cos^2\left(\frac{\theta}{2}\right)\right]\int_{0^-}^{t} C_G(t-\tau)\frac{\partial}{\partial\tau}\sigma_o(\tau)d\tau$$

As expected of a traction boundary-value problem, the stress solution is independent of material constitution, but the displacement and strain fields are not. In the context of the mechanics of fracture, the solution is commonly stated in terms of the stress intensity factor; so that, for instance: $\bar{\sigma}_{22}(s) = \frac{K_I}{\sqrt{2\pi r}}\cos\left(\frac{\theta}{2}\right)$ $\left[1 + \sin\left(\frac{\theta}{2}\right)\sin\left(3\frac{\theta}{2}\right)\right]$. According to this, the stress intensity factor, K_I, which represents the amplitude of the stress field at a crack tip, must be independent of material properties.

9.10   The Poisson's Ratio of Isotropic Viscoelastic Solids

As discussed in Sects. 9.7 and 9.8, the assumption of a constant Poisson's ratio allows any type of viscoelastic boundary-value problem to be solved directly from the solution of the same problem posed for an isotropic elastic solid. For most viscoelastic materials, however, the assumption of a constant Poisson's ratio is not supported by experimental evidence and is justified solely by the simplicity it confers upon the solution process. A more realistic assumption is that the volumetric response of many polymers, particularly rubbers, is very nearly elastic; and it is the bulk modulus which may be taken as constant in time. This observation is used here to examine the nature of the viscoelastic Poisson's ratio of an isotropic material.

Applying the correspondence principle to the elastic relation $K = E/\{3(1-2v)\}$ and rearranging, produces that: $\bar{v} = \frac{1}{2}\left(\frac{1}{s}\right) - \frac{1}{6K}\bar{E}$. Inverting this expression results in the following relationship between the uniaxial tensile modulus, $E(t)$, and the Poisson's ratio function, $v(t)$:

$$v(t) = \frac{1}{2}H(t) - \frac{1}{6K}E(t) \tag{9.34}$$

Since $E(t_2) \le E(t_1)$, whenever $t_2 \ge t_1$, [c.f. Chap. 2], expression (9.34) shows that $v(t)$ is an increasing—or at least, a non-decreasing—function of t. In other words, its glassy value, v_g, is smaller than its long term, or equilibrium value, v_∞.

This expression also shows that the long-term Poisson's ratio of a viscoelastic fluid is equal to $\frac{1}{2}$—since for such materials $E_\infty = 0$. In addition, since the long-term modulus of elastic solids never vanishes, relation (9.34) also shows that the Poisson ratio of any linear isotropic viscoelastic solid is never larger than $\frac{1}{2}$. These observations reinforce the notion that the Poisson ratio of isotropic viscoelastic materials has the time dependence shown in Fig. 8.3. Expression (9.34) also shows that the Poisson's ratio of incompressible materials, which by definition should posses an infinite bulk modulus, is also $\frac{1}{2}$. The same would be approximately true, as well, for materials which have a bulk modulus orders of magnitude larger than the uniaxial tensile or shear relaxation moduli. Such materials, like many rubber compounds, are usually referred to as nearly incompressible.

9.11   Problems

P.9.1 Use the correspondence principle to derive the equations governing the bending response of straight viscoelastic beams from the three equations for elastic beams: $V(x,t) = \frac{\partial}{\partial x}M(x,t)$, $q(x,t) = -\frac{d}{dx}V(x,t)$, and $\frac{d^2}{dx^2}y(x,t) = -\frac{M(x,t)}{EI}$; where, V, M, and q are the shearing force, bending moment, and load per unit length on the beam, respectively, and y is the deflection.

Answer:

$$V(x,t) = \frac{\partial}{\partial x}M(x,t); q(x,t) = -\frac{d}{dx}V(x,t); \frac{d^2}{dx^2}y(x,t) = -\frac{1}{I}C_E(t-\tau) * dM(x,\tau)$$

Hint:

Apply the correspondence principle to the elastic expressions and write the transformed equations for viscoelastic beams using that $s\bar{E} = 1/(sC_E)$. Invert these expressions, to arrive at the viscoelastic forms.

P.9.2 Write the auxiliary elastic constitutive equation in terms of the uniaxial tensile relaxation modulus for use in a displacement boundary-value problem of viscoelasticity.

Answer: $\sigma_{Vij}(t) = \frac{E_R}{(1+v)}\left[\frac{v}{(1-2v)}\varepsilon_{Vkk}\delta_{ij} + \varepsilon_{Vij}\right]$.

Hint:

Proceed as in Example 9.2 using the normalization (9.19) and the elastic form (9.21).

P.9.3 In Problem P8.7, we showed that the directions of principal stress and principal strain in a viscoelastic material do not coincide in general. Under what conditions would these directions coincide, if ever?

Answer: The directions of maximum principal stress and maximum principal strain will coincide for viscoelastic materials with synchronous moduli, including incompressible materials, when the boundary data are specified as proportional loading; that is, either as $\sigma_{oij}(x)f_T(t)$, or as $u_{oi}(x)f_u(t)$.

$$\theta^\sigma_{max} = \frac{2\varepsilon_{oxy}\mu * df}{(\varepsilon_{xxo} - \varepsilon_{yyo})\mu * df} = \frac{2\varepsilon_{xyo}}{(\varepsilon_{xxo} - \varepsilon_{yyo})} = \theta^\varepsilon_{max}$$

Hint: Evaluate $\theta^\sigma_{max} = \frac{2\sigma_{xy}}{(\sigma_{xx} - \sigma_{yy})}$ inserting the terms from constitutive Eq. (9.27):
$\sigma_{ij}(t) = \sigma_{oij} f_T(t) = \left[\lambda_R \varepsilon_{okk} \delta_{ij} + 2\mu_R \varepsilon_{oij}\right] m_N(t - \tau) * df_u(\tau)$, cancel the common
terms to arrive at $\theta^\sigma_{max} = \frac{2\varepsilon_{oxy} m_N * df_u}{(\varepsilon_{oxx} - \varepsilon_{oyy}) m_N * df_u} = \frac{2\varepsilon_{oxy}}{(\varepsilon_{oxx} - \varepsilon_{oyy})} \equiv \frac{2\varepsilon_{oxy} f_u(t)}{(\varepsilon_{oxx} - \varepsilon_{oyy}) f_u(t)} = \theta^\varepsilon_{max}$.

P.9.4 How would you use the finite element method to obtain the viscoelastic
solution to a traction boundary-value problem of a material with a constant
Poisson's ratio, v, if the traction data were prescribed in the form $F_o(x_i)f(t)$?

Answer:

The finite element analysis would have to be carried out using the Poisson's
ratio, v, of the viscoelastic solid, and an arbitrary shear modulus G_R, computed
such that $E_R = 2(1 + v)G_R$. The boundary data would have to correspond to
$F_o(x_i)$, or be proportional to it. This leaves the temporal part as $f_T(t) = f(t)$.
Assuming the finite element solution is given by u_{FEi}, σ_{FEij}, and ε_{FEij}, one would
use (9.23) and (9.24), to write the viscoelastic solution as:

$$\sigma_{ij}(x, y, z, t) = \sigma_{FEij}(x, y, z)f_T(t)$$

$$u_i(x, y, z, t) = u_{FEi}(x, y, z)f_u(t) = \frac{1}{\mu_R} u_{FEi}(x, y, z) \int_{0^-}^{t} m_N^{-1}(t - \tau)\frac{\partial}{\partial\tau}f_T(\tau)d\tau$$

$$\varepsilon_{ij}(x, y, z, t) = \varepsilon_{FEi}(x, y, z)f_u(t) = \frac{1}{\mu_R} \varepsilon_{FEi}(x, y, z) \int_{0^-}^{t} m_N^{-1}(t - \tau)\frac{\partial}{\partial\tau}f_T(\tau)d\tau$$

P.9.5 The radial stress in a semi-infinite elastic half space subjected to
a concentrated load, P, at the origin of coordinates, as indicated in Fig. 9.5) is
given by [5]: $\sigma^e_{rr}(t) = \frac{P(t)}{2\pi}[(1 - 2v^e)f(r, z) - g(r, z)]$ in which v^e is the Poisson's
ratio of the elastic material, and f and g are known functions of position only. (a)
What would the radial stress be in a viscoelastic half space subjected to the same
boundary values? (b) What difference would there be if the viscoelastic half space
had a constant Poisson's ratio?

Fig. 9.5 Problem 9.5

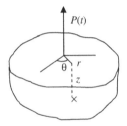

Answer:

(a) First assume the Poisson's ratio function for the half space is $v(t)$; then, invoke the correspondence principle to write the transformed solution and invert to arrive at the expression:

$$\sigma_{rr}(t) = \frac{1}{2\pi}\left\{ P(t)[f(r,z) - g(r,z)] - 2f(r,z)\int_{0^-}^{t} v(t-\tau)\frac{d}{d\tau}P(\tau)d\tau\right\}$$

(b) If the half space had a constant Poisson's ratio, the viscoelastic solution would revert to an elastic-like solution, since v would pull out from under the integral in (a).

P.9.6 Formulate the method of separation of variables for the mixed boundary-value problem in which the surface tractions and the body forces have different dependence on time.

Answer:

(a) Solution at zero body force

$$\frac{\partial}{\partial x_j}\sigma_{ij}^T = 0;\ n_j\sigma_{ji}^T(t)\Big|_{S_T} \equiv T_{oi}^o f_T^T(t);\ u_i^T(t)\Big|_{S_u} \equiv u_{oi}^o f_u^T(t)$$

$$\sigma_{ij}^T(t) = \sigma_{oij}^o f_T^T(t);\ u_i^T(t) = u_{oi}^o f_u^T(t);\ \varepsilon_{ij}^T(t) = \varepsilon_{oij}^o f_u^T(t)$$

with: $f_u^T(t) = \frac{1}{\mu_R}\int_{0^-}^{t} m_N^{-1}(t-\tau)\frac{\partial}{\partial\tau}f_T^T(\tau)d\tau$

(b) Solution at zero boundary data

$$\frac{\partial}{\partial x_j}\sigma_{ij}^B + B_i = 0;\ n_j\sigma_{ji}^B(t)\Big|_{S_T} \equiv 0;\ u_i^B(t)\Big|_{S_u} \equiv 0$$

$$\sigma_{ij}^B(t) = \sigma_{oij}^o f_T^B(t);\ u_i^B(t) = u_{oi}^o f_u^B(t);\ \varepsilon_{ij}^B(t) = \varepsilon_{oij}^o f_u^B(t)$$

with: $f_u^B(t) = \frac{1}{\mu_R}\int_{0^-}^{t} m_N^{-1}(t-\tau)\frac{\partial}{\partial\tau}f_T^B(\tau)d\tau$

Hint:

Split the solution fields into two parts: one satisfying the prescribed boundary data at zero body force, and another for which the only external force is the body force and the boundary values are identically zero. The viscoelastic solution will be the sum of the two solutions. Specifically, using the superscripts B and T, respectively, for the solutions to the problem with and without body force, write: $\sigma_{ij} = \sigma_{ij}^B + \sigma_{ij}^T$, $u_i = u_i^B + u_i^T$, and $\varepsilon_{ij} = \varepsilon_{ij}^B + \varepsilon_{ij}^T$. Split the original problem into the two parts and separate variables to arrive at the desired result.

References

1. D.C. Leigh, *Nonlinear Continuum Mechanics*, (McGraw-Hill, NY, 1968) pp. 117–138
2. E. Volterra, J.H. Gaines, *Advanced Strength of Materials*, (Prentice-Hall, NJ, 1971), pp. 12–19, pp. 177–186
3. R.M. Christensen, *Theory of Viscoelasticity*, 2nd Ed. (Dover, 1982), pp. 37–41
4. Y.C. Fung, *Foundations of Solid Mechanics*, (Prentice Hall, NJ, 1965) pp. 99–103
5. I.S. Sokolnikoff, *Mathematical Theory of Elasticity*, (McGraw-Hill, US, 1956), pp. 71–79
6. A.C. Pipkin, *Lectures on Viscoelasticity theory*, (Springer, Berlin, 1956), 2nd Ed. pp. 77–79
7. T.L. Anderson, *Fracture Mechanics, Fundamentals and Applications*, 2nd Ed. CRC, CS, 1956), pp. 51–55

Wave Propagation

10

Abstract

This chapter examines the propagation of harmonic and shock waves in viscoelastic materials of integral and differential type. For simplicity, the different topics are introduce in one dimension, presenting the balance of linear momentum across the shock front, and the jump equations in stress, strain and velocity without obscuring the subject matter. As must be expected, harmonic waves in viscoelastic media are always damped. Also shown is that shock waves travel at the glassy sonic speed of the viscoelastic material in which they occur; that is, at the speed of sound in an elastic solid with modulus of elasticity equal to the glassy modulus of the viscoelastic material at hand.

Keywords

Wave · Harmonic · Cyclic · Shock · Speed · Frequency · Front · Jump · Attenuation · Momentum · Balance

10.1 Introduction

This chapter examines in some detail the propagation of harmonic waves and shock waves in viscoelastic materials. To keep the presentation simple, only one-dimensional conditions are assumed in what follows. Harmonic waves are discussed in Sect. 10.2, and shock waves in Sect. 10.3. In both cases, the treatment covers materials of integral and differential types.

Waves develop in a body as a result of the conditions imposed on its boundaries. Harmonic waves result from cyclic boundary conditions, while shock waves are due to discontinuous boundary values. As must be expected, harmonic waves in viscoelastic media are always damped. Also, shock waves travel at the glassy sonic speed of the viscoelastic material in which they occur; in other words, at the

D. Gutierrez-Lemini, *Engineering Viscoelasticity*, DOI: 10.1007/978-1-4614-8139-3_10,   239
© Springer Science+Business Media New York 2014

speed of sound in an elastic solid having modulus of elasticity equal to the impact or glassy modulus of the viscoelastic material at hand.

10.2 Harmonic Waves

In this section, we examine the propagation of harmonic waves in a viscoelastic bar. Such waves result from cyclic boundary conditions and their speed thus depends on the forcing frequency. The treatment is restricted to longitudinal vibrations, but the same principles apply to lateral and torsional vibrations.

In longitudinal vibrations, it is assumed that bar elements can only extend or contract along the original axis of the bar; that plane cross sections—perpendicular to the original axis of the bar–remain plane during motion; and that the inertia of transverse motion is disregarded. Under these conditions, as can be inferred from Fig. 10.1, balance of linear momentum leads to the single partial differential equation:

$$\frac{\partial}{\partial x}\sigma(x,t) = \rho\frac{\partial^2}{\partial t^2}u(x,t) \tag{10.1}$$

Where σ is the axial stress, u, the axial displacement, and ρ stands for the density of the material, which may depend on position, but, for simplicity, is assumed constant in time.

Consistent with the balance laws [c.f. Appendix B], the relationship in (10.1) is valid irrespective of material constitution. The type of material at hand enters through the constitutive equation, as discussed later on.

The purpose here is to examine solutions of Eq. (10.1) associated with either displacement or traction boundary conditions which vary cyclically in time. Thus, aside from fixing one end of the bar, when needed, boundary conditions at the other end, $x = 0$, say, will be of one of the following forms:

Fig. 10.1 Balance of linear momentum in a one-dimensional bar

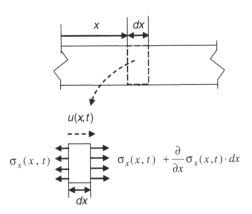

$$u(0,t) = \begin{cases} U\cos(\omega_o t) \\ \quad or: \\ U\sin(\omega_o t) \end{cases} \qquad (10.2)$$

if a harmonic displacement of amplitude U is prescribed, or:

$$\sigma(0,t) = \begin{cases} S\cos(\omega_o t) \\ \quad or: \\ S\sin(\omega_o t) \end{cases} \qquad (10.3)$$

when a harmonic surface traction of intensity S is applied.

In any event, since a frequency-domain analysis requires that both loading and response be specified in complex form—which includes cosine and sine terms—instead of (10.2) we would use:

$$u^*(0,t) = Ue^{j\omega_o t} \qquad (10.4)$$

whereas (10.3) would be replaced with:

$$\sigma^*(0,t) = Se^{j(\omega_o t + \varphi)} \qquad (10.5)$$

Clearly, the solution corresponding to a cosine boundary condition should be real part of the complex solution—the cosine component of $e^{j\omega t} = \cos\omega t + j\sin\omega t$; and, for the same reason, the imaginary part of the complex solution would represent the solution to a sine forcing function [c.f. Chap. 4].

10.2.1 Materials of Integral Type

In this case, using the strain-displacement relation $\varepsilon_{xx} = \partial u/\partial x$, converts (10.1) into:

$$E(t-\tau) * \partial\left[\frac{\partial^2}{\partial x^2} u(x,\tau)\right] = \rho\frac{\partial^2}{\partial t^2} u(x,t) \qquad (10.6)$$

Since the boundary condition is a circular function, the solution will be a combination of circular functions—that is, cosines as well as sines—of the same frequency. In addition, one would expect the intensity of the forcing function to be less at sections farther from the point of application. That is, on physical grounds, one would expect the solution to (10.6) to be a decreasing function of position. Hence, we seek a solution in the form:

$$u^*(x,t) = Ae^{j(\omega_o t - \lambda x)} \qquad (10.7)$$

In this expression, ω_o is the forcing frequency and, λ, as will be seen subsequently, is a complex quantity whose components represent the rate of amplitude decay and the wave speed. Taking (10.7) into (10.6) leads to:

$$\lambda^2 E^*(j\omega_o)u^*(x,j\omega_o t) = \omega_o^2 \rho u^*(x,j\omega_o t) \tag{a}$$

From which λ can be obtained after simplification, as:

$$\lambda = \pm\omega_o\sqrt{\frac{\rho}{E^*(j\omega_o)}} \tag{10.8}$$

This shows that there are two values of λ—one being the negative of the other—which are complex because E^* is a complex number. In what follows, we reserve the name λ for one of the roots of (10.8), the other being—λ. With that: $\lambda_1 = \lambda = \lambda' + j\lambda''$, $\lambda_2 = -\lambda$, and:

$$u^*(x,j\omega_o t) = A_1 e^{-\lambda''x}e^{j(\omega_o t-\lambda'x)} + A_2 e^{+\lambda''x}e^{j(\omega_o t+\lambda'x)} \tag{10.9}$$

The physical interpretation of this form is that under a cyclic forcing function there will always be two waves running in opposite directions along the bar. The first term in (10.9) represents a wave running in the direction of increasing x, and the second term is a wave moving in the opposite direction. The solution (10.9) also shows that each wave decreases exponentially as it advances; the imaginary part, λ'', of λ is responsible for the wave's rate of decay; and, by dimensional homogeneity, λ' must be inversely proportional to the wave speed, as will be shown later. Finally, in the case of an infinitely-long bar, the solution can contain only the first term, since the second term must be omitted for regularity, to keep the solution from blowing up.

10.2.2 Materials of Differential Type

In this case, the stress–strain law is introduced by applying the differential operator P to the equation of motion (10.1), thus:

$$P\left[\frac{\partial}{\partial x}\sigma_{xx}(x,t)\right] = P\left[\rho\frac{\partial^2}{\partial t^2}u(x,t)\right]$$

Using that $P(\sigma) = Q(\varepsilon)$, and $\varepsilon_{xx} = \partial u/\partial x$, and assuming the material's density to be time independent, transforms the dynamic equation into:

$$Q\left[\frac{\partial^2}{\partial x^2}u(x,t)\right] = \rho P\left[\frac{\partial^2}{\partial t^2}u(x,t)\right] \tag{10.10}$$

On the same arguments as before, the solution of (10.10) is sought in the form (10.7); so that:

$$\lambda^2 Q[j\omega_o]u^*(x,j\omega_o t) = \omega_o^2 \rho P[j\omega_o]u^*(x,j\omega_o t) \tag{c}$$

Or, since $E^*(j\omega_o) = Q(j\omega_o)/P(j\omega_o) = Q^*/P^*$ [c.f. Chap. 4]:

$$\lambda = \pm\omega_o\sqrt{\rho\frac{P[j\omega_o]}{Q[j\omega_o]}} == \pm\omega_o\sqrt{\frac{\rho}{E^*(j\omega_o)}} \tag{10.11}$$

This expression is exactly the same as that in (10.8), which was derived for a material with constitutive equation of hereditary integral type. Hence, irrespective of the form of the material's stress–strain law, once the complex modulus $E^*(j\omega_o)$ is made available, λ can be evaluated from either of the equivalent forms (10.8) or (10.11).

In particular, since $E^*(j\omega_o) = E'(\omega_o) + jE''(\omega_o)$, with $\tan\delta = E''/E'$ [c.f. Chap. 4], multiplying and dividing the quantity under the radical sign in (10.11) by the complex conjugate of E^* leads, after some algebraic manipulations, to the following form, which is more adequate for numerical computations[1]:

$$\lambda = \omega_o\sqrt{\frac{\rho}{||E^*(j\omega_o)||}}e^{-j(\delta+2k\pi)/2} \equiv \pm\frac{\omega_o}{\sqrt{\frac{||E^*(j\omega_o)||}{\rho}}}e^{-j\delta/2} \tag{10.12}$$

In this expression, the quantity under the radical sign on the far right has the dimensions of speed. In analogy with the elastic case [1], the complex sonic speed is defined as follows:

$$c^*(j\omega_o) = \sqrt{\frac{E^*(j\omega_o)}{\rho}} \equiv ||c^*(j\omega_o)||e^{j\delta/2} \tag{10.13}$$

This gives the amplitude $||c^*(j\omega_o)||$ of the complex sonic speed, as:

$$||c^*(j\omega_o)|| = \sqrt{\frac{||E^*(j\omega_o)||}{\rho}} \tag{10.14}$$

With the definitions introduced in (10.12) and (10.13), λ, in (10.11), may be put in the form:

$$\lambda = \frac{\omega_o}{||c^*(j\omega_o)||}e^{j\delta/2} \tag{10.15}$$

---

[1] The root of index n of a complex number, $z = re^{j(\theta+2k\pi)}$, is given by: $z^{1/n} = r^{1/n}e^{j(\theta+2k\pi)/n}$, as shown in Appendix A.

which serves to express λ', and λ'' as:

$$\lambda' = \frac{\omega_o}{||c^*(j\omega_o)||}\cos(\delta/2); \quad \lambda'' = \frac{\omega_o}{||c^*(j\omega_o)||}\sin(\delta/2) \tag{10.16}$$

Two limiting cases of the complex sonic speed amplitude, (10.14), which are of interest, pertain to forcing functions of very low and very high frequencies. Using the limit properties of E^* [c.f. Chap. 4], the following results are obtained:

$$\lim_{\omega_o \to 0} \sqrt{\frac{E^*(j\omega_o)}{\rho}} = \sqrt{\frac{E'_\infty}{\rho}} \equiv \sqrt{\frac{E_\infty}{\rho}} = c_\infty \tag{10.17}$$

and:

$$\lim_{\omega_o \to \infty} \sqrt{\frac{E^*(j\omega_o)}{\rho}} = \sqrt{\frac{E'_o}{\rho}} \equiv \sqrt{\frac{E_g}{\rho}} = c_g \tag{10.18}$$

These expressions indicate that waves induced by harmonic forcing functions with very low or very high frequencies propagate as if the material were linearly elastic. Harmonic waves of very low forcing frequency will travel at the long-term sonic speed of the viscoelastic material at hand, while those of very high frequency will travel with the material's glassy sonic speed.

The preceding discussion shows that, irrespective of whether the constitutive equations are of integral- or differential-operator type, harmonic waves in a viscoelastic substance will always be a superposition of decaying sine and cosine waves of the form (10.7)—or in more detail, (10.9) and that the wave speed and the rate of decay depend on the material's density and on the forcing frequency— through the complex modulus.

Example 10.1 Find the magnitude of the complex sonic speed of an elastomer whose complex modulus at the frequency of interest is $E^*(j\omega) = 10 + 3.5\,j$ MPa, if its mass density is 1,660 kg/m^3.

Solution: The sonic speed depends on the magnitude of the complex modulus, $||E^*|| \approx 10.6$ Mpa, and the mass density: $\rho = 1,660$ kg/m^3; thus, using (10.14), it follows that: $||c^*|| \approx 79.9$ m/s.

Example 10.2 A bar of a viscoelastic material with tensile relaxation modulus $E(t) = 70 + 35e^{-t/0.1}$ kPa, and mass density $= 1,800$ kg/m^3 is subjected to the following boundary conditions[2]:

$$u(0,t) = U \cdot \cos(\omega_o t); \quad \omega_o = 25 \text{ rad/s}$$
$$u(L,t) = 0$$

Determine the bar's response at $\omega_o t = 0, \pi/4, \pi/2, 3\pi/4, \pi$, if the length of the bar is such that $L = 7\pi/\lambda'$ (≈ 6.5 m).

Solution: First, the boundary conditions are expressed in the complex form $U e^{j(\omega_o t)}$. Noting that the actual condition, $U \cdot cos(\omega_o t)$, corresponds to the real part of $U \cdot e e^{j(\omega_o t)}$, the solution will be given by the real part of (10.9). Putting the boundary conditions in it leads to a system of equations in the unknown amplitudes, A_1 and A_2:

$$U = A_1 + A_2$$
$$0 = A_1 e^{-\lambda'' L} + A_2 e^{+\lambda'' L}$$

Solving for A_1 and A_2 and plugging the results back into 10.9), leads, after some manipulation, to the complex form [cf. Appendix A]:

$$u^*(x, t) = U \frac{\sin \lambda (L - x)}{\sin \lambda L} e^{j\omega t}$$

Since λ is a complex number, to decipher this expression requires the circular functions of a complex variable [c.f. Appendix A]: $\sin z = \sin(x+jy) = \sin(x) \cdot \cosh(y) + j\cos(x) \cdot \sinh(y)$ and similarly for $\cos(z)$. After some complex and trigonometric algebra, there results:

$$
\begin{aligned}
u^*(x, t) = {} & \frac{U}{\cosh 2\lambda'' L - \cos 2\lambda' L} \langle [\cos \lambda' x \cosh \lambda''(2L - x) \\
& - \cosh \lambda'' x \cos \lambda'(2L - x)] \cos(\omega_o t) \\
& - [\sin \lambda' x \sinh \lambda''(2l - x) - \sinh \lambda'' x \sin \lambda'(2L - x)] \sin(\omega_o t) \\
& + j\{[\cos \lambda' x \cosh \lambda''(2L - x) - \cosh \lambda'' x \cos \lambda'(2L - x)] \sin(\omega_o t) \\
& + [\sin \lambda' x \sinh \lambda''(2l - x) - \sinh \lambda'' x \sin \lambda'(2L - x)] \cos(\omega_o t)\}\rangle
\end{aligned}
$$

The solution corresponding to the boundary values of the present problem (which pertain to a cosine forcing function) is the real part of the previous expression. The parameters of this problem: $E^* = 100.17 + 12.07$ j MPa; $\delta \approx 0.120$ rad; $\|c^*\| \approx 7.487$ m/s; $\lambda' \approx 3.333$; and $\lambda'' \approx -0.200$, yield the solution, which is displayed in Fig. 10.2 at a few selected times.

Also shown by this solution is that although the wave amplitudes decrease from the left to right, they increase upon first entering the bar. This indicates that the bar is at or near resonance. In fact, an elastic bar of the same density and modulus $M = E'(\omega_o)$ would be in resonance under the prescribed boundary conditions and would exhibit standing waves of infinite amplitude [2].

Fig. 10.2 Example 10.2:
harmonic waves in given
viscoelastic bar

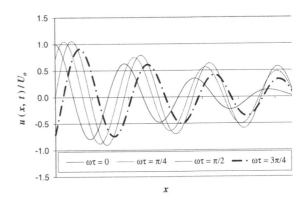

10.3   Shock Waves

This section examines the propagation of shock waves in a bar made of a linearly
viscoelastic material. When the governing equations—the equations of motion in
the present case—are of hyperbolic type, any abruptness in the boundary values
show up as discontinuities in the first derivatives of the displacement field. These
discontinuities travel along surfaces, which are known as the characteristics of the
differential equation, and which are in themselves smooth functions of the spatial
coordinates and time. In continuum mechanics, these discontinuities are commonly referred to as shock waves.

A rigorous theory admits the propagation of shock waves only in unbounded
media. In a finite body, multiple reflections of waves from the boundaries give rise
to dispersion and blur the shock front. To avoid these complications and still use a
one-dimensional model in the study of shock waves, we assume a bar in which
plane sections remain plane, there is no transverse inertia, nor lateral motion of any
sort, not even due to Poisson's effect; and most of all, such that waves do not
bounce off its lateral surfaces. These assumptions allow the study of shock waves
in an idealized bar in virtually the same way as in an infinite medium.

Per the previous definitions, a shock wave through the one-dimensional bar
model will be a smooth function, $x = Y(t)$, in the (x,t) plane. The first partial
derivative of the shock wave with respect to time, which exists by the smoothness
assumption, represents the wave speed:

$$\frac{d}{dt}x = \frac{d}{dt}Y(t) = c(t) \tag{10.19}$$

As long as the bar does not come apart, the displacement field, $u(x,t)$, will be
continuous throughout. However, the first partial derivatives of the displacement
field with respect to time and space will exhibit jump discontinuities across the
shock front.

The jump of a function, f, at a point, is denoted by the symbol $[f]$ and defined as the difference between the values of the function at points infinitely close to the point in question, but situated on opposite sides of it. More explicitly, the magnitude of the jump of a function f at $x = y$ is the difference between the right- and left-hand side limits of the function at $x = y$:

$$[f] = f^+ - f^- \tag{10.20}$$

where the one-sided limits are defined as follows [4]:

$$f^- \equiv \lim_{\substack{x \to y \\ x < y}} f(x); \quad f^+ \equiv \lim_{\substack{x \to y \\ x > y}} f(x) \tag{10.21}$$

Once in motion, a shock wave travels forward from one end of the bar to the other; and upon reaching it, reflects and travels back to the point of origin. The wave fronts or characteristics of the differential equation of motion can be represented by straight lines in the $x - t$ plane:

$$\xi = t - \frac{x}{c}; \quad \eta = t + \frac{x}{c} \tag{10.22}$$

The equation $\xi = \text{constant}$, represents a wave moving in the direction of increasing x, and $\eta = \text{constant}$ corresponds to a wave front moving backwards, toward the origin, as illustrated in Fig. 10.3.

As seen in Fig. 10.3, on the forward wave $\xi = \text{constant}$, $d\xi = 0$; which implies that $dt = dx/c$. Likewise, on the return wave $\eta = \text{constant}$, $d\eta = 0$, and so, $dt = -dx/c$. Hence:

$$\frac{\partial}{\partial x} = \begin{cases} \frac{1}{c} \frac{\partial}{\partial t}; & \text{on } \xi = const \\[2mm] -\frac{1}{c} \frac{\partial}{\partial t}; & \text{on } \eta = const \end{cases} \tag{10.23}$$

Since the total differential of a function $f\,(x,t)$ is $df = \frac{\partial f}{\partial x} dx + \frac{\partial f}{\partial t} dt$, (10.23) implies that:

Fig. 10.3 Graphical representations of shock waves in the x-t plane

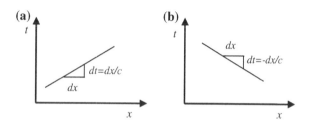

$$\frac{Df}{Dx} = \begin{cases} \frac{\partial f}{\partial x} + \frac{1}{c}\frac{\partial f}{\partial t}; & on\ \xi = const \\[2mm] \frac{\partial f}{\partial x} - \frac{1}{c}\frac{\partial f}{\partial t}; & on\ \eta = const \end{cases} \qquad (10.24)$$

And also that:

$$\frac{Df}{Dt} = \begin{cases} \frac{\partial f}{\partial t} + c\frac{\partial f}{\partial x}; & on\ \xi = const \\[2mm] \frac{\partial f}{\partial t} - c\frac{\partial f}{\partial x}; & on\ \eta = const \end{cases} \qquad (10.25)$$

Additionally, (10.24) and (10.25) lead to:

$$\frac{D}{Dt} = \begin{cases} c\frac{D}{Dx}; & on\ \xi = const \\[2mm] -c\frac{D}{Dx}; & on\ \eta = const \end{cases} \qquad (10.26)$$

The differential equation of motion (10.1) applies on either side of the front, but not on the front itself. For ease of reference that equation is repeated here, with a change in notation from displacement to velocity: $\partial^2 u \backslash \partial t^2 \equiv \partial v \backslash \partial t$; thus:

$$\frac{\partial}{\partial x}\sigma(x,t) = \rho\frac{\partial}{\partial t}v(x,t) \qquad (10.27)$$

To see what happens at the front, the balance of momentum equation is used in its global form. To this end, consider a portion of the bar which straddles the wave front, as shown in Fig. 10.4, and let the size of the element of bar approach zero. Eq. (10.27) then leads to:

$$-A\sigma(X_1(t),t) + A\sigma(X_2(t),t) = A\frac{\partial}{\partial t}\left[\int_{X_1}^{Y^-(t)} \rho v(x(t),t)dx + \int_{Y^+(t)}^{X_2} \rho v(x(t),t)dx\right] \quad (a)$$

To cast this equation in a more useful form, cancel out the common factor A, and carry out the integration using Leibnitz' rule [c.f. Appendix A]. Let $x_1 \to Y^-$ from the left, and $x_2 \to Y^+$ from the right, and, recognizing the continuity of the

Fig. 10.4 Balance of linear momentum across the wave front

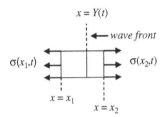

shock wave (which requires that: $dY^-/dt = dY^+/dt = c$), re-order terms and arrive at the balance of linear momentum across the front:

$$[\sigma] = -\rho c[v] \tag{10.28}$$

The continuity condition at the shock front is obtained by differentiating the jump of the displacement. Since the displacement is continuous across the front, $[u] = 0$ there; and so must be, its total derivative, Du/D. Using (10.24): $\frac{Du}{Dx} = 0 \equiv \frac{\partial u}{\partial x} + \frac{1}{c}\frac{\partial u}{\partial t}$, to evaluate the jump leads to:

$$[\frac{\partial u}{\partial x}] = -\frac{1}{c}[\frac{\partial u}{\partial t}]; \quad or : \quad [\varepsilon] = -\frac{1}{c}[v] \tag{10.29}$$

Combining the last two expressions leads to a relation between the jumps in stress and strain across the front, which seems independent of the constitutive equation.

$$[\sigma] = \rho c^2 [\varepsilon] \tag{10.30}$$

Expressions (10.28–10.30) relate the jumps in stress, strain, and velocity. Material type or constitution enters directly through the density, per (10.28), and the stress—strain relations, through (10.30).

10.3.1 Materials of Integral Type

Without loss of generality, the stress-strain law in this case may be represented in the integral form [c.f. Chap. 2]:

$$\sigma(x, t) = E_g \varepsilon(x, t) - \int_0^t \frac{\partial E(t - \tau)}{\partial \tau} \varepsilon(x, \tau) d\tau \tag{10.31}$$

Since the relaxation function $E(\cdot)$ is continuous across the front, and the strain history, $\varepsilon(x, \tau)$, is continuous for all τ except $\tau = t$, the integral vanishes across the jump. Thus:

$$[\sigma] = E_g [\varepsilon] \tag{10.32}$$

This, together with (10.30) yields the expression for the speed of the shock wave, which is of the same form as for shock waves in a linearly elastic bar [1]:

$$c = \sqrt{\frac{E_g}{\rho}} \tag{10.33}$$

As indicated before, shock waves in a viscoelastic bar travel with the material's glassy sonic speed.

Example 10.3 Obtain the speed of longitudinal shock waves in a bar of visco-elastic material with tensile relaxation modulus $E(t) = 4.2 + 39.3e^{-t/0.25}$ MPa and density 1,800 kg/m^3.

Solution:The speed of the shock front is given by (10.33). The glassy modulus of the material in this case is 43.5 MPa—the sum of the equilibrium and the transient components of the tensile relaxation modulus. Hence, $c = (43.5 \cdot 10^6/ 1{,}800)^{1/2} \approx 155.5$ m/s.

A shock wave moving in a viscoelastic material would change its amplitude both as the wave progresses through the medium, and as time goes by. This may be proven by constructing a differential equation—in the spatial coordinate or in time, respectively—out of the stress-velocity jump equation and examining its solution, as follows.

Differentiating the stress-velocity jump equation (10.28) across the front and using (10.24), leads to:

$$\frac{D}{Dx}[\sigma] = -\rho c \frac{D}{Dx}[v] \tag{a}$$

Or:

$$\left[\frac{\partial \sigma}{\partial x}\right] + \frac{1}{c}\left[\frac{\partial \sigma}{\partial t}\right] = -\rho c \frac{D}{Dx}[v] \tag{b}$$

The second term on the left-hand side of this expression is evaluated from the stress-strain law, (10.31), using Leibnitz' rule to differentiate under the integral sign, to get:

$$\frac{1}{c}\left[\frac{\partial \sigma}{\partial t}\right] = \frac{1}{c}\left\{E_g\left[\frac{\partial \varepsilon}{\partial t}\right] - \left[\int_0^t \frac{\partial^2 E(t-\tau)}{\partial t \partial \tau}\varepsilon(x,\tau)d\tau\right] - \left.\frac{\partial E(t-\tau)}{\partial \tau}\right|_{\tau=t}[\varepsilon(x,t)]\right\} \tag{c}$$

Since the integrand in the second term on the right is continuous for all τ, except $\tau = t$, its jump is zero. Therefore:

$$\frac{1}{c}\left[\frac{\partial \sigma}{\partial t}\right] = \frac{1}{c}E_g\left[\frac{\partial \varepsilon}{\partial t}\right] - \left(\frac{1}{c}\right)\left.\frac{\partial E(t-\tau)}{\partial \tau}\right|_{\tau=t}[\varepsilon(x,t)] \tag{d}$$

Upon using that: $\partial \varepsilon/\partial t = \partial v/\partial x$, and (10.33): $E_g = \rho c^2$, on the first term of the right-hand side of the last expression; together with (10.29): $[\varepsilon] = -[v]/c$, on the second term, there results:

$$\frac{1}{c}\left[\frac{\partial\sigma}{\partial t}\right] = \rho c\left[\frac{\partial v}{\partial x}\right] + \left(\frac{1}{c}\right)\frac{\partial E(t-\tau)}{\partial\tau}\bigg|_{\tau=t}\left(\frac{1}{c}\right)[v] \tag{e}$$

Taking this result into (b), using the equation of motion (10.27) to replace $\partial\sigma/\partial x$, collecting terms and recalling the definition (10.24) of the derivative, D/Dx, across the front, leads to:

$$\frac{D}{Dx}[v] + \frac{1}{2c}\Gamma(0)[v] = 0; \quad \Gamma(0) \equiv \frac{1}{E_g}\frac{\partial}{\partial\tau}E(t-\tau)\big|_{\tau=t} \equiv -\frac{1}{E_g}\frac{d}{dt}E(t)\big|_{t=0} \tag{10.34a}$$

This is a linear differential equation of first order with constant coefficients. It thus has an integrating factor $e^{\Gamma(0)x/(2c)}$[c.f. Appendix A]. This allows casting it as $d\left[v \cdot e^{\Gamma(0)x/(2c)}\right] = 0$; from which follows the solution:

$$[v] = Ae^{-\frac{\Gamma(0)}{2c}x} \tag{10.35a}$$

Since $\Gamma(0) > 0$, this relationship shows that the jump in velocity decreases as the wave moves in the direction of increasing x.

Example 10.4 Prove that $\Gamma(0)$, as defined in (10.34a) is always positive for relaxation functions expressed in Dirichlet-Prony series, as sums of exponentials. Solution: Use (10.34a): $\Gamma(t-\tau) \equiv (1/E_g)\frac{\partial}{\partial\tau}E(t-\tau)$, and the Prony-series form of the relaxation modulus:$E(t) = E_g - \sum_{i=1}^{N} E_i(1 - e^{-t/\tau_i})$ to compute: $\frac{d}{d\tau}E(t-\tau) =$

$\sum_{i=1}^{N}\frac{E_i}{\tau_i}e^{-(t-\tau)/\tau_i}$ and arrive at the result: $\Gamma(\tau=t) \equiv \Gamma(0) = \frac{1}{E_g}\sum_{i=1}^{N}\frac{E_i}{\tau_i}$. Use that since all the E_i, and $\tau_i > 0$, it follows that $\Gamma(0) > 0$.

The attenuation of shock waves with time may be examined using Eq. (10.26): $D/Dx = (1/c)D/Dt$; to write (10.34a) as an equation in time:

$$\frac{D}{Dt}[v] + \frac{1}{2}\Gamma(0)[v] = 0; \quad \Gamma(0) \equiv \frac{1}{E_g}\frac{d}{d\tau}E(t-\tau)\big|_{\tau=t} \equiv -\frac{1}{E_g}\frac{d}{dt}E(t)\bigg|_{t=0} \tag{10.34b}$$

Proceeding as before, the solution in time becomes:

$$[v] = Be^{-\frac{\Gamma(0)}{2}t} \tag{10.35b}$$

This result indicates that the amplitude of a shock wave decreases with time, as expected it should in viscoelastic materials. Using (10.28) with either of Eq. (10.35a, b) establishes the stress jump:

$$[\sigma] = \begin{cases} -\rho c B e^{-\frac{1}{2}\Gamma(0)t} \\ \quad or \\ -\rho c A e^{-\frac{1}{2c}\Gamma(0)x} \end{cases} \tag{10.36}$$

Similarly, (10.29) and (10.35a, b) produce the strain jump:

$$[\varepsilon] = \begin{cases} -\frac{1}{c} B e^{-\frac{1}{2}\Gamma(0)t} \\ \quad or \\ -\frac{1}{c} A e^{-\frac{1}{2c}\Gamma(0)x} \end{cases} \tag{10.37}$$

A wave front may also move from right to left, on the characteristic $\eta = $ constant. In that case, the wave velocity will be negative, and the previous equations apply as well.

Additional information about the response of the bar can be gained solving (10.27). To this end, insert into it the constitutive equation $\sigma(t) = E(t - \tau) * d\frac{\partial}{\partial x} u(\tau)$ [c.f. Chap. 2] writing:

$$\int_0^t E(t - \tau) \frac{\partial}{\partial s} \left(\frac{\partial^2}{\partial x^2} u \right) d\tau = \rho \frac{\partial^2}{\partial t^2} u \tag{f}$$

Now take the Laplace transform, assuming at-rest initial conditions to get [c.f. Appendix A]:

$$\overline{E}(s) s \overline{u''}(x, s) - \rho s^2 \overline{u}(x, s) = 0 \tag{g}$$

Or:

$$\overline{u''}(x, s) - \frac{\rho s}{\overline{E}(s)} \overline{u}(x, s) = 0 \tag{10.38}$$

This is a partial differential equation in x for the transformed function $\overline{u}(x, s)$. Because only derivatives with respect to x enter this equation, its solution may be obtained by methods for ordinary differential equations. Employing differential operators leads to a characteristic equation with roots [c.f. Appendix A]:

$$\lambda = \pm \sqrt{\frac{\rho s}{\overline{E}(s)}} \equiv \pm s \sqrt{\frac{\rho}{s\overline{E}(s)}} \tag{10.39}$$

These roots imply a general solution of the form:

$$\overline{u}(x, s) = A_1(s) e^{+\lambda x} + A_2(s) e^{-\lambda x} \tag{10.40a}$$

Inserting the Laplace-transformed strain, $\overline{\varepsilon}(x, s) = \partial \overline{u}(x, s)/\partial x$, leads to the transformed stress:

$$\bar{\sigma}(x, s) = \lambda[A_1(s)e^{\lambda(s)x} - A_2(s)e^{-\lambda(s)x}] \tag{10.41a}$$

From here, the solution in physical space-time would be obtained, at least in principle, by Laplace-transform inversion. The mathematical manipulations required to accomplish this, however, become intractable for all but the simplest of constitutive models.

Lastly, for a semi-infinite bar, the root with positive sign in (10.40a) has to be discarded.[2] It represents a wave that starts traveling from the right at time zero. This is inconsistent with at-rest initial conditions. Thus, the solution for a semi-infinite bar in transformed space is simply:

$$\bar{u}(x, s) = A(s)e^{-\lambda(s)x} \equiv A(s)e^{-s\sqrt{\rho/s\bar{E}(s)}x} \tag{10.40b}$$

$$\bar{\sigma}(x, s) = -\lambda(s)A(s)e^{-\lambda(s)x} \tag{10.41b}$$

Example 10.5 Find the Laplace-transformed response of a semi-infinite bar of a viscoelastic material subjected to a discontinuous velocity of amplitude V at its end.

Solution:First, note that $v = \partial u/\partial t$, so that: $\bar{v}(x, s) = s\bar{u}(x, s)$. The transform of the boundary value $v(0,t) = V \cdot H(t)$ is $\bar{v}(0, s) = V/s$; $\bar{u}(0, s) = V/s^2$. Evaluating (10.41b) at $x = 0$ establishes $A(s) = V/s^2$. Thus, the Laplace-transformed solution is $\bar{u}(x, s) = \frac{V}{s^2}e^{-\lambda x}$. Note that, per (10.39), λ is a function of the transformed modulus, so the inverse transform of $\bar{u}(x, s)$ is not simply the transform of $1/s^2$ times $Ve^{-\lambda x}$.

Example 10.6 Find the Laplace transform stress response of the semi-infinite bar of Example 10.5 if the bar's material is of Maxwell type, with modulus E_g and viscosity η.

Solution:Using that $\bar{\sigma} = s\bar{E} \cdot \bar{\varepsilon} \equiv s\bar{E} \cdot \partial \bar{u}/\partial x$ together with the results of Example 10.5, leads to the transformed stress: $\bar{\sigma} = -\bar{E}\frac{V}{s}\lambda e^{-\lambda x}$. The relaxation modulus for the Maxwell model is given by [c.f. Chap. 3]: $E(t) = E_g e^{-t/\tau}$, with relaxation time $\tau = \eta/E_g$, so that: $\bar{E} = \frac{E_g}{s+1/\tau}$ and hence, from (10.39), using that $E_g = \rho c^2$, $\lambda = \sqrt{\frac{\rho}{E_g}s(s + 1/\tau)} \equiv \frac{1}{c}\sqrt{s(s + 1/\tau)}$. And the transformed stress becomes:[3] $\bar{\sigma} = -\rho Vc\sqrt{s(s + 1/\tau)}e^{-\sqrt{s(s+1/\tau)}(x/c)}$.

---

[2] In mathematical terms, this has to be so because every Laplace transform must vanish at $s \to \infty$, per the limit theorems [c.f. Appendix A].

[3] In this case, the solution in physical space can be established from a table of Laplace transforms and it is [3] $\sigma(x, t) = \rho cVe^{-t/(2\tau)}I_o\left(\frac{1}{2\tau}(t^2 - x^2/c^2)\right)H(t - x/c)$, where $I_o(\cdot)$ is the "modified Bessel function of order zero".

10.3.2 Materials of Differential Type

The general form of the stress–strain law in this case may be expressed as: $P(\sigma) = Q(\varepsilon)$ [Cf. Chap. 3], where P and Q are differential operators of order m, and n, respectively, such that either $m = n - 1$, or $m = n$:

$$\sum_{k=1}^{m} p_k \frac{d^k \sigma}{dt^k} = \sum_{l=1}^{n} q_l \frac{d^l \varepsilon}{dt^l} \tag{10.42}$$

Where, for a homogeneous material, all the p_k's and q_l's are constant.

The differential equation (10.42) can be integrated with respect to time, t, starting from a fixed but arbitrary point $x = X_1$, located before the shock front, and ending at a variable point $x = X_2(t)$, before or after the front. After the first integration, the generic term on either side of (10.42) becomes:

$$\int_{X_1}^{X_2} \frac{\partial^k}{\partial t^k}(\cdot) dt = \int_{X_1}^{X_2(t)} \frac{\partial}{\partial t} \frac{\partial^{k-1}}{\partial t^{k-1}}(\cdot) dt = \frac{\partial^{k-1}}{\partial t^{k-1}}\Big|_{X_1}^{X_2(t)} \tag{h}$$

With this, any term of the sum in (10.42) becomes a function of the upper limit and can be integrated again. Repeating this integration a total of n times produces an integral whose limits lie on both sides of the shock front. At this stage, most of the terms that result from the integrations would be continuous functions of time, in such a manner that their jumps across the front would be zero. The exception being the two terms with the highest order derivative. If we now let X_1 and X_2 approach each other and the shock front at $x = X^*$, the following is obtained after n integrations:

$$\lim_{\substack{X_1 \to X_2(t) \\ X_1 \leq X^{*-}(t)}} \left\{ p_n \int_{X_1}^{X^{*-}(t)} \frac{\partial}{\partial t} \sigma dt \right\} + \lim_{\substack{X_2 \to X_1 \\ X_2 > X^{*+}(t)}} \left\{ p_n \int_{X^{*+}(t)}^{X_2(t)} \frac{\partial}{\partial t} \sigma dt \right\} \tag{i}$$

and similarly:

$$\lim_{X_1 \to X_2(t)} = q_n \{\varepsilon(X^{*-}) - \varepsilon(X^{*+})\} + q_n \{\varepsilon(X^*) - \varepsilon(X^*)\} \equiv -q_n[|\varepsilon|] \tag{j}$$

These expressions lead to the stress-strain law in jump form for a viscoelastic substance of the differential-operator type, as:

$$[\sigma] = E_g[\varepsilon]; \quad E_g \equiv \frac{q_n}{p_n} \tag{10.43}$$

This expression is the exact analog of (10.32). Also, since (10.28–10.30) apply irrespective of the form of the stress–strain law, combining (10.30) and (10.43) yields the same equation for the wave speed, c, as Eq. (10.33):

$$c = \sqrt{\frac{E_g}{\rho}}; \quad E_g \equiv \frac{q_n}{p_n} \tag{10.44}$$

Hence, the jump equations: (10.28–10.30) and (10.32) or (10.43), as well as (10.33) or (10.44), for the sonic speed, are the same for all viscoelastic substances.

Example 10.7 Obtain the speed of longitudinal shock waves in a bar made of a Maxwell material with modulus E_o, viscosityη, and density ρ.

   Solution:The constitutive equation for this Maxwell material is [c.f. Chap. 3]: $\sigma + \frac{\eta}{E_o}\frac{d\sigma}{dt} = \eta\frac{d\varepsilon}{dt}$. Using (10.44) with $n = 1$, $p_n = \eta/E_o$, $q_n = \eta$ yields: $c = \sqrt{E_o/\rho}$.

   The differential equations governing the jumps in strain, stress, and velocity for the materials of differential type are obtained in the same way as for viscoelastic materials of integral type: by taking the total derivative, on the shock front, of the stress-velocity jump equation (10.28), evaluating the partial derivative of the stress–strain law with respect to time, and using expressions (10.28–10.30) in the process. Specifically, taking the derivative of (10.28) on the shock front leads to:

$$\frac{D}{Dt}[\sigma] = -\rho c\frac{D}{Dt}[v] \tag{k}$$

Using (10.25) on the left-hand side of this expression converts it to:

$$\left[\frac{\partial\sigma}{\partial t}\right] + c\left[\frac{\partial\sigma}{\partial x}\right] = -\rho c\frac{D}{Dt}[v] \tag{l}$$

To evaluate the first term on the left-hand side of (*l*), first differentiate the stress–strain law once with respect to time. Considering the case $m = n$ only, this leads to:

$$p_0\frac{\partial\sigma}{\partial t} + \cdots p_{n-1}\frac{\partial^n\sigma}{\partial t^n} + p_n\frac{\partial^{n+1}\sigma}{\partial t^{n+1}} = q_0\frac{\partial\varepsilon}{\partial t} + \cdots q_{n-1}\frac{\partial^n\varepsilon}{\partial t^n} + q_n\frac{\partial^{n+1}\varepsilon}{\partial t^{n+1}} \tag{m}$$

Integrating this expression n times with respect to t, following the same approach that led to Eq. (10.43)—starting at a fixed but arbitrary point below the shock front and ending at a variable point above it and then taking the limit as the two points approach the front—produces the jump equation:

$$\left[\frac{\partial\sigma}{\partial t}\right] = \frac{q_n}{p_n}\left[\frac{\partial\varepsilon}{\partial t}\right] + \frac{q_{n-1}}{p_n}[\varepsilon] - \frac{p_{n-1}}{p_n}[\sigma] \tag{n}$$

Using that $q_n/p_n = E_g = \rho c^2$, together with $[\partial\varepsilon/\partial t] = [|\partial v/\partial x|]$, $[\varepsilon] = -[v]/c$, and $[\sigma] = -\rho c[v]$, converts this expression into the following:

$$\left[\frac{\partial\sigma}{\partial t}\right] = \rho c^2\left[\frac{\partial v}{\partial x}\right] - \frac{q_{n-1}}{p_n}\rho c[v] + \frac{p_{n-1}}{p_n}\rho c[v] \tag{o}$$

Inserting (o) and the jump of the equation of motion across the front: $[\partial\sigma/\partial x] = \rho[\partial v/\partial t]$, into ($l$), yields, after collecting and regrouping terms:

$$\rho c\left\{\left[\frac{\partial v}{\partial t}\right] + c\left[\frac{\partial v}{\partial x}\right]\right\} + \left(\frac{p_{n-1}}{p_n} - \frac{q_{n-1}}{q_n}\right)\rho c[v] = -\rho c\frac{D}{Dt}[v] \tag{p}$$

Since, by virtue of (10.25), the expression in braces is the total time derivative, $D[v]/Dt$, of $[v]$ across the front, the final result is:

$$\frac{D}{Dt}[v] + \frac{1}{2}\left(\frac{p_{n-1}}{p_n} - \frac{q_{n-1}}{q_n}\right)[v] = 0 \tag{10.45}$$

This first-order differential equation is the exact analog of (10.34b), derived for materials of relaxation integral type. The solution of this equation can therefore be cast in the same form as (10.35b):

$$[|v|] = Ae^{-\frac{1}{2}\Gamma(0)t} ; \quad \Gamma(0) \equiv \left(\frac{p_{n-1}}{p_n} - \frac{q_{n-1}}{q_n}\right) \tag{10.46a}$$

As was the case with materials of integral type, using (10.8): $D/Dt = cD/Dx$ converts (10.45) into an equation in x with solution:

$$[v] = Ae^{-\frac{1}{2c}\Gamma(0)x} ; \quad \Gamma(0) \equiv \left(\frac{p_{n-1}}{p_n} - \frac{q_{n-1}}{q_n}\right) \tag{10.46b}$$

This seems to be the first time the general expression for $\Gamma(0)$ listed in (10.46a, b) appears in the literature. In lieu of this expression, the procedure to estimate the rate of decay of shock waves in viscoelastic substances of differential type seems to have use (10.34a), which requires the relaxation modulus—a more cumbersome process.

Example 10.8 Determine the rate of decay of shock waves in a bar made of the standard linear solid with the material properties shown in Fig 10.5.

Solution:The rate of decay of shock waves in viscoelastic materials of differential type is given by (10.46a, 10.46b). To compute it requires the coefficients of the constitutive equation for this material. Using [c.f. Chap. 3]: $\frac{E_0+E_1}{\eta_1}\sigma +$ $\frac{\partial\sigma}{\partial t} = \frac{E_0 E_1}{\eta_1}\varepsilon + E_0\frac{\partial\varepsilon}{\partial t}$, identifies $p_o = \frac{E_0+E_1}{\eta_1}$, $p_1 = 1$, $q_o = \frac{E_0 E_1}{\eta_1}$, and $q_1 = E_o$. Taking

Fig. 10.5 Standard solid of
example 10.8

this into (10.46a, 10.46b) gives the rate of decay of shock waves for the
standard linear solid as: $\Gamma(0) = \left(\frac{p_{n-1}}{p_n} - \frac{q_{n-1}}{q_n}\right) \equiv \left(\frac{E_0+E_1}{\eta_1} - \frac{E_1}{\eta_1}\right) = \left(\frac{E_0}{\eta_1}\right).$

10.4   Problems

P.10.1 Determine the complex sonic speed at $\omega = 50$ rad/s for the three-parameter
solid of Example. 10.6 if $E_o = 140$ kPa, $E_1 = 140$ kPa, $\eta = 276$ kPa-s,
$\rho = 1{,}600$ kg/m^3.

  Answer: $c^* = 9.35e^{-0.0101j} = 9.353 + 0.0474\,j$

  Hint: Use the constitutive polynomials P and Q [c.f. Chap. 3] to compute the
complex modulus $E^*(j\omega) = Q^*(j\omega)/P^*(j\omega) \approx 139.97 + 1.42\,j$ kPa. From it,
determine the loss angle $\delta = tan^{-1}(E''/E') \approx 0.0101$ *rad*, and use Eq. (10.13) to
obtain c^*.

P.10.2 The equilibrium modulus of an elastomeric compound is 145 psi; its mass
density is 0.065 lbm/in^3; and the transient part of its tensile relaxation modulus, in
Prony-series form, has the following coefficients and time constants at the temperature of interest:

E_k (MPa)	9.665	20.69	3.448	5.517	1.379	1.379	0.2410	0.2070
τ_k (sec)	$4.5 \cdot 10^{-5}$	$6.5 \cdot 10^{-4}$	$5.0 \cdot 10^{-3}$	$2.0 \cdot 10^{-2}$	$2.5 \cdot 10^{-1}$	$3.5 \cdot 10^{0}$	$1.5 \cdot 10^{1}$	$4.0 \cdot 10^{2}$

Determine the complex sonic speed of the material at a forcing frequency of
50 Hz.

  Answer: $c^* \approx 74.2 + 13.0\,j$ m/s.

  Hint: Compute E^* per Chap. 3, as: $E^* \approx 9.64 + 3.345\,j$ MPa, together with its
amplitude, $\|E^*\| \approx 10.20$ MPa, and phase angle $\delta = tan^{-1}(3.345/9.64) \approx$
0.347 rad. Using $\|E^*\|$ and $\rho = 1{,}600$ kg, in (10.14), gives $\|c^*\| \approx 75.3$ m/s. The
complex sonic speed at the given frequency is obtained from (10.13).

P.10.3 Determine the response of a 30 inch long bar of the same material and
under the same conditions as the bar in Example 10.2. What would the response of
the bar be if its material were such that $E'' = 0$?

  Answer: The solution is displayed in Fig. 10.6 for the viscoelastic bar and in
Fig. 10.7, for the bar without damping ($E'' = 0$), which would then be elastic. As
seen in the figures, the viscoelastic and elastic bars respond quite similarly.

Fig. 10.6 Problem 10.3: harmonic waves in given viscoelastic bar

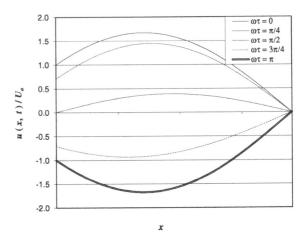

Hint: In this case, E^*, ω, and λ, are the same as for the bar in Example 10.2. Hence, the form of the solution will be the same, and the difference is due to the change in length. Note the symmetry of the response for the case with no damping ($E'' = 0$).

P.10.4 Prove that: $[|v|] = A \cdot e^{-\frac{t}{2E_g c}\frac{dE}{dt}|_{t=0}} \equiv A \cdot e^{-\frac{(E'(0)/E_g)}{2c}t}$

Hint: Use that: $\Gamma(0) \equiv \frac{1}{E_g}\frac{d}{ds}E(t-s)\Big|_{t=0} = -\frac{1}{E_g}qE(q)\Big|_{q=0} = -\frac{E'(0)}{E_g}$, to put (10.35b) into the desired form.

P.10.5 Find the stress response in transformed space, of a semi-infinite bar of Maxwell material subjected to a stress of amplitude Σ, suddenly applied at its end.

Answer: $\bar{\sigma}(x,s) = \frac{\Sigma}{s}e^{-(x/c)\sqrt{s(s+1/\tau)}}$

Hint: First, obtain the transform of $\sigma(0,t) = \Sigma\cdot H(t)$, as: $\bar{\sigma}(0,s) = \Sigma/s$. Now use the stress-strain law (3.23a) in transformed space, to get: $A(S) - A = -\Sigma/(\lambda s)$. The result follows from this and the expression for $\bar{\sigma}(x,s)$.

P.10.6 Find the Laplace-transformed solution of a bar of length L, fixed at its end $x = L$ and subjected to a discontinuous change in velocity of amplitude V at $x = 0$.

Fig. 10.7 Problem 10.3: harmonic waves in given elastic bar ($E'' = 0$)

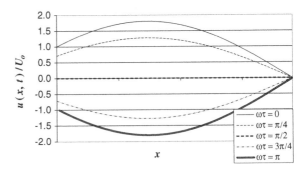

Answer: $\bar{v}(x, s) = \frac{V}{s} e^{\lambda(s)L} \frac{\sinh \lambda(s)x}{\sinh \lambda(sL)}$

Hint: Use that $v(x,t) = \partial u(x,t)/\partial t$ to write: $\bar{v}(x, s) = s\bar{u}(x, s)$, and thus: $\bar{v}(0, s) = V/s$. Use this, and the fact that $v(L,t) = 0$, $\bar{v}(x, s) = D_1 e^{\lambda x} + D_2 e^{-\lambda x}$ to evaluate the functions $D_1(s)$ and $D_2(s)$ and arrive at the result.

P.10.7 Determine the rate of decay of shock waves in the standard linear solid of Example 10.5, using the material's relaxation modulus and expression (10.34a).

Answer: $\Gamma(0) = E_0/\eta_1$

Hint: The rate of decay of shock waves in terms of the relaxation modulus given in (10.34a) requires the material's glassy modulus as well as the first derivative of the relaxation modulus at $t = 0$. The glassy modulus of this materials is $E_g = E_0$; and its relaxation modulus [c.f. Chap. 3]: $E(t) = \frac{E_0 E_1}{E_0 + E_1} + \left(E_0 - \frac{E_0 E_1}{E_0 + E_1} \right) e^{-\left(\frac{E_0 + E_1}{\eta_1} \right) t}$.

From this, it follows that $\frac{d}{dt} E(t)\big|_{t=0} = \frac{E_0^2}{\eta_1}$; and, from (10.34a): $\Gamma(0) = E_0/\eta_1$; the same as in Ex. 10.8, which used Eqs. (10.46a, b), pertinent for materials of differential type.

References

1. H. Kolsky, *Stress Waves in Solids* (Dover Publications, NY, 1963), pp. 41–53
2. W. Flügge, *Viscoelasticity*, 2nd edn. (Springer, Berlin, 1975), pp. 121–140
3. R.M. Christensen, *Theory of Viscoelasticity*, 2nd edn. (Dover, NY, 1982), pp. 110–116
4. N. Cristescu, I. Suliciu, *Viscoplasticity* (Martinus Nijhoff, The Hague, 1982)

Variational Principles and Energy Theorems

11

Abstract

This chapter introduces the subject of the variation of a functional and develops variational principles of instantaneous type which are the equivalent of Castigliano's theorems of elasticity for computing the generalized force associated with a generalized displacement and vice versa, by means of partial derivatives of the potential energy and the complementary potential energy functionals, respectively. A natural consequence of the variational principle of instantaneous type is that the constitutive potentials of viscoelastic materials are not unique. Any dissipative term can be added to them without changing the stress strain law. The viscoelastic versions of the unit load theorem of elasticity, and the theorems of Betti and Maxwell for elastic bodies, are also developed in detail.

Keywords

Variation · Variational · Functional · Potential · Stationary · Admissible · Field · Castigliano · Unit load · Reciprocal

11.1 Introduction

The calculus of variations is concerned with the development of methods to establish the maxima and minima of a special class of functions, called functionals. Loosely speaking, a functional is a function whose arguments are themselves functions. Functionals occur naturally in viscoelasticity through the constitutive equations, in which the arguments are the position coordinates, the current stress or strain, and the history of the stress or strain.

D. Gutierrez-Lemini, *Engineering Viscoelasticity*, DOI: 10.1007/978-1-4614-8139-3_11, 261
© Springer Science+Business Media New York 2014

Many variational theorems have been developed for the theory of elasticity; among them are the principles of minimum potential energy and minimum complementary potential energy, the principle of virtual work or virtual displacements, and the theorems of Hu–Washizu, Hellinger–Reiner, Castigliano, and Lagrange [1]. Not surprisingly, variational theorems also exist which are the viscoelastic counterparts of those for elasticity.

Variational principles provide a rigorous means of establishing the field equations of the theory of viscoelasticity without recourse to simplifying assumptions. The purpose of the present chapter, however, is to use variational principles to develop computational tools that facilitate solving certain types of viscoelastic boundary-value problems. To do this, the subject matter is developed in five parts, as follows.

Section 2 treats viscoelastic functionals as functions of two time-dependent arguments: the present values and the past values of the stress or strain. Depending on whether the current values or the past values of the arguments are varied, two types of variations are identified. Instantaneous variations result when only the current values of the arguments (stress or strain) are varied, and history variations are concerned only with variations in the past values of the arguments. This is used in Sect. 3 to develop two variational principles of instantaneous type, which are the equivalent of Castigliano theorems of elasticity for computing the generalized force associated with a generalized displacement and vice versa, by means of derivatives of the potential energy and the complementary potential energy, respectively. The viscoelastic version of the unit load theorem of elastic solids, which is used to obtain the deflection at an arbitrary point in a structure, is developed in Sect. 4. The reciprocal theorems of viscoelasticity which correspond to the theorems of Betti and Maxwell for elastic bodies [2] are developed in Sect. 5.

11.2 Variation of a Functional

Aside from the position coordinates, x_k, two types of time-dependent arguments occur in viscoelastic functionals. To account for memory effects, the response is made to depend on all past values of the stress or strain. To account for instantaneous response, the current values of the stress or strain are also included. Simply put then, a viscoelastic functional may be considered as a function of two variables: a "history" variable and a variable that describes the current state. Therefore, in examining the variation in a given viscoelastic functional, one is at liberty to consider variations in the current values of its arguments or to consider variations in the past history of the arguments. The former type of variation leads to instantaneous variational principles and the latter to hereditary variational principles. Both types of variations produce equivalent principles, but only variational principles of instantaneous type are discussed in the present chapter.[1]

---

[1] Viscoelastic variational principles concerned with variations of the histories of the arguments will be included in subsequent editions of the text.

With this, a functional, F, of the past history, $\varepsilon(t-s)$, and current value, $\varepsilon(t)$, of a variable may be represented as follows [3]:

$$F = \underset{\tau=0}{\overset{\infty}{\psi}} \left[\varepsilon_{ij}(t-\tau), \varepsilon_{ij}(t)\right] \tag{11.1}$$

Assuming that all pertinent smoothness and continuity requirements are satisfied by all the functions involved and considering the past history $\varepsilon_{ij}(t-\tau)$ fixed, the first instantaneous variation, or total differential, δF, of the functional is defined as the limit[2]:

$$\delta F = \lim_{h\to 0} \frac{1}{h} \left\{ \underset{s=0}{\overset{\infty}{\psi}}\left[\varepsilon_{ij}(t-s), \varepsilon_{ij}(t) + h \cdot \delta\varepsilon_{ij}(t)\right] - \underset{s=0}{\overset{\infty}{\psi}}\left[\varepsilon_{ij}(t-s), \varepsilon_{ij}(t)\right] \right\}\Bigg|_{h=0} \tag{a}$$

As for ordinary scalar-valued scalar functions, this expression is the derivative of ψ with respect to h, evaluated at $h = 0$ [c.f. Appendix A]. Therefore, the first variation of the functional F with respect to the current value of its arguments, which is also called first instantaneous variation of F, is given by

$$\delta F = \left\{ \frac{d}{dh} \underset{s=0}{\overset{\infty}{\psi}}\left[\varepsilon_{ij}(t-s), \varepsilon_{ij}(t) + h \cdot \delta\varepsilon_{ij}(t)\right] \right\}\Bigg|_{h=0} \tag{11.2}$$

Using the chain rule of differentiation to evaluate it, results in

$$\delta F = \frac{\partial}{\partial\varepsilon_{kl}(t)} \overset{\infty}{\underset{0}{\psi}}\left[\varepsilon_{ij}(t-s), \varepsilon_{ij}(t)\right] \cdot \delta\varepsilon_{kl}(t) \tag{11.3}$$

Clearly, entirely similar expressions would be obtained if the roles of the stress and strain tensors were interchanged.

11.3 Variational Principles of Instantaneous Type

These principles assume variation in the current, instantaneous values of the variables σ_{ij} and ε_{ij}, while their histories determine the state of the material at the current time. In this case, as indicated by 11.3, the first variation of a functional is obtained by taking its partial derivative with respect to its instantaneous argument, regarding the hereditary component as constant.

---

[2] In keeping with standard practice, in this chapter only, we use the Greek letter, δ, to denote variation of a function or functional; and as such, is not to be mistaken for the unit impulse or Dirac Delta function, nor for the Kronecker delta, used to represent the unit tensor.

For the present treatment, the three-dimensional forms of the constitutive laws presented in 2.34a and b will be used.[3] Thus, as discussed in Chp. 8, for a general anisotropic material

$$\sigma_{ij}(t) = M_{gijkl}\varepsilon_{kl}(t) - \int_0^t \frac{\partial}{\partial \tau} M_{ijkl}(t-\tau)\varepsilon_{kl}(\tau)d\tau \qquad (11.4a)$$

$$\varepsilon_{ij}(t) = C_{gijkl}\varepsilon_{kl}(t) - \int_0^t \frac{\partial}{\partial \tau} C_{ijkl}(t-\tau)\sigma_{kl}(\tau)d\tau \qquad (11.5a)$$

which, according to 11.3, have the following instantaneous variations:

$$\delta\sigma_{ij}(t) = M_{gijkl}\delta\varepsilon_{kl}(t) \qquad (11.4b)$$

$$\delta\varepsilon_{ij}(t) = C_{gijkl}\delta\sigma_{kl}(t) \qquad (11.5b)$$

Different viscoelastic potential functionals and associated variational principles can be constructed, which have their counterparts in the theory of elasticity. Prominent among these are a Hellinger–Reissner-type functional and a Castigliano-type functional that is similar to the potential energy functional of elasticity. The Castigliano-type functional is equivalent to the elastic complementary potential energy functional. Like their elastic counterparts, these viscoelastic functionals acquire stationary values and give rise to principles which can be used to derive the field equations of viscoelasticity and various computational methods.

11.3.1 First Castigliano-Type Principle

This is the instantaneous variational counterpart of the theorem of minimum potential energy of elasticity. The corresponding functional which, for reference only, is called instantaneous viscoelastic potential energy, ϕ_o, in this text, is defined for all kinematically admissible displacement fields, $u_i'(t)$. A displacement field u_i' is said to be kinetically admissible if it is continuously differentiable up to third order[4] inside the region occupied by the body in question and, in addition, identically satisfies both the equations of strain compatibility and the displacement boundary conditions on the part of the boundary of the body where they are prescribed. Clearly, to a kinematically admissible displacement field, there correspond kinematically admissible strains $\varepsilon_{ij}' = \frac{1}{2}(u_{i,j}' + u_{j,i}')$, defined by the strain

---

[3] As in previous chapters, the subscript g stands for "glassy", to indicate the value of the corresponding material property function at $t = 0$.

[4] The requirement of continuity up to the third order derivatives of the displacement field stems from the fact that the equations of compatibility involve second derivatives of the strains, which are defined in terms of the first derivatives of the displacement field.

displacement relationships (9.5), and stresses, σ'_{ij}, defined through constitutive equations (11.4a).

Similar to elastic solids, the instantaneous viscoelastic potential function, ϕ_o, is defined for all kinematically admissible displacement fields, $u'_i(t)$, including the actual field, $u_i(t)$, in the form:

$$\varphi_o(u'_i) = \int_V U_o\left(\varepsilon'_{ij}\right) dV - \int_V F_i(t) u'_i(t)\, dV - \int_{S_T} T_i^o(t) u'_i(t)\, ds \qquad (11.6)$$

In this expression, the functional U_o is of the same form as the internal energy of elasticity theory and, just like it, is assumed to be an (instantaneous) internal energy potential, from which the stress–strain relations of viscoelasticity can be obtained by differentiation [c.f. Appendix B]. Similar to the elastic strain energy density, and using the actual fields σ_{ij} and ε_{ij}, for generality, the instantaneous potential energy density functional, U_o, is defined as

$$\sigma_{ij} = \frac{\partial}{\partial \varepsilon_{ij}} U_o(\varepsilon_{mn}) \qquad (11.7)$$

As a potential function, U_o (ε) can be formally constructed from 11.7 by integration. To do this, the stress–strain constitutive Eq. (11.4a) is put on the left-hand side of 11.7 and integration is carried out holding the hereditary part constant, in accordance with the present type of variation. This leads to

$$U_o[\varepsilon_{mn}(t)] = \frac{1}{2}\varepsilon_{ij}(t) M_{gijkl}\varepsilon_{kl}(t) - \varepsilon_{ij}(t) \int_0^t \frac{\partial}{\partial s} M_{ijkl}(t-s)\varepsilon_{kl}(s)\,ds + \overset{\infty}{\underset{s=0}{D}}[\varepsilon_{ij}(t-s)] \qquad (a)$$

In this expression, D $[(\varepsilon_{ij}(t-s)]$ is a functional of the strain history and represents a purely dissipative contribution. The physical interpretation of this is that

"The constitutive potentials of viscoelastic substances are not unique. Any dissipative term can be added to them without changing the stress strain law."

Thus, without any loss of generality, the functional D is assumed to be identically zero, so that:

$$U_o(\varepsilon_{mn}) \equiv \frac{1}{2}\varepsilon_{ij}(t) M_{gijkl}\varepsilon_{kl}(t) - \varepsilon_{ij}(t) \int_0^t \frac{\partial}{\partial \tau} M_{ijkl}(t-\tau)\varepsilon_{kl}(\tau)\,d\tau \qquad (11.8)$$

By its construction, as the integral of 11.7 with respect to the current strains, it should be clear that the first instantaneous variation in this expression produces the constitutive equations.

Example 11.1 Check that the first instantaneous variation in U_o with respect to current strains yields the constitutive equations of viscoelasticity in stress–strain form.

Solution:

Take the first instantaneous variation in U_o as

$\delta U_o \equiv \frac{1}{2} \left[\delta \varepsilon_{ij} M_{gijkl} \varepsilon_{kl} + \varepsilon_{ij} M_{gijkl} \delta \varepsilon_{kl} \right]$ and use the symmetry of the material property tensor: $M_{gijkl} = M_{gklij}$ to collect terms and arrive at constitutive Eq. (11.4a).

For future reference, U_o is split into an elastic part, U_{go}, and a hereditary part, U_{to}:

$$U_o \equiv U_{go} - U_{to} \tag{11.9a}$$

$$U_{go} \equiv \frac{1}{2} \varepsilon_{ij}(t) M_{gijkl} \varepsilon_{kl}(t) \tag{11.9b}$$

$$U_{to} \equiv \varepsilon_{ij}(t) \int_0^t \frac{\partial}{\partial \tau} M_{ijkl}(t - \tau) \varepsilon_{kl}(\tau) d\tau \tag{11.9c}$$

Multiplying (11.4) by $\varepsilon_{ij}(t)$, and using the previous definitions, yields the expression:

$$\sigma_{ij}(t) \varepsilon_{ij}(t) \equiv 2U_{go} - U_{to} \tag{11.10}$$

Returning now to the instantaneous potential ϕ_o, introduced in 11.6, the other integrals in it correspond to the instantaneous work of the body forces, $F_i(t)$, and the specified surface tractions, $T_i^o(t)$. Also, the stress and strain fields will in general be functions of position, but such spatial dependence is omitted for clarity.

The conditions that 11.6 must fulfill at its stationary points are obtained by setting its first variation to zero.

$$\delta \varphi_o(u_i') = \int_V \delta U_o(\varepsilon_{ij}') \, dV - \int_V F_i(t) \delta u_i'(t) \, dV - \int_{S_T} T_i^o(t) \delta u_i'(t) \, dS = 0 \tag{b}$$

Adding and subtracting $\int_{S_T} T_i \delta u_i' \, dS \equiv \int_{S_T} n_j \sigma_{ji} \delta u_i'$:

$$\delta \varphi_o(u_i') = \int_V \delta U_o(\varepsilon_{ij}') \, dV - \int_V F_i \delta u_i' \, dV + \int_{S_T} (T_i - T_i^o) \delta u_i' \, dS - \int_{S_T} n_j \sigma_{ji} \delta u_i' \, dS$$
$$= 0$$

$$\tag{c}$$

Extending the integral on the far right to the whole surface of the body $S_T + S_u$, which is valid because $\delta u' = 0$ on S_u, and invoking Gauss' theorem to convert that integral into a volume integral [c.f. Appendix A], use the strain–displacement relations, and collect terms to write:

$$\delta\varphi_o(u_i') = \int_V [\sigma_{ij} - \frac{\partial}{\partial\varepsilon_{ij}}U_o(\varepsilon_{kl}')]\delta\varepsilon_{ij}\, dV - \int_V [\sigma_{ij,j} + F_i]\delta u_i'\, dV + \int_{S_T} (T_i - T_i^o)\delta u_i'\, dS$$

$$= 0$$

$$(d)$$

So that, for arbitrary variations in the kinematically admissible displacement field and its associated strain field, the first variation in the instantaneous potential is stationary if

$$\sigma_{ij}(t) = \frac{\partial U_o(\varepsilon_{ij})}{\partial\varepsilon_{ij}(t)} \equiv M_{gijkl}\varepsilon_{kl}(t) - \int_0^t \frac{\partial}{\partial\tau}M_{ijkl}(s)\varepsilon_{kl}(t-\tau)\, d\tau$$

$$\sigma_{ij,j}(t) + F_i(t) = 0$$

$$T_i(t) = T_i^o(t); on \quad S_T$$

$$(11.11)$$

Conversely, if these conditions are fulfilled, one can construct the potential ϕ_o. For this reason, the first variation in the instantaneous potential ϕ_o has a stationary value, if and only if the constitutive and equilibrium equations as well as the traction boundary conditions are satisfied. Clearly, by definition of the admissible fields, the displacement boundary conditions are also identically satisfied. These are the exact same requirements on the potential energy functional of elasticity.

That the stationary value of the instantaneous potential ϕ_o corresponds to a minimum may be proven in simple terms. Indeed, the symmetry of the stress and strain tensors implies that the glassy modulus tensor, M_{gijkl} is symmetric. Because of this, the glassy term of the instantaneous potential ϕ_o, which is quadratic in $\varepsilon_{ij}(t)$, must be positive definite. As in ordinary differential calculus, the second variation—or second derivative—determines the character of the stationary points of any given functional. Also, both the current and hereditary terms of the instantaneous potential, ϕ_o, and the external work are linear in $\varepsilon_{ij}(t)$. Hence, it is the quadratic term in $\varepsilon_{ij}(t)$ that defines the second variation. By the positive definiteness of this term, the instantaneous functional ϕ_o will attain a minimum for all kinematically and statically admissible displacement fields.

The theorem of minimum instantaneous potential ϕ_o may be used to develop a method to determine the generalized force required to maintain a set of displacements in a body. This is so when the instantaneous work of the external forces can be expressed as a linear combination of the work of a set of generalized forces, $P_r(t)$, as they move through their corresponding generalized displacements, $d_r'(t)$, that is, if

$$\int_V F_i(t)u_i'(t)\, dV + \int_{S_T} T_i^o(t)u_i'(t)\, dS = \sum_r P_r d_r'$$

$$(11.12)$$

In this expression, the P_r's are known generalized forces, while the d'_r's are kinematically admissible, but otherwise arbitrary generalized displacements.

The displacement field, $u'_i(t)$, and with it the corresponding strain field, $\varepsilon'_{ij}(t)$, can then be expressed as linear combinations of the d'_r. Therefore, the instantaneous potential, U_o, and through it, ϕ_o, become quadratic functions of the d_r, as

$$\varphi_o = \int_V U_o(d'_q)\, dV - \sum_r P_r d'_r \tag{11.13}$$

Invoking the stationarity of ϕ_o, and recalling the definition of instantaneous variation, one can write $\delta\varphi_o = \frac{\partial}{\partial d'_s}\left[\int_V U_o(d'_q)\, dV - \sum_r P_r d'_r\right]\delta d'_s = 0$; which yields the relationship sought:

$$P_s = \frac{\partial}{\partial d'_s}\int_V U_o(d'_q)\, dV \tag{11.14}$$

11.3.2 Second Castigliano-Type Principle

This is the instantaneous viscoelastic variational counterpart of the theorem of minimum complementary potential energy of elasticity. The viscoelastic functional, which may be thought of as an instantaneous complementary potential energy density, ψ_o, is defined for all statically admissible displacement fields u". A displacement field, u", with strains $\varepsilon''_{ij} = \frac{1}{2}(u''_{i,j} + u''_{j,i})$, and stresses, $\sigma''_{ij} = M_{ijkl} * d\varepsilon''_{kl}$, is statically admissible if it is continuous and continuously differentiable inside the region occupied by the body and identically satisfies the equations of equilibrium inside the body, as well as the traction boundary conditions, where prescribed.

In analogy with elastic materials, an instantaneous complementary potential functional, ψ_o, is defined for all statically admissible fields u''_i, ε''_{ij}, and σ''_{ij}—which include the actual field $u_i(t)$—in the form:

$$\psi_o\left(\sigma''_{ij}\right) = \int_V Y_o\left(\sigma''_{ij}\right) dV - \int_{S_u} T''_i(t) u^o_i(t)\, ds \tag{11.15}$$

In this expression, the functional Y_o is of the same form as the complementary potential energy density of elasticity theory and, just like it, is assumed to be an (instantaneous) energy potential, from which the strain–stress relations of viscoelasticity can be obtained by differentiation [c.f. Appendix B]. Similar to the elastic case, and using the actual fields σ_{ij} and ε_{ij}, for generality, the instantaneous complementary potential energy functional, Y_o, is assumed such that

$$\varepsilon_{ij} = \frac{\partial}{\partial \sigma_{ij}} Y_o(\sigma_{kl}) \tag{11.16}$$

As a potential function, Y_o (σ) can be formally constructed from 11.16 by integration. To do this, the strain–stress constitutive Eq. (11.5a) is put on the left-hand side of 11.16 and integration is carried out holding the hereditary argument constant, in accordance with the present type of variation. This leads to

$$Y_o[\sigma_{mn}(t)] = \frac{1}{2}\sigma_{ij}(t)C_{gijkl}\sigma_{kl}(t) - \sigma_{ij}(t)\int_0^t \frac{\partial}{\partial\tau}C_{ijkl}(t-\tau)\sigma_{kl}(\tau)d\tau + \overset{\infty}{\underset{s=0}{Q}}[\sigma_{mn}(t-s)]$$

$$(11.17)$$

This expression has the following instantaneous variation:

$$\delta Y_o(\sigma_{mn}) = \delta\sigma_{mn}(t)\left[C_{gijkl}\sigma_{kl}(t) - \int_0^t \frac{\partial}{\partial s}C_{ijkl}(t-s)\sigma_{kl}(\tau)d\tau\right] \qquad (a)$$

In general, although the term Q, above, may be a functional of the stress history, as far as the instantaneous variation is concerned, it does not contribute at all to the strain–stress equations and may thus be taken as zero without loss of generality. In other words, the complementary energy potential functional for a viscoelastic substance is not uniquely defined, since one can add to it an arbitrary functional of the stress history without altering the strain–stress relations. In other words, the functional Q represents a purely dissipative contribution. Hence, in the sequel, Y_o is taken simply, as:

$$Y_o(\sigma_{mn}) \equiv \frac{1}{2}\sigma_{ij}(t)C_{gijkl}\sigma_{kl}(t) - \sigma_{ij}(t)\int_0^t \frac{\partial}{\partial\tau}C_{ijkl}(t-\tau)\sigma_{kl}(\tau)d\tau \qquad (11.18)$$

Also, for future reference, Y_o is separated into an instantaneous component, Y_{go}, and a transient part, Y_{to}, as follows[5]:

$$Y_o \equiv Y_{go} - Y_{to} \qquad (11.19a)$$

$$Y_{go} \equiv \frac{1}{2}\sigma_{ij}(t)C_{gijkl}\sigma_{kl}(t) \qquad (11.19b)$$

$$Y_{to} \equiv \sigma_{ij}(t)\int_0^t \frac{\partial}{\partial\tau}C_{ijkl(t-\tau)}\sigma_{kl}(\tau)\, d\tau \qquad (11.19c)$$

Premultiplying (11.5a) by $\sigma_{ij}(t)$, using the above decomposition of Y_o, and collecting terms produce:

$$\sigma_{ij}\varepsilon_{ij} = 2Y_{go} - Y_{to} \qquad (11.20)$$

---

[5] The negative sign to define Y_{to} is only used for mathematical convenience. .

In addition, replacing Y_{to} with the aid of 11.19a and rearranging

$$Y_o = \sigma_{ij}\varepsilon_{ij} - Y_{go} \tag{11.21}$$

Returning to the instantaneous potential ψ_o, introduced in 11.5a and b, in terms of the arbitrary but statically admissible fields u_i'', ε_{ij}'', and σ_{ij}'', the second integral in it is the work of the arbitrary but statically admissible surfaced tractions, T''', acting on the actual displacement field $u_i^o(t)$.

The stationarity conditions of Y_o are derived by setting the first variation in ψ_o to zero:

$$\delta\psi_o(\sigma_i'') = \int_V \delta Y_o(\sigma_{ij}'')dV - \int_{S_u} \delta T_i''(t)u_i^o(t)dS = 0 \tag{b}$$

Adding and subtracting $\int_{S_u} \delta T_i'' u_i\, dS$ and using $\int_{S_T} \delta T_i'' u_i dS \equiv \int_{S_u} n_j \delta\sigma_{ij}'' u_i dS$ result in:

$$\delta\psi_o(\sigma_i'') = \int_V \delta Y_o(\sigma_{ij}'')\, dV - \int_{Su} \delta T_i''(u_i^o - u_i'')\, dS + \int_{S_T+S_u} n_j\delta\sigma_{ij}''u_i S = 0 \tag{c}$$

Before proceeding, the third integral on the right of this expression is converted into a volume integral by means of the Gauss theorem [c.f. Appendix A]:

$$\int_S n_j\delta\sigma_{ij}''u_i dS \equiv \int_V \delta\sigma_{ij,j}''u_i dV + \int_V \delta\sigma_{ij}''u_{i,j}dV \tag{d}$$

The first term on the right vanishes because the equations of equilibrium are identically satisfied by the statically admissible field, and the instantaneous variation in the actual body forces is identically zero. Now, by the symmetry of the stress tensor, the second integral may be expressed in the form:

$$\int_V \delta\sigma_{ij}''u_{i,j}dV \equiv \int_V \frac{1}{2}(u_{i,j} + u_{j,i})\delta\sigma_{ij}''dV \tag{e}$$

Using these results and rearranging, $\delta\psi_o$ becomes

$$\delta\psi_o(\sigma_{mn}'') = \int_V [\frac{\partial}{\partial\sigma_{ij}''}Y_o(\sigma_{kl}'') - \frac{1}{2}(u_{i,j} + u_{j,i})]\delta\sigma_{ij}dV + \int_{S_u} (u_i'' - u_i^o)\delta T_i''dS = 0 \tag{f}$$

So that, for arbitrary variations in the statically admissible fields, the first variation of the instantaneous complementary energy potential is stationary if

$$\varepsilon_{ij}''(t) = \frac{1}{2}\left(u_{i,j}'' + u_{j,i}''\right); \; in \, V \tag{11.22}$$

$$u_i''(t) = u_i^o; on \, S_u$$

Clearly, if these conditions are fulfilled, one can also construct the potential ψ_o. For this reason, the first variation in the instantaneous complementary potential has a stationary value, if and only if the strain–displacement relationships and the displacement boundary conditions are satisfied. By the static admissibility of the fields involved, the equations of equilibrium are satisfied as well. As stated before, these are precisely the requirements put on the potential energy functional of elasticity.

The proof that the stationary value of the instantaneous complementary potential energy corresponds to a minimum follows from the simple fact that ψ_o is quadratic in the instantaneous stresses and only linear in its hereditary part. By the symmetry of the tensor of glassy compliances, C_{gijkl}, the quadratic term of ψ_o is positive definite. Since the character of a stationary point is determined by the second variation in a functional, it being positive for ψ_o due to the positive definiteness of its quadratic term, the stationary points of ψ_o must correspond to a minimum.

The theorem of minimum instantaneous complementary potential ψ_o may be applied to derive the viscoelastic counterpart of a second Castigliano-type theorem of elasticity. This principle is used to determine the generalized displacement required to maintain a set of specified tractions in a structure. In so doing, it is assumed that the work term appearing in 11.15 may be expressed as discrete sum of products of arbitrary but statically admissible generalized forces,[6] P_r'', and actual generalized displacements d_r as:

$$\int\limits_{S_u} T_i''(t)u_i^o(t)dS \equiv \sum_r P_r'' d_r \tag{11.23}$$

In this case, the stress field may be expressed as a linear combination of the generalized forces; consequently, the instantaneous complementary potential energy functional, Y_o, is quadratic in the P_r''. Using this and 11.19a, b, and c allow one to express 11.23 as:

$$\psi_o = \int\limits_V Y_o(P_q'')dV - \sum_r P_r'' d_r \tag{11.24}$$

---

[6] The term generalized force is used to denote either a concentrated force or a concentrated moment, while the term generalized displacement denotes either a linear or an angular displacement. A generalized force and a generalized displacement are work-conjugate if the work done by the former acting on the later can be correctly calculated from their product.

Stationarity of ψ_o requires that:

$$\delta\psi_o = \frac{\partial}{\partial P''_s}\left[\int_V Y_o(P''_q)dV - \sum_r P''_r d_r\right]\delta P''_r = 0 \tag{g}$$

That is

$$d_s = \frac{\partial}{\partial P''_s}\int_V Y_o(P''_q)dV \tag{11.25}$$

This expression shows that the generalized displacement corresponding to a generalized force may be obtained as the first partial derivative of the instantaneous complementary potential function with respect to that force. This is the analog of Castigliano's second theorem for linear elastic structures.

Example 11.2 A beam of a viscoelastic material with tensile relaxation modulus $E(t)$ is subjected to a concentrated load $P(t)$, as shown Fig 11.1. The length of the beam is L and its cross-section have second moment of area of magnitude I. Find the deflection of the beam under the point of load application .
Solution:
The deflection, Δ, of the beam under the load P may be readily obtained by means of Castigliano's theorem (11.25). Hence, the following expression needs to be evaluated: $\Delta = \frac{\partial}{\partial P}\int_V Y_o(P)dV$. For the beam in question [c.f. Chap. 5],

$\sigma = \frac{My}{I} \equiv \frac{P(t)xy}{2I}$. Using the strain–stress relation

$$\varepsilon(t) = C_g\sigma(t) - \partial_t C(t-\tau) * \sigma(\tau) \equiv C_g\sigma(t) - K(t-\tau) * \sigma(\tau)$$

and Eq. (11.21): $\gamma o = \sigma\varepsilon - \gamma_{go}$ leads to the following expression for the complementary potential function:

$$\int_V Y_o dv \equiv \int_V \left[\frac{1}{2}\sigma(t)C_g\sigma(t) - \sigma(t)K(t-\tau) * \sigma(\tau)\right]dv$$

$$= \frac{P^2(t)C_g L^3}{96I} - \frac{P(t)C_g K(t-\tau) * P(\tau)}{48I}$$

Fig. 11.1 Example 11.2

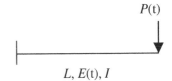

$P(t)$

$L, E(t), I$

Applying the theorem leads to $\Delta(t) = \frac{\partial}{\partial P} \int_V V_o dv \equiv \frac{P(t)C_g L^3}{48I} - \frac{L^3 C_g K(t-\tau)*P(\tau)}{48I}$; or:

$\Delta(t) = \frac{L^3}{48I}[C_g P(t) - \int_0^t \frac{\partial}{\partial \tau} C(t - \tau)P(\tau)d\tau]$. This is the viscoelastic analog of the elastic form $PL^3/(48E_g I)$ and could, of course, have been obtained by the methods of Chap. 5.

Castigliano's theorem may appear to be of limited value because its form (11.25) seems to imply that one can determine generalized displacements only at the exact locations where generalized forces act. The unit theorem presented subsequently is derived from Castigliano's theorem and extends its applicability to locations in a structure where no actual generalized forces need to be acting.

11.3.3 Unit Load Theorem

The unit load theorem for viscoelastic structures is derived from Castigliano's principle in a simple fashion, as follows. A generalized fictitious load P is applied to the structure at the location and in the direction in which the deflection is desired.[7] The stresses and strains resulting from P are then determined and used to construct the complementary potential functional $Y_o(P)$. The generalized displacement, d, is obtained by applying Castigliano's theorem (11.25), differentiating Y_o with respect to P and, since P is fictitious, evaluating the result at $P = 0$.

Indeed, let $\sigma_{ij}(t)$ be the stress field induced by the loads acting on the structure and $\sigma_{uij}(t)$ that due to a generalized load of unit magnitude. With this, the stress field induced in a linear viscoelastic structure by a generalized load of magnitude P will be $P \cdot \sigma_{uij}(t)$, and the combined action of the actual loads and P will yield the field: $\sigma_{ij}(t) + P \cdot \sigma_{uij}(t)$. Under this system, using 11.19a, b, and c results in the following instantaneous potential functional, Y_o:

$$Y_o = \frac{1}{2}[\sigma_{ij}(t) + P\sigma_{uij}(t)][C_{gijkl}]\{\sigma_{kl}(t) + P\sigma_{ukl}(t)\}+$$

$$- [\sigma_{ij}(t) + P\sigma_{uij}(t)] \int_0^t [\frac{\partial}{\partial \tau}C_{tijkl}(t - \tau)]\{\sigma_{kl}(\tau) + P\sigma_{ukl}(\tau)\}d\tau \tag{11.26}$$

Applying the second Castigliano-type theorem, (11.25) produces the expression sought:

---

[7] For simplicity of exposition, the prime notation used to distinguish statically admissible fields is dropped.

$$u|_P = \frac{\partial}{\partial P} Y_0|_{P=0}$$

$$= \int_V \sigma_{uij}(t) C_{gijkl} \sigma_{kl}(t)\, dV - \int_V \sigma_{uij}(t) \left[\int_0^t \frac{\partial}{\partial \tau} C_{tijkl}(t-\tau)\sigma_{kl}(\tau)\, d\tau \right] dV$$

$$(11.27)$$

This form is easily applied to obtain the deflection of linearly viscoelastic bodies made of structural members such as beams, columns, and bars, for which the most general state of stress, σ, is made up of a normal component and two shear stress components. The direct or normal stress is due to the normal force and bending moment—possibly two of them—acting on the cross-section. The shear stresses are induced by the shear forces and, if present, the torsional moment. As it turns out, the use of principal centroidal axes in (11.27) eliminates all terms containing mixed mechanical elements.[8]

Example 11.3 Derive the viscoelastic unit load theorem for prismatic structural members accounting only for bending moment effects.
Solution:

Start with the normal stress $\sigma(t) = \frac{M(x)\,y}{I}$ produced in a viscoelastic beam by a bending moment M (t) [c.f. Chap. 5] and note that $\sigma_u(t) = \frac{M_u(x)\,y}{I}$ would be the stress induced by a generalized load of unit magnitude acting along the generalized direction at the desired location. Then, insert these two expressions on the right-hand side of (11.26) to obtain

$$u|_P = \int_L \int_A \frac{M_u(x,t)}{I} y C_g \frac{M(x,t)}{I} y\, dA\, dx$$

$$- \int_L \int_A \frac{M_u(x,t)}{I} y \left[\int_0^t \frac{\partial}{\partial \tau} C(t-\tau) \frac{M(x,\tau)}{I} y\, d\tau \right] dA\, dx$$

Carry out the integral over the cross-sectional area, using that $I \equiv \int_A y^2 dA$ and simplify to get:

$$u|_P = \int_L \frac{M_u(x,t) C_g M(x,t)}{I}\, dx - \int_L \left\{ \frac{M_u(x,t)}{I} \int_0^t \frac{\partial C(t-\tau)}{\partial \tau} M(x,\tau)\, d\tau \right\} dx$$

$$(11.28)$$

---

[8] This is so, because mixed terms involve integrals of the form: $\int_A y\, dA$; which are identically zero by the assumption that the reference axes are centroidal.

This is the analog of the linear elastic expression: $u|_P = \int_L \frac{M_u(t)M(t)}{EI} dx$, only when bending effects are included. For convenience, Eq. (11.28) will be simplified by introducing an auxiliary operator, $\hat{C}$:

$$A(t) \cdot \hat{C} * B(t) \equiv A(t)C_g B(t) - A(t)\int_0^t \frac{\partial}{\partial \tau} C(t - \tau)B(\tau)d\tau \qquad (11.29)$$

So, that, operationally

$$\hat{C}* \equiv [C_g - \int_0^t \frac{\partial}{\partial \tau} C(t - \tau)*] \qquad (11.30)$$

With this definition, Eq. (11.28) takes the symbolic form:

$$u|_P = \int_L \frac{M_u(x,t) \cdot \hat{C} * M(x,t)}{I} dx \qquad (11.31)$$

Example 11.3 Determine the mid-span deflection of the uniformly loaded, simply supported viscoelastic beam shown in Fig. 11.2, where q, L, I, and E represent, respectively, the uniform load on the beam, reckoned per unit length of the beam, the beam's length, the second moment of area of the beams cross-section, and the tensile relaxation modulus.

Solution:
  The solution is obtained by evaluating Eq. (11.31), in which M is the moment induced by the actual loading q and M_u is the moment due to a unit concentrated load acting downward.[9] To do this, obtain the required mechanical elements by establish equilibrium of the beam under the unit, generalized load and the actual load.
  To establishing equilibrium of the beam under a unit load, use Fig. 11.3:

Fig. 11.2 Example 11.3

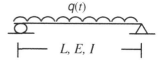

$q(t)$

L, E, I

Fig. 11.3 Example 11.3:
Equilibrium under a unit load

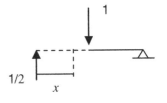

Fig. 11.4 Example 11.3:
Equilibrium under actual load

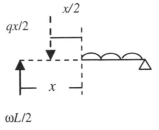

From it, write that:

$$M_u = \begin{cases} \frac{1}{2}x; & 0 \leq x \leq \frac{L}{2} \\ \frac{L}{2} - \frac{x}{2}; & \frac{L}{2} < x \leq L \end{cases}$$

For equilibrium under the actual load, use Fig. 11.4 to write

$$M = \frac{qL}{2}x - \frac{q}{2}x^2; \quad 0 \leq x \leq L \tag{b}$$

Insert   these   expressions   into   Eq.   (11.31)   to   obtain

$$u(x = \tfrac{L}{2}) = \tfrac{2}{I} \int\limits_0^{L/2} \left[\tfrac{1}{2}x \cdot \hat{C}_E * \left(\tfrac{qL}{2}x - \tfrac{q}{2}x^2 \right) \right] dx$$

Here, use has been made of the symmetry of M_u and M with respect to the center of the beam to perform the integration over one half of the beam only. Integrating this relation, noting that $\hat{C}$ is independent of x because the material is homogeneous and using the $C_g = 1/E_g$ yields: $\quad u(x = \tfrac{L}{2}) = \tfrac{5L^4}{384I} \hat{C} * q(t)$

$$\equiv \tfrac{5q(t)L^4}{384E_g I} - \tfrac{5L^4}{384I} \int\limits_0^t \tfrac{\partial C_E(t-\tau)}{\partial \tau} q(\tau)\, d\tau;$$ in which the first part on far right is the deflection

at the center of an elastic beam of Young modulus E_g.

Expressions (11.31) can be generalized to cover the case of multiple mechanical elements—axial force N_z, shear forces V_x and V_y, bending moments M_x and M_y, and torsional moment M_z—acting simultaneously on a structural member. Indeed, let the mechanical elements in a structural member be defined in a coordinate system with its z-axis passing through the centroid of the member's

Fig. 11.5 Vectorial sign convention for structural mechanical elements

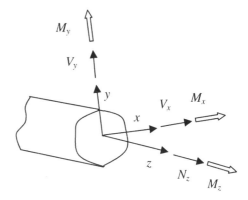

cross-sections and located along the tangent to the principal centroidal rectangular coordinate system.[10] For definiteness, the vector sign convention is used. According to this convention, a mechanical element is positive if as a vector it acts in the positive direction of the corresponding coordinate axis, as illustrated in Fig. 11.5.

Under these assumptions, the most general state of stress in any given structural member would include the following components [4]:

$$\sigma_{xx}(t) = \sigma_{xy}(t) = \sigma_{yy}(t) = 0 \tag{11.32a}$$

$$\sigma_{zx}(t) = \frac{V_x Q_x}{I_x b_y} - \frac{M_z(t)x}{J_z}$$

$$\sigma_{zy}(t) = \frac{V_y Q_y}{I_x b_x} + \frac{M_z(t)y}{J_z} \tag{11.32b}$$

$$\sigma_{zz}(t) = \frac{N_z(t)}{A} + \frac{M_x(t)y}{I_x} - \frac{M_y(t)x}{I_y} \tag{11.32c}$$

In these expressions, A is the area of the cross-section; I_x and I_y are the second rectangular moments of area of the cross-section with respect to principal and

---

[10] A centroidal and principal coordinate system is located at the center of area –or centroid– of a cross section, with its axes coinciding with the cross section's principal axes of inertia. In such a system, the following relations hold:

$$\int_A x dA = \int_A y dA = \int_A r dA = 0; I_{\overline{xy}} \equiv \int_A xy dA = 0; I_{\overline{x}} \equiv \int_A y^2 dA; I_{\overline{y}} \equiv \int_A x^2 dA; J_{\overline{z}} \equiv \int_A r^2 dA$$

centroidal x and y axes, respectively; and $J_z = I_x + I_y$ is the polar moment of area of the cross-section with respect to the z-axis. Also, $b_x(y)$ is the length of the fiber at which the shear stress is being evaluated, which is located at a distance y from the centroidal x-axis, and $Q_x(y)$ is the first moment of area with respect to x of that portion of the cross-section delimited by this fiber and the outermost fiber of the cross-section; $b_y(x)$ and $Q_y(x)$ have completely analogous interpretations.

Using Eq. (11.32a, b, c) and the corresponding expressions for the P-multiplied stress field with Eq. (11.27) leads to:

$$
\begin{aligned}
u|_P = \frac{\partial}{\partial P} Y_o|_{P=0} = & \int_L \int_A \sigma_{uzz}(t) C_g \sigma_{zz}(t)\, dV - \int_L \int_A \sigma_{uzz}(t) \left[\int_0^t \frac{\partial}{\partial \tau} C(t-\tau) \sigma_{zz}(\tau)\, d\tau \right] dV \\
& + \int_L \int_A \sigma_{uzx}(t) C_g \sigma_{zx}(t)\, dV - \int_L \int_A \sigma_{uzx}(t) \left[\int_0^t \frac{\partial}{\partial \tau} C(t-\tau) \sigma_{zx}(\tau)\, d\tau \right] dV \\
& + \int_L \int_A \sigma_{uzy}(t) C_g \sigma_{zy}(t)\, dV - \int_L \int_A \sigma_{uzy}(t) \left[\int_0^t \frac{\partial}{\partial \tau} C(t-\tau) \sigma_{zy}(\tau)\, d\tau \right] dV
\end{aligned}
$$

$$(11.33)$$

Example 11.4 Derive the expression for the unit load theorem for a prismatic beam loaded only on its principal plane of bending $z-y$, by axial and transverse forces, disregarding the effect the shear forces might have on the deflection of the beam.

Solution:

In this case, $M_y = M_z = V_x = 0$; and $\sigma_{xx} = \sigma_{xy} = \sigma_{yy} = \sigma_{zx} = 0$, and $\sigma_{zz} = \frac{N_z}{A} - \frac{M_x y}{I_x}$. Also, although $\sigma_{zy} = \frac{V_y Q_y}{I_y b_y}$, its effect on the deflection will be neglected. Inserting this field in 11.33 together with the corresponding stresses $\sigma_{uzz} = \frac{N_{uz}}{A} - \frac{M_{ux} y}{I_x}$, due to the unit load, and using (11.29) to simplify notation, produce the expression:

$u|_P = \int_L \left(\frac{N_{uz}}{A} - \frac{M_{ux} y}{I_x} \right) \hat{C} * \int_A \left(\frac{N_z}{A} - \frac{M_x y}{I_x} \right) dA\, dx$. Carrying out the operations and

noting that $\int_A y\, dA = 0$, because the reference axes are centroidal, and also that

$\int_A y^2\, dA = I_x$ lead to $u|_P = \int_L \frac{N_{uz} \hat{C} * N_z}{A}\, dx + \int_L \frac{M_{ux} \hat{C} * M_x}{I_x}\, dx\tau$.

Just as for elastic systems, the unit load theorem can also be used to determine reactions in statically indeterminate structures. This and other related topics will be taken up in a future edition of the text.

11.4 Reciprocal Theorems

Consider a viscoelastic body under the action of two separate sets of body forces, surface tractions and boundary displacements, $\{\overline{F}^1, \overline{T}_o^1, \overline{u}_o^1\}$ and $\{\overline{F}^2, \overline{T}_o^2, \overline{u}_o^2\}$, to which correspond stress and strain solution fields $\{\underline{\sigma}^1, \underline{\varepsilon}^1\}$ and $\{\underline{\sigma}^2, \underline{\varepsilon}^2\}$, respectively. It can then be shown that under certain additional restrictions, any two such systems enjoy several reciprocity relations.

11.4.1 Static Conditions

Here, viscoelastic bodies are considered which have been at rest prior to the start of the observation and are loaded so that their accelerations remain negligible. Such systems have boundary tractions and displacements that satisfy a reciprocity relation analogous to the theorem of Betti for elastostatics [see for instance A. Ghali, A.M. Neville, cited].

$$\int_S T_i^1 * du_i^2 \, dS + \int_V F_i^1 * du_i^2 \, dV = \int_S T_i^2 * du_i^1 \, dS + \int_V F_i^2 * du_i^1 \, dV \quad (11.34a)$$

Or written out in full:

$$\int_{S_T} \int_0^t T_{oi}^1(t-\tau) \frac{\partial}{\partial\tau} u_i^2(\tau) \, d\tau \, dS + \int_V \int_0^t F_i^1(t-\tau) \frac{\partial}{\partial\tau} u_i^2(\tau) \, d\tau \, dV = {}^2|_1 \quad (11.34b)$$

where dependence on the position coordinates is omitted for clarity. Also, to avoid repetition, the symbol $^p|_q$ is meant to indicate that the expression on the right-hand side is identical to that on the left-hand side, but with superscript p taking the role of superscript q and vice versa [1].

The present reciprocity relation is proven by showing that starting with one side of it, say the left, one can produce its other side. To do this, first transform the surface integral on the left-hand side of (11.34a, b) to a volume integral, replacing the surface traction in it by the scalar product of the surface normal and stress tensor and applying Gauss' divergence theorem [c.f. Appendix A]:

$$\int_S T_i^1 * du_i^2 \, dS = \int_S n_j \sigma_{ij}^1 * du_i^2 \, dS = \int_V (\sigma_{ij}^1 * du_i^2)_{,j} \, dV$$

$$= \int_V (\sigma_{ij,j}^1 * du_i^2 + \sigma_{ij}^1 * du_{i,j}^2) \, dS \quad (a)$$

$$= -\int_V F^1 * du_i^2 \, dV + \int_V \sigma_{ij}^1 * du_{i,j}^2 \, dV$$

In the last step, the equilibrium equations, $\sigma^1_{ij,j} + F^1_i = 0$, were invoked. Operating now on the second integral on the right-hand side of (a), using the symmetry of the stress tensor

$$\int_V \sigma^1_{ij} * du^2_{i,j}\, dV \equiv \int_V \sigma^1_{ij} * \frac{1}{2} d(u^2_{i,j} + u^2_{j,i})\, dV = \int_V \sigma^1_{ij} * d\varepsilon^2_{ij}\, dV \qquad \text{(b)}$$

Use of (a) and (b) transforms the left-hand side of (11.34a) into

$$\int_S T^1_i * du^2_i\, dS + \int_V F^1_i * du^2_i\, dV = \int_V \sigma^1_{ij} * d\varepsilon^2_{ij}\, dV \qquad \text{(c)}$$

Consequently, also

$$\int_S T^2_i * du^1_i\, dS + \int_V F^2_i * du^1_i\, dV = \int_V \sigma^2_{ij} * d\varepsilon^1_{ij}\, dV \qquad \text{(d)}$$

Now, in view of the stress–strain constitutive equations, $\sigma^1_{IJ} = M_{ijkl} * d\varepsilon^1_{kl}$, together with the associative and commutative properties of the Stieltjes convolution [c.f. Appendix A]:

$$\int_V \sigma^1_{ij} * d\varepsilon^2_{ij}\, dV \equiv \int_V (M_{ijkl} * d\varepsilon^1_{kl}) * d\varepsilon^2_{ij}\, dV = \int_V (M_{ijkl} * d(\varepsilon^1_{kl} * d\varepsilon^2_{ij})\, dV$$

$$= \int_V M_{ijkl} * d(\varepsilon^2_{kl} * d\varepsilon^1_{ij})\, dV = \int_V (M_{ijkl} * d\varepsilon^2_{kl}) * d\varepsilon^1_{ij}\, dV \qquad \text{(e)}$$

$$= \int_V \sigma^2_{ij} * d\varepsilon^1_{ij}\, dV$$

Expressions (c) and (e) prove the reciprocity relations (11.34a, b).

11.4.2 Dynamic Conditions

Using the Laplace transformation and assuming at-rest initial conditions, a reciprocity relationship for dynamics may be derived, which is applicable to both homogeneous and inhomogeneous elastic and viscoelastic materials of arbitrary density [1].

The reciprocity relations for dynamics, which are entirely analogous to those of the quasi-static case—including that no time derivatives appear in the convolution integrals—take the form:

$$\int_S T^1_i * u^2_i\, dS + \int_V F^1_i * u^2_i\, dV = \int_S T^2_i * u^1_i\, dS + \int_V F^2_i * u^1_i\, dV \qquad \text{(11.35a)}$$

Or, explicitly

$$\int_{S_T} \int_0^t T_{oi}^1(t-\tau) u_i^2(\tau)\, d\tau\, dS + \int_V \int_0^t F_i^1(t-\tau) u_i^2(\tau)\, d\tau\, dV = {}^2|_1 \qquad (11.35b)$$

Indeed, starting with the equations of motion for the body under the first loading system $\sigma_{ij,j}^1 + F_i^1 = \rho \ddot{u}_i$, convolving it with the dot product of the displacement vector of the second system, u_i^2, and integrating over the volume, manipulating the spatial derivative:

$$\int_V (\sigma_{ij,j}^1 + F_i^1) * u_i^2\, dV \equiv \int_V (\sigma_{ij}^1 * u_i^2)_{,j}\, dV - \int_V \sigma_{ij}^1 * u_{i,j}^2\, dV + \int_V F_i^1 * u_i^2\, dS$$
$$= \int_V \rho \frac{d^2 u_i^1}{dt^2} * u_i^2\, dV$$

Applying now the divergence theorem to the first integral on the right of the identity sign and using the strain–displacement relations and symmetry of the stress and strain tensors in the second integral; and rearranging:

$$\int_S T_i^1 * u_i^2\, dS + \int_V F_i^1 * u_i^2\, dV = \int_V \sigma_{ij}^1 * \varepsilon_{ij}^2\, dV + \int_V \rho \frac{d^2 u_i^1}{dt^2} u_i^2\, dV$$

On applying the Laplace transform to this expression, assuming at-rest initial conditions:

$$\int_S \overline{T}_i^1 \overline{u}_i^2\, dS + \int_V \overline{F}_i^1 \overline{u}_i^2\, dV = \int_V \overline{\sigma}_{ij}^1 \overline{\varepsilon}_{ij}^2\, dV + \int_V \rho s^2 \overline{u}_i^1 \overline{u}_i^2\, dV$$

And, upon inserting the Laplace transform, $\overline{\sigma}_{ij}^1 = s\overline{M}_{ijkl}\overline{\varepsilon}_{kl}^1$, of the constitutive relations:

$$\int_S \overline{T}_i \overline{u}_i^2\, dS + \int_V \overline{F}_i^1 \overline{u}_i^2\, dV = \int_V s\overline{M}_{ijkl}\overline{\varepsilon}_{kl}^1\overline{\varepsilon}_{ij}^2\, dV + \int_V \rho s^2 \overline{u}_i^1\overline{u}_i^2\, dV \qquad \text{(a)}$$

Similarly, reversing the roles of solution fields 1 and 2:

$$\int_S \overline{T}_i^2 \overline{u}_i^1\, dS + \int_V \overline{F}_i^2 \overline{u}_i^1\, dV = \int_V s\overline{M}_{ijkl}\overline{\varepsilon}_{kl}^2\overline{\varepsilon}_{ij}^1\, dV + \int_V \rho s^2 \overline{u}_i^2\overline{u}_i^1\, dV \qquad \text{(b)}$$

Equating (a) and (b) yields:

$$\int_S \overline{T}_i^1 \overline{u}_i^2\, dS + \int_V \overline{F}_i^1 \overline{u}_i^2\, dV = \int_S \overline{T}_i^2 \overline{u}_i^1\, dS + \int_V \overline{F}_i^2 \overline{u}_i^1\, dV \qquad \text{(c)}$$

Fig. 11.6 Problem 11.2

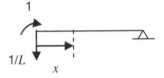

The inverse Laplace transform of this expression recovers the general reciprocity relation for elastokinetics presented in Eqs. (11.35a, b).

11.5 Problems

P.11.1 derives explicit expressions for the instantaneous potentials U_{go} and U_{to}, of a linear isotropic viscoelastic solid of constant Poisson's ratio, v, and uniaxial tensile relaxation modulus $E(t)$.

Answer:

$$U_{go} = \frac{E_g}{2(1+v)}\left\{\varepsilon_{ij}(t)\varepsilon_{ij}(t) + \frac{v}{1-2v}[\varepsilon_{kk}(t)]^2\right\}$$

$$U_{to} = \frac{1}{(1+v)}\left[\frac{1}{(1-2v)}\varepsilon_{kk}(t)\int_0^t \frac{\partial}{\partial\tau}E(t-\tau)\varepsilon_{kk}(\tau)d\tau + \varepsilon_{ij}(t)\int_0^t \frac{\partial}{\partial\tau}E(t-\tau)\varepsilon_{ij}(\tau)d\tau\right]$$

Hint:

Use constitutive Eq. (8.26) and relations (9.21) for isotropic viscoelastic materials with constant Poisson's ratio, together with Eq. (11.9b, c), and carry out the indicated operations.

P.11.2 Use the unit load theorem to determine the rotation at the left support of the beam in Example 11.3.

Answer: $\theta(x = 0, t) = \frac{L^2}{16I}\left[\frac{P(t)}{E_g} - \int_0^t \frac{\partial}{\partial\tau}C_E(t-\tau)P(\tau)d\tau\right]$

Hint:

Proceed as in Example 11.3, using a unit moment as generalized load, applied at the left support, where the rotation, as generalized deflection is wanted. Thus, as seen in Fig. 11.6, equilibrium under the unit generalized load leads to $M_u = 1 - x/L$. Insert this and $M = \frac{qL}{2}x - \frac{q}{2}x^2$; $0 \le x \le L$, into (11.31) and carry out the operations to arrive at the result.

References

1. Y.C. Fung, Foundations of Solid Mechanics, Prentice Hall, Inc., p 429–433 (1965)
2. A. Ghali, A.M. Neville, Structural Analysis. A Unified Classical and Matrix Approach, International Textbook Co., p. 90–110 (1972)
3. R.M. Christensen, Theory of Viscoelasticity, 2nd Ed., Dover, p. 3–9 (1982)
4. T. Oden, Mechanics of Elastic Structures, McGraw-Hill, pp. 76–78, 89–93 (1967)

Errata to: Engineering Viscoelasticity

12

Danton Gutierrez-Lemini

Errata to:
D. Gutierrez-Lemini, *Engineering Viscoelasticity*,
DOI 10.1007/978-1-4614-8139-3

In the below table, text content given in "Reads" column should be corrected, and the corrected text content have been given in "Should read as" column:

The online version of the original book can be found under 10.1007/978-1-4614-8139-3

D. Gutierrez-Lemini (✉)
Special Products Division, Oil States Industries, Inc., Commercial Blvd. N. 1031,
Arlington, TX 76001, USA
e-mail: dantonglemini@tx.rr.com

D. Gutierrez-Lemini, *Engineering Viscoelasticity*, DOI: 10.1007/978-1-4614-8139-3_12,
© Springer Science+Business Media New York 2014

Page numbers	Location	Reads	Should read as
Chapter 1			
4	Section 1.3, second paragraph	"…and let σ stands…"	"…and let σ stand…"
11	Equtaion (l)	$\sigma(t) = R$	$\sigma(t) = \eta R$
Chapter 2			
24	Last sentence	This topic of great…	This topic is of great…
26	Equation (c)	$\sigma(t) = \begin{cases} 0, & for * \tau < \tau_\phi \\ M(t-t_1)\varepsilon_0, & for * \tau \geq \tau_\phi \end{cases}$	$\sigma(t) = \begin{cases} 0, & for\ t < t_1 \\ M(t-t_1)\varepsilon_0, & for\ t \geq t_1 \end{cases}$
26	Last paragraph	t_{k+1}	t_k
27	Second paragraph	"…and using the properties of the Heaviside step function,"	The phrase: "…and using the properties of the Heaviside step function," should be deleted
34	Last sentence of first paragraph	timescale	time scale
39	Last paragraph	ϕ and ψ	ϕ and $d\psi/dt$
41	Last sentence of first paragraph	"…and the applied action…"	"…and the derivative of the applied action…"
43	First paragraph, second line	$\overline{\sigma} D_{ij} = 2s\overline{G} \overline{\varepsilon} D_{ij}$	$\overline{\sigma}_{Dij} = 2s\overline{G}\overline{\varepsilon}_{Dij}$
43	First paragraph, fourth line	$\sigma D_{ij} = 2G * d\varepsilon D_{ij}$	$\sigma_{Dij} = 2G * d\varepsilon_{Dij}$
43	First paragraph, after "Solution"	According to (2.31), the creep compliance…then invert it…	The phrase: "then invert it" should be deleted

(continued)

(continued)

Page numbers	Location	Reads	Should read as
43	First paragraph, after "Solution"	"…there results: "$s^2 M(s) = \ldots$""	"…there results: "$1/s^2 M(s) = \ldots$""
45	Equation (2.39)	$\sigma(t) = M_g\left\{\varepsilon_{rel}(t) + \int_0^t \frac{\partial}{\partial \tau} m(t-\tau)\varepsilon_{rel}(\tau)d\tau\right\}$	$\sigma(t) = M_g\left\{\varepsilon_{rel}(t,0) + \int_0^t \frac{\partial}{\partial \tau} m(t-\tau)\varepsilon_{rel}(t,\tau)d\tau\right\}$
47	First paragraph	"…given as $\varepsilon = u/l$, we cast…"	"…given as $\varepsilon = u/l$, and $\sigma = F/A$, we cast…"
50	Answer to P.2.5	Answer: $\sigma(t) = \left[E_e + \frac{E_1(\omega\tau)^2}{1+(\omega\tau)^2}\right]\varepsilon_o \cos(\omega t) - \frac{E_1(\omega\tau)}{1+(\omega\tau)^2}\varepsilon_o \cos(\omega t)$	Answer: $\sigma(t) = \left[E_e + \frac{E_1(\omega\tau)^2}{1+(\omega\tau)^2}\right]\varepsilon_o \cos(\omega t) - \frac{E_1(\omega\tau)}{1+(\omega\tau)^2}\varepsilon_o \sin(\omega t)$

Chapter 3

Page numbers	Location	Reads	Should read as
53	Midway down Abstract	"…it turns out that the constitutive…"	"…the constitutive…"
59	First paragraph	$f(t) \equiv f(t) \equiv \varepsilon_0$	$f(t) \equiv \sigma_0$
59	First paragraph	"and performing the required integration:"	", performing the required integration, and dividing by σ_o:"
59	Equation (3.6a)	$C(t-t_0) = \frac{1}{\eta}\int_{t_0}^t \varepsilon_0 ds \equiv \frac{(t-t_0)}{\eta} H(t-t_0)$	$C(t-t_0) = \frac{\varepsilon(t)}{\sigma_0} \equiv \frac{1}{\eta}\int_{t_0}^t ds \equiv \frac{(t-t_0)}{\eta} H(t-t_0)$
59	Last paragraph	(3.1a, b)	(3.1a) and (3.4b)
60	First line, second bullet point	$E\varepsilon(t) + \eta\partial_t\varepsilon(t) = [E + \eta\partial_t]\varepsilon(t) = [\eta\partial_t + E J\varepsilon(t) = \eta\partial_t\varepsilon(t) + E\varepsilon(t)$;	$E\varepsilon(t) + \eta\partial_t\varepsilon(t) = [E + \eta\partial_t]\varepsilon(t) = [\eta\partial_t + E]\varepsilon(t) = \eta\partial_t\varepsilon(t) + E\varepsilon(t)$;
62	Second bullet point	$M(\infty) = \frac{p_0}{q_0} = \cdots$	$M(\infty) = \frac{q_0}{p_0} = \cdots$
66	Equation (h)	$_{n-1}\sigma(t_0^+)$	$p_{n-1}\sigma(t_0^+)$
66	Last paragraph	"When stress is…"	"(a) When stress is…"

(continued)

(continued)

Page numbers	Location	Reads	Should read as
66	Equation (l), third term on right hand side	$\ldots \dfrac{q_n}{p_n}\dfrac{d}{dt}\varepsilon(t_0^+)\cdots$	$\ldots \dfrac{q_n}{p_n}\dfrac{d^2}{dt^2}\varepsilon(t_0^+)\cdots$
67	First paragraph	"Similarly, when…"	"(b) Similarly, when…"
67	Equation (m), third term on right hand side	$\ldots \dfrac{p_n}{q_n}\dfrac{d}{dt}\sigma(t_0^+)\cdots$	$\ldots \dfrac{p_n}{q_n}\dfrac{d^2}{dt^2}\sigma(t_0^+)\cdots$
73	Third paragraph	"…setting $t_o = 0$,…"	"…setting $f(t) = \sigma(t)$, and $t_o = 0$,…"
75	Item 2	"2. he overall…"	"2. The overall…"
76	First paragraph	$u = e^{\int dt/\tau_r} = e^{1/\tau_r}$	$u = e^{\int dt/\tau_r} = e^{1/\tau_r}$
76	Equation (c), first term on right hand side	$\ldots H(t - t_o)\int_{t_o}^t Ee^{\frac{\pm}{\tau_r}}\dfrac{df}{dt'}ds\ldots$	$\ldots H(t - t_o)\int_{t_o}^t Ee^{\frac{\pm}{\tau_r}}\dfrac{df}{ds}ds\ldots$
78	Paragraph after Eq. (k)	σ and $(t) = \sigma_o H(t - t_o)$ should not be split, but should be $\sigma(t) = \sigma_o H(t - t_o)$	
80	Equation (3.27a)	$\displaystyle\sum_{i=1}^{n}\varepsilon_i = \cdots$	$\displaystyle\sum_{i=0}^{n}\varepsilon_i = \cdots$
80	Paragraph after Eq. (3.27a)	"…Kelvin unit, switching…"	"…Kelvin unit, and switching…"
80	Equation (3.27B)	$\displaystyle\sum_{i=0}^{n}\prod_{\substack{j=0\\j\neq i}}^{n}(E_i + \eta_i\partial_t)\sigma = \prod_{i=0}^{n}(E_i + \eta_i\partial_t)\varepsilon$	$\displaystyle\sum_{i=0}^{n}\prod_{\substack{j=0\\j\neq i}}^{n}(E_j + \eta_j\partial_t)\sigma = \prod_{j=0}^{n}(E_j + \eta_j\partial_t)\varepsilon$

(continued)

(continued)

Page numbers	Location	Reads	Should read as
81	Paragraph after Eq. (3.29)	τ_i, 1/ E_i should not be split, but should be (τ_i, 1/E_i)	"...we write these expressions in stress..."
82	List item 2	"...we express it in stress..."	"...we write these expressions in stress..."
83	End of first paragraph	"...equation (3.31a) reveals..."	"...equation (3.28) reveals..."
83	End of list item (a)	"will always exhibits..."	"will always exhibit..."
85	Second paragraph	"...$M_o = q_1/p_1 \equiv M_g$..."	"$M_o = q_1/p_1 = E_o \equiv M_g$..."
85	Equation (f)	Two occurrences of "e_τ^{t}" should read as "$e^{t/\tau}$" also, ds should read dt' and $f(t)$ should be replaced with $f(t')$	
86	Equation (j), first term on right hand side	The coefficient σ_o should be deleted from the term	
87	Second paragraph	"...and strain becomes..."	"...and stress becomes..."
88	Section 3.7.2, second paragraph	"...equation of the Maxwell..."	"...equation of the generalized Maxwell..."
90	Problem P3.6	"...Fig. 3.19 adjacent sketch, show..."	"...Fig. 3.19, show..."
Chapter 4			
98	Example 4.2, Solution	$M^* = E\dfrac{j\omega t}{1+j\omega t} = E\dfrac{(\omega t)^2 + j(\omega t)}{1+(\ )^2}$	$M^* = E\dfrac{j\omega t}{1+j\omega t} = E\dfrac{(\omega t)^2 + j(\omega t)}{1+(\omega t)^2}$
98	Example 4.2, Solution	"...and (4.7a), as:..."	"...and (4.7c), as:..."

(continued)

(continued)

Page numbers	Location	Reads	Should read as
98	First paragraph after Eq. (4.11)	"…this into (2.33a) and…"	"this into (2.33b) and…"
99	Second paragraph	"…(4.7a) and (4.13)…"	"…(4.7a) and (4.14)…"
99	Equation (4.19a)	$\|C^*\|(j\omega) = \|C^*\|e^{-j\delta}$	$C^*(j\omega) = \|C^*\|e^{-j\delta}$
99	Equation (4.19b)	$\|C^*\| = \sqrt{(C')^2 + (c'')^2}$	$\|C^*\| = \sqrt{(C')^2 + (C'')^2}$
100	First paragraph after Eq. (4.22)	"…such as constitutive equation."	"…such as constitutive equations."
100	Last paragraph	"…equations (2.32)."	"…equation (2.1b)."
104	Example 4.5, Solution	$\dfrac{d\sigma}{dt} + \dfrac{1}{\tau_r}\sigma = E\dfrac{d}{dt}f(t)$	$\dfrac{d\sigma}{dt} + \dfrac{1}{\tau_r}\sigma = E\dfrac{d}{dt}\varepsilon(t)$
104	Example 4.5, Solution	Replace all occurrences of the Greek letter τ in the expression for C^* with the subscripted Greek letter τ_r	
105	End of second paragraph	"…by parts:"	"…by parts, using $u = M(t)$ and $dv = e^{-j\omega t}dt$:"
107	Second paragraph	"…whereas those of a fluid…"	"…whereas those of the three-parameter fluid…"
108	Figure 4.5, caption	"… $= M''(\infty) = \infty C'(0) = C_e$,…"	$= M''(\infty) = \infty$, $C'(0) = C_e$,…"
109	Phrase after Eq. (4.43)	The correct expression should be $\sigma(t) = \sigma_0 \sin(\omega t + \delta)$; and it should not be split	

(continued)

(continued)

Page numbers	Location	Reads	Should read as
109	First line of last paragraph	"…substituting $_o = \|C^*\|\sigma_o\ldots$"	"…substituting $\varepsilon_o = \|C^*\|\sigma_o\ldots$"
109	Last line	"…in (4.19):"	"…in (4.19d):"
111	Problem P4.5	Question mark appearing at end of problem statement should be removed	
112	Hint to Problem P.4.9	$\varepsilon_k^* = C_k^* \varepsilon_k^*$	$\varepsilon_k^* = C_k^* \sigma_k^*$
112	Problem 4.10, first line	"fork=1,2,…"	"…for k = 1,2,…"
Chapter 5			
115	Figure 5.1, caption	"…**d** Force resulant at x"	"…**d** Force resulants at x"
116	End of first paragraph	The term "tan $\varphi \approx \ldots$" should not be split	
117	Second line of second paragraph	"…neglecting higher order terms yields:"	"…in the limit as $\Delta x \to 0$, yields:"
119	Figure 5.3, part (a)	$P(t) = Po$	$P(t) = F_o f(t)$
119	Example 5.1, statement	"…in Fig. 5.2…"	"…in Fig. 5.3…"
121	Equation (5.17) left hand side	$\tau(r, xt) =$	$\tau(r,x,t) =$
121	Paragraph, after Eq. (5.18)	"…at station, z."	"…at station x."

(continued)

(continued)

Page numbers	Location	Reads	Should read as
121	Equation (5.19a)	$\theta(x,t) \equiv \dfrac{\partial}{\partial z}\psi(x,t) \equiv \cdots$	$\theta(x,t) \equiv \dfrac{\partial}{\partial x}\psi(x,t) \equiv \cdots$
122	Second paragraph	"...(5.18)..."	"...(5.17)..."
122	First paragraph of 5.3.2	"...(5.15). Thus"	"...(5.13). Thus"
123	First paragraph of 5.3.3	...(5.19a)...	...(5.19a,b)...
123	Second paragraph of 5.3.3	$T(t) = JG(t-\tau) * d\dfrac{d\psi}{dx}(x,\tau)$	$T(t) = JG(t-\tau) * d\dfrac{\partial\psi}{\partial x}(x,\tau)$
125	Paragraph after Eq. (5.25)	"...this take the form:"	"...this takes the form:"
126	Second paragraph. after Eq. (5.32)	"...is that as follows:"	"...is as follows:"
133	Paragraph. after Eq. (5.42b)	"...time of as in Chap. 2..."	"...time, as in Chap. 2..."
135	Equation (g)	$\overline{T}(s) = \dfrac{J\overline{Q}(s)/\overline{P}(s)}{L}\overline{\psi}(s) \equiv \overline{K}_T(s)\overline{\psi}(s)$	$\overline{T}(s) = \dfrac{J\overline{Q}(s)/\overline{P}(s)}{L}\overline{\psi}(s) \equiv \overline{K}_T(s)\overline{\psi}(s)$
135	Equation (5.46a)	"...E^*..."	"...G^*..."
137	Second line, last paragraph	"...elastic relationships..."	"...viscoelastic relationships..."

(continued)

(continued)

Page numbers	Location	Reads	Should read as
138	Fourth paragraph	"...except that as follows:"	"...except as follows:"
138	Bullet point b	"...angular acceleration, $\ddot{\psi}^*$, or rotational acceleration, $\ddot{\psi}^*$,"	"...bending acceleration, $\ddot{\phi}^*$, or torsional acceleration, $\ddot{\psi}^*$,"
138	Footnote	"...the complex-controlled variable..."	"...the complex, controlled variable..."
139	Paragraph after Eq. (5.48)	"the component K' and ..."	"the components K' and ..."
141	Fifth paragraph	"...the magnitude, F^* is..."	"...the magnitude of F^* is..."
144	Paragraph after Eq. (5.59)	"...Chap. 4 developed ..."	"...Chap. 4, developed ..."
145	Problem P.5.1, Answer	"...$= q(x,t)$"	"...$= -q(x,t)$"
146	Problem P.5.3, Answer	$Q[v(L,t)] = -\frac{L^3}{3J}P[F(t)] \equiv -\frac{F_0 L^3}{3J}P[F(t)]$	$Q[v(L,t)] = -\frac{L^3}{3J}P[F(t)] \equiv -\frac{F_0 L^3}{3J}P[f(t)]$
147	Problem P.5.9, Hint	"...Apply (2.37)..."	"...Apply (2.45)..."
Chapter 6			
150	Last paragraph	"...dealt within this..."	"...dealt with in this..."
152	Second paragraph	"...where it is not for..."	"...were it not for"
153	Last paragraph	"...time t, and T...."	"...time t, and temperature T...."
154	Second paragraph after Eq. (6.2b)	"...to $t/aT(T,T_r)$ units..."	"...to $t/a_T(T,T_r)$ units..."

(continued)

(continued)

Page numbers	Location	Reads	Should read as
154	Last paragraph	"…curve at $t/a_T(T_r)$…"	"…curve at $t/a_T(T,T_r)$…"
159	Second line of footnote	"…when $T_r = T_g$."	"…when $T_r \neq T_g$."
160	Fourth paragraph, second line	"…inserting into…"	"…inserting (6.18) into…"
160	Fourth paragraph, third line	$\varepsilon^M(t) = \cdots = C(t-\tau)^* d\sigma(\tau)$	$\varepsilon^M(t) = \cdots = C(t-\tau)^* d\sigma(\tau)$
160	Fourth paragraph, fourth line	$M_* dC = H(t)$	$M * dC = H(t)$
160	Last line of fourth paragraph	"…with (2.20) or (2.21) yields"	"…with (2.28) yields"
161	Last paragraph	$C_2(°F) = 9/5 \cdot C_2(°C)\_134.9\ °F$ $T_r(°F) = 9/5 \cdot T_r(°F) + 32\_73.0\ °F$	$C_2(°F) = 9/5 \cdot C_2(°C) \approx 134.9\ °F$ $T_r(°F) = 9/5 \cdot T_r(°C) + 32 \approx 73.0\ °F$
163	Last reference	"W.G. Knauss, I. Emri…"	"7. W.G. Knauss, I. Emri…"
Chapter 7			
165	Abstract, second to last line	"…function in a Prony…"	"…functions in Prony…"
166	Third paragraph	"…using Eq. (2.32a)…"	"…using Eq. (2.1b)…"

(continued)

(continued)

Page numbers	Location	Reads	Should read as
169	First paragraph of 7.2.2	"...per Eq. (2.2)..."	"...per Eq. (2.2b)..."
169	Last paragraph, 3rd line	"...to the loading is..."	"...to the loading..."
169	Last paragraph	"...temperature, by calculating..."	"...temperature. By calculating..."
176	Equation (7.24)	$M_g - \sum_{i=1}^{N} \frac{M_r \beta_r}{\beta_r - \alpha_i} = 0; \quad r = 1, \ldots, N$	$M_g + \sum_{i=1}^{N} \frac{M_i \alpha_i}{\beta_r - \alpha_i} = 0; \quad r = 1, \ldots, N$
176	Equation (7.25)	$C_e - \sum_{i=1}^{N} \frac{C_i \beta_i}{\beta_r - \alpha_i} = 0; \quad i = 1, \ldots, N$	$C_e + \sum_{r=1}^{N} \frac{C_i \alpha_i}{\beta_i - \alpha_i} = 0; \quad i = 1, \ldots, N$
177	Equation (7.26)	$\Phi(\beta) = M_g - \sum_{i=1}^{N} \frac{M_i \beta}{\beta - \alpha_i} = 0$	$\Phi(\beta) = M_g + \sum_{i=1}^{N} \frac{M_i \alpha_i}{\beta - \alpha_i} = 0$
177	Third paragraph	"These considerations reveal that Φ tends to M_g and M_e as β approaches 0 and ∞, respectively. In addition, Φ tends to $-\infty$ and $+\infty$ as β approaches α_i from above ..."	These considerations reveal that Φ tends to M_e and M_g as β approaches 0 and ∞, respectively. In addition, Φ tends to $+\infty$ and $-\infty$ as β approaches α_i from above ...
177	Equation (7.27)	$\alpha_i < \beta_i < \alpha_{i+1}; for\ i = 1, N-1; and\ \beta_N > \alpha_N$	$\beta_1 < \alpha_1; \alpha_i < \beta_{i+1} < \alpha_{i+1}; for\ i = 1, N-1$
177	Paragraph after Eq. (7.27)	"...are shorter than the corresponding relaxation times"	"...are longer than the corresponding relaxation times"

(continued)

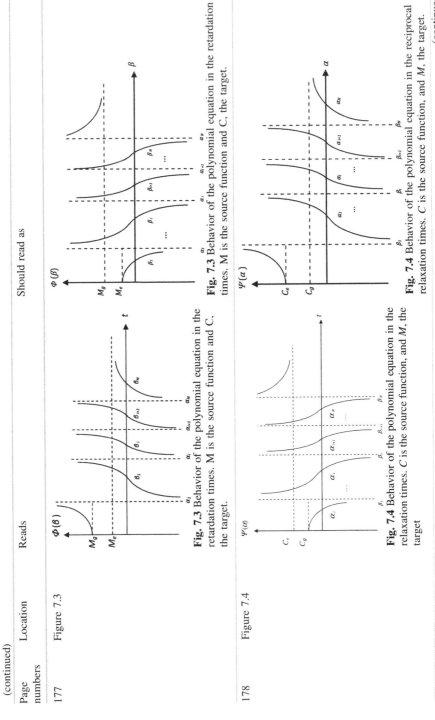

Fig. 7.3 Behavior of the polynomial equation in the retardation times. M is the source function and C, the target.

Fig. 7.3 Behavior of the polynomial equation in the retardation times. M is the source function and C, the target.

Fig. 7.4 Behavior of the polynomial equation in the relaxation times. C is the source function, and M, the target

Fig. 7.4 Behavior of the polynomial equation in the reciprocal relaxation times. C is the source function, and M, the target.

(continued)

(continued)

Page numbers	Location	Reads	Should read as
178	Second paragraph	$\Psi(\alpha=0)=C_g$ and $\Psi(\alpha=\infty)=C_e$, respectively. Also, Ψ tends to $+\infty$ and $-\infty$ as α approaches…"	$\Psi(\alpha=0)=C_e$ and $\Psi(\alpha=\infty)=C_g$, respectively. Also, Ψ tends to $-\infty$ and $+\infty$ as α approaches…"
180	Paragraph after bullet "c."	"…to scheme…"	"…to the scheme…"
180	Bullet point "4."	"…relative to curve…"	"…relative to the curve…"
181	Paragraph before Eq. (7.32)	"…of temperature shift values…"	"…of temperature and shift values…"
181	Table 7.3 heading	Column headings "i" and "t (sec)" should appear at same level as column heading $\log(t/a_{TK})$
183	First line	(a)	7.5.2.1
184	Middle of page	(b)	7.5.2.2
188	First bullet (•), second line	"…to assign correct sign …"	"…to assign the correct sign…"
188	Second bullet (•)	"…and And…"	"…and…"
190	First paragraph	"…annotations are only permissible…"	"…annotations are permissible…"
190	Last paragraph	"…(1 = modulus, 0 = compliance)"	"…(1 = modulus, −1 = compliance)"
Chapter 8			
194	Last paragraph	"…tensors of the same order can be multiplied…"	"…tensors can be multiplied…"
195	Second paragraph (expression split at index)	"$..A_{ij}\delta_{jk}=A_{i1}\delta_{1k}+A_{i2}\delta_2,$, $k+A_{i3}\delta_{3k}\cdots$	"$..A_{ij}\delta_{jk}=A_{i1}\delta_{1k}+A_{i2}\delta_{2k}+A_{i3}\delta_{3k}$

(continued)

(continued)

Page numbers	Location	Reads	Should read as
195	Last paragraph	"…parallel to the axis…"	"…parallel to the axes…"
198	Paragraph after Eg. (8.6)	$\{\varepsilon_{ij}\}^T \equiv \{\varepsilon\}^T \equiv \{\varepsilon_{11},\ \varepsilon_{22},\ \varepsilon_{33},\ \varepsilon_{12},\ \varepsilon_{13}\}$	$\{\varepsilon_{ij}\}^T \equiv \{\varepsilon\}^T \equiv \{\varepsilon_{11},\ \varepsilon_{22},\ \varepsilon_{33},\ \varepsilon_{12},\ \varepsilon_{13},\ \varepsilon_{23}\}$
199	Section 8.4, 1st paragraph	"…linear anisotropic…"	"…linear orthotropic…"
201	Last paragraph	"…each Poisson's ratio functions…"	"…each Poisson's ratio function…"
203	Example 8.1, Solution	"…of the complex stress…"	"…of the response to the complex stress…"
204	Equation (8.20)	$M_{ijkl}(t) = \lambda(t) \cdot (\delta_{ij}\delta_{kl} + \delta_{il}\delta_{jk}) + \mu(t) \cdots$	$M_{ijkl}(t) = \lambda(t) \cdot \delta_{ij}\delta_{kl} + \mu(t) \cdots$
204	Equation (8.23)	$M_{ijkl}(t) = \lambda_M(t)\left(\delta_{ij}\delta_{kl} + \delta_{ik}\delta_{jl}\right) + \mu_M(t) \cdots$	$M_{ijkl}(t) = \lambda_M(t) \cdot \delta_{ij}\delta_{kl} + \mu_M(t) \cdots$
205	Equation (8.24)	$C_{ijkl}(t) = \lambda_C(t)\left(\delta_{ij}\delta_{kl} + \delta_{ik}\delta_{jl}\right) + \mu_C(t) \cdots$	$C_{ijkl}(t) = \lambda_C(t)\delta_{ij}\delta_{kl} + \mu_C(t) \cdots$
206	First paragraph	"…$(1 + s\bar{v})(1 - 2s\bar{v})$ $s\bar{\lambda} = s\bar{v}s\bar{E}$…"	$(1 + s\bar{v})(1 - 2s\bar{v})s\bar{\lambda} = s\bar{v}s\bar{E}$
206	Second paragraph	"As will become apparent later shortly …"	"As will become apparent shortly …"
206	Fourth paragraph	"…$A_{sij} \equiv (A_{kk}/3)\delta_{ij} = 1/3(A_{11} + A_{22} + A_{33})\delta_{ij}\cdots$"	$A_{sij} \equiv (A_{kk}/3)\delta_{ij} = 1/3(A_{11} + A_{22} + A_{33})\delta_{ij}$
206	Fourth paragraph	$A_{Di\ j} \equiv \cdots$	$A_{Dij} \equiv \cdots$
207	Equation (8.36b)	$\sigma_{Dij}(t) = 2\mu_M * d\varepsilon_{Dij}$	$\varepsilon_{Dij}(t) = 2\mu_C * d\sigma_{Dij}$
208	Third line of text	"Likewise take the convolutions of (8.33b) to write"	"Likewise, from (8.33b)"

(continued)

(continued)

Page numbers	Location	Reads	Should read as
209	Equation (8.40)	$\sigma_{ij}^*(j\omega) = \lambda_M^*(j\omega) \cdot \varepsilon_{kk}^*(j\omega) \cdot \delta_{ij} + 2\mu_M^*(j\omega) \cdot \varepsilon_{ij}^*(j\omega)$	$\bar{\sigma}_{kl}(j\omega) = (j\omega)\bar{\lambda}_M(j\omega) \cdot \bar{\varepsilon}_{mm}(j\omega) \cdot \delta_{kl} + 2(j\omega)\bar{\mu}_M(j\omega) \cdot \bar{\varepsilon}_{kl}(j\omega)$
211	Second paragraph	Using the notation: $j\omega \cdot \bar{\mu} \equiv \mu^*, \bar{p} \equiv p^*; \bar{\sigma}_{ij} \equiv \sigma_{ij}^*, \bar{\varepsilon}_{ij} \equiv \varepsilon_{ij}^*$, as is customary, (43-a) may expressed alternatively as	The steady-state response σ_{ij}^* and p^* to the complex strain ε_{ij}^* may obtained from (8.41a), proceeding as in Chap. 4, using the notation $j\omega \cdot \bar{\mu} \equiv \mu^*$ as:
214	Paragraph in first Hint	$\sigma_{D22} = -1/3\sigma_{11} + 2/3\sigma_{22}$ $\varepsilon_{kk} = (\sigma_{11} + \sigma_{22})/3$	$\sigma_{D22} = -1/3\sigma_{11} + 2/3\sigma_{22}$ $\varepsilon_{kk} = (\sigma_{11} + \sigma_{22})/3K$
214	Problem P.8.5, Hint	"The constitutive applicable constitutive…" "…In addition, $-p(t) = \sigma_o H(t)/3…$"	"The applicable constitutive…" "…In addition, $-p(t) = \sigma_o H(t)/3…$"
214	Problem P.8.6, Hint	$\sigma_{kk} = (1/3)\sigma_{11}, \varepsilon_{kk} = (\varepsilon_{11} + 2\,\varepsilon_{22})/3$	$\sigma_{kk} = \sigma_{11}, \varepsilon_{kk} = (\varepsilon_{11} + 2\,\varepsilon_{22})$
215	Problem P.8.8, Answer	$\varepsilon_{zz}(t) = -p_o\cdots;\ \tau \equiv \eta\left(\dfrac{4G + 3K}{3KG}\right)$	$\varepsilon_{zz}(t) = -p_o\cdots;$ $\tau \equiv \eta\left(\dfrac{4G + 3K}{3KG}\right)$
216	Figure 8.6, caption	"Fig. 8.6 Problem 8.10"	"Fig. 8.6 Problem 8.11"
Chapter 9			
219	Abstract, last sentence	"…elastic ones and as a consequence…"	"…elastic ones and is a consequence…"
222	Third paragraph	"…(as listed in [2])…"	"…(as listed in Appendix B)…"

(continued)

(continued)

Page numbers	Location	Reads	Should read as
222	Third paragraph last line	"...respectively, and listed here:"	"...respectively, are listed here:"
226	Paragraph after Eq. (9.14)	"...auxiliary displacement u_{Vi}..."	"...auxiliary displacements, u_{Vi}..."
228	Equation (9.21)	$\lambda_R = \dfrac{v}{(1+v)(1-2v)}; E_R = \dfrac{2v}{1-2v}\,G_R$	$\lambda_R = \dfrac{v}{(1+v)(1-2v)}\;E_R = \dfrac{2v}{1-2v}\,G_R$
231	Equation (9.30)	$\overline{\sigma}_{ij}(s) = s\overline{\lambda}_M(s)\overline{\varepsilon}_{kk}(\tau)\delta_{ij} + \dots)$	$\overline{\sigma}_{ij}(s) = s\overline{\lambda}_M(s)\overline{\varepsilon}_{kk}(s)\delta_{ij} + \dots$
Chapter 10			
239	Abstract, third line	"...are introduce in..."	"...are introduced in..."
242	Paragraph after Eq. (10.8)	"...other being $-\lambda$..."	"...other being $-\lambda$..."
243	Equation (10.11)	$\lambda = \pm\omega_o\sqrt{\rho\dfrac{P|_{j\omega_o}}{Q|_{j\omega_o}}} == \pm\omega_o\sqrt{\dfrac{\rho}{E'(j\omega_o)}}$	$\lambda = \pm\omega_o\sqrt{\rho\dfrac{P|_{j\omega_o}}{Q|_{j\omega_o}}} = \pm\omega_o\sqrt{\dfrac{\rho}{E'(j\omega_o)}}$
244	Last paragraph	Footnote 2 is missing	(2) A parametric treatment of a problem of this type, for a three-parameter solid, may be found in W. Flügge, Viscoelasticity, pp. 135–140, Springer-Verlag, 1975.
245	Second paragraph	"...real part of $U \cdot ee^{j(\omega_o t)}$, the..."	"...real part of $U \cdot e^{j(\omega_o t)}$, the..."
245	Third paragraph	"...back into 10.9)..."	"...back into (10.9)..."
245	Last paragraph	"...decrease from the left to right..."	"...decrease from left to right..."
252	Equation (f)	$\int_o^t E(t-\tau)\dfrac{\partial}{\partial s}\left(\dfrac{\partial^2}{\partial x^2}u\right)d\tau = \dots$	$\int_o^t E(t-\tau)\dfrac{\partial}{\partial \tau}\left(\dfrac{\partial^2 u}{\partial x^2}\right)d\tau = \dots$

(continued)

(continued)

Page numbers	Location	Reads	Should read as
253	Example 10.5, Solution	(10.41b)	(10.40b)
253	Example 10.6, Solution	$\bar{\sigma} = -\rho Vc\sqrt{s(s+1/\tau)}\,e^{-\sqrt{s(s+1/\tau)(\frac{x}{c})}}$	$\bar{\sigma} = -\rho Vce^{-\sqrt{s(s+1/\tau)(\frac{x}{c})}}/\sqrt{s(s+1/\tau)}$
256	Second paragraph	The term $[\partial\sigma/\partial x]$ in expression $[\partial\sigma/\partial x] = p[\partial v/\partial x]$ should not be split	
256	Paragraph after Eq. (10.46a)	"...using (10.8)..."	"...using (10.26)..."
256	Paragraph after Eq. (10.46a)	The term D/Dt in expression D/Dt = cD/Dx should not be split	
256	Equation (10.46b)	"$[v] = Ae^{-}$..."	"$[v] = Be^{-}$..."
257	P.10.1	"...Example. 10.6)..."	"...Example (10.8)..."
257	P.10.2, Statement	"145 psi and mass density 0.065 lbm/in³"	"1 MPa and mass density 1600 kg/m³"
257	P.10.2, Hint	"...$\rho = 1600$ kg..."	"...$\rho = 1600$ kg/m³..."
258	P.10.4, equation in Hint	$\Gamma(0) \equiv \frac{1}{E_g}\frac{d}{ds}E(t-s)\big|_{t=0} = -\frac{1}{E_g}qE(q)\big|_{q=0} = -\frac{E'(0)}{E_g}$	$\Gamma(0) \equiv \frac{1}{E_g}\frac{d}{ds}E(t-s)\big|_{t=s} = -\frac{1}{E_g}\frac{d}{dt}E(t)\big|_{t=0}$
258	P.10.5, Hint, second line	"...(3.23a)..."	"...(10.41b)..."
258	P.10.5, Hint, second line	"...A(S) – A = ..."	"A(s) = ..."
259	P.10.7, Statement	"of Example 10.5..."	"of Example 10.8..."

(continued)

(continued)

Page numbers	Location	Reads	Should read as
259	Problem P.10.7, Hint	$\frac{d}{dt}E(t)|_{t=0} = \frac{E_0^2}{\eta_1}$	$\frac{d}{dt}E(t)|_{t=0} = -\frac{E_0^2}{\eta_1}$
Chapter 11			
266	Paragraph after Eq. (11.9c)	"Multiplying (11.4)…"	"Multiplying by (11.4a)…"
268	Second paragraph	"…of the d_r"	"…of the d_r…"
268	Section 11.3.2, First paragraph	… fields u". A displacement field, u" …	… fields u''. A displacement field, u'' …
270	Second paragraph	"…introduced in 11.5…"	"…introduced in 11.15…"
270	Line after Equation (c)	"…and using $\int_{S_T} \delta T_i'' u_i dS \equiv \int_{S_u} n_j \delta\sigma_{ij}'' u_i dS$	"…also subtracting $\int_{S_T} \delta T_i'' u_i dS = 0$, and using $\int_{S_T+S_u} \delta T_i'' u_i dS \equiv \int_{S_T+Su} n_j \delta\sigma_{ij}'' u_i dS$
270	Equation (c)	$\delta\psi_o(\sigma_{ij}'') = \int_V \delta Y_o(\sigma_{ij}'')\,dV$ $-\int_{Su}\delta T_i''(u_i^o - u_i'')\,dS + \int_{S_T+S_u} n_j\delta\sigma_{ij}''u_i S = 0$	$\delta\psi_o(\sigma_{ij}'') = \int_V \delta Y_o(\sigma_{ij}'')\,dV$ $-\int_{Su}\delta T_i''(u_i^o - u_i)\,dS - \int_{S_T+S_u} n_j\delta\sigma_{ij}''u_i\,dS = 0$
271	Equation (11.22)	$\varepsilon_{ij}''(t) = \frac{1}{2}\left(u_{i,j}'' + u_{j,i}''\right);\ in\ V$	$\varepsilon_{ij}''(t) = \frac{1}{2}\left(u_{i,j} + u_{j,i}\right);\ in\ V$

(continued)

(continued)

Page numbers	Location	Reads	Should read as
271	Last paragraph	"…express 11.23 as:"	"…express 11.15 as:"
272	Example 11.2, Statement	"…its cross section have second…"	"…its cross section has second…"
272	Example 11.2, Solution	"…and Eq. 11.21: $\gamma_o = \sigma\varepsilon - \gamma_{go}$"	"…and Eq. 11.21: $Y_o = \sigma\varepsilon - Y_{go}$…"
273	First line	"…leads to $\Delta(t) = \int_V V_o \, dv \equiv \cdots$"	"…leads to $\Delta(t) = \int_V Y_o \, dV \equiv \cdots$"
275	First line	"…only when bending effects are included…"	"…when only bending effects are included…"
275	Example 11.3, Solution	"…by establish equilibrium…"	"…by establishing equilibrium…"
276	Figure 11.4, inset	"$\omega L/2$"	"$qL/2$"
276	Next-to-last paragraph	"…the first part on far right is…"	"…the first part on the far right is…"
278	Example 11.4, Solution, last equation	$u|_P = \int_L \dfrac{N_{uz}\hat{C} * N_z}{A}\,dx + \int_L \dfrac{M_{ux}\hat{C} * M_x}{I_x}\,dx\tau$	$u|_P = \int_L \dfrac{N_{uz}\hat{C} * N_z}{A}\,dx + \int_L \dfrac{M_{ux}\hat{C} * M_x}{I_x}\,dx$
279	Equation (11.34b)	$\int_{S_T}\int_0^t T_{0i}^1(t-\tau)\dfrac{\partial}{\partial\tau}u_i^2(\tau)d\tau dS \cdots = {}^2|_1$	$\int_{S_T} T_i^1(t-\tau)\dfrac{\partial}{\partial\tau}u_i^2(\tau)d\tau dS \cdots = {}^2|_1$
280	Paragraph after Eq. (d)	$\sigma_{IJ}^1 = M_{ijkl} * d\varepsilon_{kl}^1$	$\sigma_{ij}^1 = M_{ijkl} * d\varepsilon_{kl}^1$

(continued)

(continued)

Page numbers	Location	Reads	Should read as
281	Equation (11.35b)	$\int_{S_T}\int_0^t T_{0i}^1(t-\tau)u_i^2(\tau)d\tau dS + \cdots = \ ^2|_1$	$\int_{S_T}\int_0^t T_i^1(t-\tau)u_i^2(\tau)d\tau dS + \cdots = \ ^2|_1$
281	Paragraph after Eq. (11.35b)	"...system $\sigma_{ij,j}^1 + F_i^1 = \rho\ddot{u}_i^1 \ldots$"	"...system $\sigma_{ij,j}^1 + F_i^1 = \rho\ddot{u}_i^1 \ldots$"

Appendix A
Mathematical Background

A.1 Average Value of a Function

Simply put, an integrable function is one whose integral exists. The average value of an integrable function, f, over a finite interval $[a,b]$ is defined as the quotient of the value of the integral divided by the amplitude of the integration interval.[1]

$$A(f) = \frac{1}{(b-a)} \int_a^b f(s)ds \qquad (A.1)$$

Stated another way:

$$\int_a^b f(s)ds = (b-a)A(f) \qquad (A.2)$$

In other words, the average value $A(f)$ of the function f may be thought of as the height of a rectangle with base equal to the length of the interval over which the average is taken. This is indicated schematically in Fig. A.1.

When f is continuous, its average value is equal to the value of f at some point in the interval $[a,b]$. This is the mean value theorem for integrals, proven next.

A.2 Mean Value Theorem for Integrals

If f is continuous on $[a,b]$, then for some c ε $[a,b]$:

$$\int_a^b f(s)ds = (b-a) \cdot f(c) \qquad (A.3-a)$$

---

[1] Reference material for sections A.1 to A.5 may be found in T.M. Apostol, Calculus, 2nd Edition, Xerox College Publishing, (1967), pp. 154, 184–186.

D. Gutierrez-Lemini, *Engineering Viscoelasticity*, DOI: 10.1007/978-1-4614-8139-3, © Springer Science+Business Media New York 2014

Fig. A.1 Geometric interpretation of mean value theorem

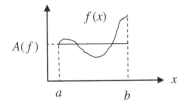

Proof Let m and M denote, respectively, the minimum and maximum values of $f(x)$ on $[a,b]$. Then, $m \leq f(x) \leq M$. Integrating this system of inequalities in the given interval, dividing through by $(b-a)$, and using the average value of a function, there results

$$m \leq \frac{1}{(b-a)} \int_a^b f(s)ds = A(f) \leq M \qquad \text{(a)}$$

In addition, the intermediate value theorem for continuous functions tells us that f takes on every value between $f(a)$ and $f(b)$, somewhere in the interval $[a,b]$. Thus, $A(f) = f(c)$, for some $c \; \varepsilon \; [a,b]$; completing the proof. The theorem is alternatively expressed as

$$\int_a^b f(s)ds = (b-a) \cdot f[a + \lambda(b-a)], \quad 0 < \lambda < 1 \qquad \text{(A.3–b)}$$

A.3 Weighted Mean Value Theorem for Integrals

Let f and g be continuous on $[a,b]$. If g never changes sign on $[a,b]$, then for some c in $[a,b]$:

$$\int_a^b f(s)g(s)ds = f(c) \int_a^b g(s)ds \qquad \text{(A.4)}$$

Proof The proof proceeds as that of the mean value theorem for integrals. Since g never changes sign in $[a,b]$, it is always nonnegative or always non-positive. Assuming g is nonnegative, we multiply the system of inequalities $m \leq f(x) \leq M$ by $g(x)$, to yield $m\,g(x) \leq f(x)g(x) \leq M\,g(x)$, and integrate it between the limits of the given interval, so that

$$m \int_a^b g(s)ds \le \int_a^b f(s)g(s)ds \le M \int_a^b g(s)ds \qquad (a)$$

If the integral of g is zero, the theorem is trivially satisfied; since in that event, the integral of $f \cdot g$ is also zero, and both members of the theorem are zero, for any c in $[a,b]$. Otherwise, the integral of g would be positive. This would allow division of the above system of inequalities by the integral of g and apply the intermediate value theorem—as we did for the mean value theorem—to complete the proof.

This theorem is useful to get an estimate of the integral of a product of two functions, especially when one of the functions is easy to integrate.

Example A.1 Prove the inequality: $\frac{1}{10\sqrt{2}} \le \int_0^1 \frac{x^9}{\sqrt{1+x}} dx \le \frac{1}{10}$

Solution

First note that both, x^9 and $1/\sqrt{1+x}$ are continuous and nonnegative in the interval of integration. Then, apply the weighted mean value theorem for integrals, to write $\int_0^1 \frac{x^9}{\sqrt{1+x}} dx = \frac{1}{\sqrt{1+\xi}} \int_0^1 x^9 dx \equiv \frac{1}{10\sqrt{1+\xi}}$; $\xi \in [0, 1]$. Now, since $\frac{1}{\sqrt{2}} \le \frac{1}{\sqrt{1+\xi}} \le 1$, for all ξ in $[0,1]$, it follows that $\frac{1}{10\sqrt{2}} \le \int_0^1 \frac{x^9}{\sqrt{1+x}} dx \le \frac{1}{10}$.

A.4 Rolle's Theorem

For any function, f, which is continuous in $[a,b]$, and differentiable in (a,b), and such that $f(a) = f(b)$, there exist at least one point $c\ \varepsilon\ (a,b)$ where its derivative vanishes $f'(c) = 0$.

Proof This theorem is proven assuming that $f'(x) \ne 0$ everywhere in (a,b), and noting that such assumption leads to a contradiction, which then implies that $f'(x) = 0$ for at least one x in (a,b). Indeed, because f is continuous in $[a,b]$, it acquires its minimum and maximum values, m, and M, respectively, somewhere in $[a,b]$. Also, since a necessary condition for the existence of an extreme value (maximum or minimum) of a function is that its derivative vanishes at the critical point, it follows that $f'(x) = 0$ for some x in $[a,b]$. However, by assumption, $f'(x) \ne 0$ everywhere in the interval (a,b). This leaves $x = a$ and $x = b$ as the only critical points of f in $[a,b]$. That is $f'(a) = 0$ and $f'(b) = 0$. This, and the property that $f(a) = f(b)$, would require that $m = M$; in which case, f would be constant in $[a,b]$, and $f'(x) = 0$, everywhere in $[a,b]$. This contradicts the hypothesis that $f'(x) \ne 0$ everywhere in (a,b), and hence, $f'(x) = 0$ somewhere in (a,b); completing the proof.

A.5 Mean Value Theorem for Derivatives

If f is continuous in $[a, b]$, and differentiable everywhere in (a,b), then there is a point c in (a,b) for which

$$f'(c) = \frac{f(b) - f(a)}{(b - a)} \tag{A.5}$$

In other words, as is depicted in Fig. A.2, there is—at least—a point in the interval (a, b), where the tangent to the curve has the same slope as the segment joining the end points of the interval.

Proof The proof is based on Rolle's theorem, which applies to the same type of function as the present theorem, but with the added requirement that $f(a) = f(b)$. To apply Rolle's theorem, we construct the function: $h(x) = f(x) \cdot (b - a) - x \cdot [f(b) - f(a)]$. By construction, $h(x)$ is continuous in $[a,b]$, differentiable in (a,b), and such that $h(a) = h(b)$. Under these conditions, Rolle's theorem applies to $h(x)$; thus, $h'(c) = 0$ for some c in (a,b). Hence, $h'(c) = 0 = f'(c) \cdot (b - a) - 1 \cdot (b - a)$; proving the theorem.

A.6 The Total Derivative

If $y = f(x)$ represents a real-valued function of a real variable, its derivative, $f'(x)$, is defined by

$$f'(x) = \lim_{h \to 0} \frac{1}{h} [f(x + h) - f(x)] \tag{A.6}$$

Now consider the following limit, denoted by $f'(x; z)$, or $\delta f(x; z)$ and called the total derivative or first variation of f with increment z^2:

$$\delta f(x; z) \equiv \lim_{h \to 0} \frac{1}{h} [f(x + hz) - f(x)] = \frac{d}{dh} f(x + hz)|_{h=0} \tag{A.7}$$

As will be shown subsequently, this limit is linear in z, so that

$$\delta f(x; z) \equiv f'(x; z) = f'(x)z \tag{A.8}$$

This establishes the relation between the derivative $f'(x)$ and its total derivative or first variation, $f'(x; z)$, or $\delta f(x; z)$.

The linearity of $\delta f(x; z)$ on z follows from its homogeneity and additivity. To prove that $\delta f(x; z)$ is homogeneous of degree one in z, consider the variation $\delta f(x; kz)$; and using its definition in (A.7), multiply and divide its right-hand side by k and invoke that the product of the limit is the limit of the product, to write

---

[2] D.C. Leigh, Nonlinear Continuum Mechanics, McGraw-Hill (1968), pp. 46–49.

Fig. A.2 Mean value theorem for derivatives

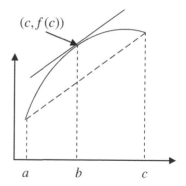

$$\delta f(x; kz) = \lim_{h \to 0} \frac{1}{h} [f(x + hkz) - f(x)] = \lim_{h \to 0} \frac{k}{kh} [f(x + hkz) - f(x)] \qquad \text{(a)}$$

Now set $l = kh$ and arrive at the following expression, which proves homogeneity:

$$\delta f(x; kz) = k \lim_{l \to 0} \frac{1}{l} [f(x + lz) - f(x)] = k \delta f(x; z) \qquad \text{(b)}$$

To prove additivity, evaluate $\delta f(x; z + w)$ using (A.7). Then, split the argument $x + h(z + w)$ into $(x + hz)$ and hw, and add and subtract the function $f(x + hz)$ to express the result as

$$\begin{aligned}
\delta f(x; z + w) &= \lim_{h \to 0} \frac{1}{h} [f\{x + h(z + w)\} - f\{x\}] \\
&= \lim_{h \to 0} \frac{1}{h} [f\{(x + hz) + hw\} - f(x + hz)] \\
&\quad + \lim_{h \to 0} \frac{1}{h} [f\{(x + hz)\} - f\{x\}] \\
&= \delta f(x; z) + \delta f(x; w)
\end{aligned} \qquad \text{(c)}$$

A.7 Differentiation Under the Integral Sign

Differentiation under the integral sign is carried out by means of Leibnitz's rule, that

If $u_1(\alpha)$ and $u_2(\alpha)$ are differentiable functions in the closed interval $[a,b]$, and $f(x,\alpha)$ and $\partial f(x,\alpha)/\partial \alpha$ are continuous in the region of the x-α plane delimited by $a \le \alpha \le b$, $u_1(\alpha) \le x \le u_2(\alpha)$, then

$$\frac{d}{d\alpha} \int_{u_1(\alpha)}^{u_2(\alpha)} f(x, \alpha) dx = \int_{u_1(\alpha)}^{u_2(\alpha)} \frac{\partial}{\partial \alpha} f(x, \alpha) dx$$

$$+ f(u_2(\alpha), \alpha) \frac{d}{d\alpha} u_2(\alpha) - f(u_1(\alpha), \alpha) \frac{d}{d\alpha} u_1(\alpha) \quad \text{(A.9)}$$

Proof Letting: $\Phi(\alpha) \equiv \int_{u_1(\alpha)}^{u_2(\alpha)} f(x, \alpha) dx$, we proceed to compute its derivative using the "four-step rule." We thus evaluate $\Phi(\alpha + \Delta\alpha)$, $\Delta\Phi$, $\Delta\Phi/\Delta\alpha$ and $\lim_{\Delta\alpha \to 0}(\Delta\Phi/\Delta\alpha)$:

$$\Delta\Phi(\alpha) \equiv \Phi(\alpha + \Delta\alpha) - \Phi(\alpha) = \int_{u_1(\alpha+\Delta\alpha)}^{u_2(\alpha+\Delta\alpha)} f(x, \alpha + \Delta\alpha) dx - \int_{u_1(\alpha)}^{u_2(\alpha)} f(x, \alpha) dx \quad \text{(a)}$$

We now split the limits of integration of the first integral on the right-hand side into three subintervals; the first, from $u_1(\alpha + \Delta\alpha)$ to $u_1(\alpha)$; the second, from $u_1(\alpha)$ to $u_2(\alpha)$; and the third, from $u_2(\alpha)$ to $u_2(\alpha + \Delta\alpha)$. We then combine the second integral on the right-hand side of the above expression with the split integral of the same limits, to arrive at

$$\Delta\Phi(\alpha) \equiv \int_{u_1(\alpha+\Delta\alpha)}^{u_1(\alpha)} f(x, \alpha + \Delta\alpha) dx + \int_{u_1(\alpha)}^{u_2(\alpha)} [f(x, \alpha + \Delta\alpha) - f(x, \alpha)] dx$$

$$+ \int_{u_2(\alpha)}^{u_2(\alpha+\Delta\alpha)} f(x, \alpha + \Delta\alpha) dx \quad \text{(b)}$$

Rewrite the first integral in this expression by switching its limits of integration–which picks up a negative sign:

$$\Phi_1(\alpha) \equiv \int_{u_1(\alpha+\Delta\alpha)}^{u_1(\alpha)} f(x, \alpha + \Delta\alpha) dx = - \int_{u_1(\alpha)}^{u_1(\alpha+\Delta\alpha)} f(x, \alpha + \Delta\alpha) dx \quad \text{(c)}$$

Apply the mean value theorem for integrals to this expression to get

$$\Phi_1(\alpha) = -f(\xi_1, \alpha + \Delta\alpha) \cdot [u_1(\alpha + \Delta\alpha) - u_1(\alpha)] \quad \text{(d)}$$

That is

$$\Phi_1(\alpha) = -f(\xi_1, \alpha + \Delta\alpha) \cdot \Delta u_1(\alpha); \ u_1(\alpha) < \xi_1 < u_1(\alpha + \Delta\alpha) \quad \text{(e)}$$

In similar fashion, invoking the mean value theorem for integrals, the third integral in the expression for Φ may be cast in the form:

$$\Phi_2(\alpha) \equiv \int_{u_2(\alpha)}^{u_2(\alpha+\Delta\alpha)} f(x, \alpha + \Delta\alpha) dx = f(\xi_2, \alpha + \Delta\alpha) \cdot [u_2(\alpha + \Delta\alpha) - u_2(\alpha)] \quad \text{(f)}$$

In other words,

$$\Phi_2(\alpha) = f(\xi_2, \alpha + \Delta\alpha) \cdot \Delta u_2(\alpha); \; u_2(\alpha) < \xi_2 < u_2(\alpha + \Delta\alpha) \tag{g}$$

Using the expressions for Φ_1 and Φ_2 to replace the corresponding integrals appearing in $\Delta\Phi$, leads to

$$\frac{\Delta\Phi(\alpha)}{\Delta\alpha} \equiv \int_{u_1(\alpha)}^{u_2(\alpha)} \frac{[f(x, \alpha + \Delta\alpha) - f(x, \alpha)]}{\Delta\alpha} dx - f(\xi_1, \alpha + \Delta\alpha) \cdot \frac{\Delta u_1(\alpha)}{\Delta\alpha}$$

$$+ f(\xi_2, \alpha + \Delta\alpha) \cdot \frac{\Delta u_2(\alpha)}{\Delta\alpha} \tag{h}$$

Taking then limit as $\alpha \to 0$, noting that $\xi_1 \to u_1(\alpha)$ and $\xi_2 \to u_2(\alpha)$ as $\alpha \to 0$, and using the definition of the partial derivative of a function, completes the proof.

Example A.2 Obtain the derivative with respect to t, of the function $q(t) = \int_0^t e^{-\frac{t-s}{\alpha}} p(s) ds$

Solution

Direct application of Leibnitz's rule yields

$$\frac{d}{dt} q(t) = \int_0^t \frac{\partial}{\partial t} e^{-\frac{t-s}{\alpha}} p(s) ds + e^{-\frac{t-t}{\alpha}} p(t) \frac{dt}{dt} - e^{-\frac{t-0}{\alpha}} p(0) \frac{d0}{dt}$$

That is

$$\frac{d}{dt} q(t) = -\frac{1}{\alpha} \int_0^t e^{-\frac{t-s}{\alpha}} p(s) ds + p(t) \equiv -\frac{1}{\alpha} q(t) + p(t)$$

Expressions of the form $\frac{d}{dt} q(t) = -\frac{1}{\alpha} \int_0^t e^{-\frac{t-s}{\alpha}} p(s) ds + p(t) \equiv -\frac{1}{\alpha} q(t) + p(t)$, occur naturally in viscoelasticity and are solved for the state variable q (stress or strain) in terms of the source function, p (strain or stress, respectively) using the procedure described in the next section.

A.8 Linear Differential Equation of First Order

The general linear ordinary differential equation of first order may be written as

$$\frac{d}{dx} y(x) + p(x) \cdot y(x) = q(x) \tag{A.10}$$

The solution to this equation is obtained by rewriting it in terms of differentials, and multiplying it by a continuous function, $u(x)$[3]:

$$u(x) \cdot dy(x) + u(x) \cdot p(x) \cdot y(x) \cdot dx = u(x) \cdot q(x) \cdot dx \qquad \text{(a)}$$

With this, the left-hand side of the equation becomes an exact differential of the product $u(x) \cdot y(x)$. That is, look for a (non-zero) function $u(x)$ such that

$$d[u(x) \cdot y(x)] = u(x) \cdot q(x) \cdot dx \qquad \text{(b)}$$

The solution, $y(x)$, of the general linear differential equation posed, is obtained integrating this expression. This can be accomplished either by indefinite or by definite integration.

Using indefinite integration and the fact that the indefinite integral of the total differential of a function is the function itself, results in:

$$u(x) \cdot y(x) = \int_x u(s) \cdot q(s) \cdot ds + C \qquad \text{(A.11)}$$

The dummy variable of integration of the indefinite integral on the right-hand side was changed for clarity. The constant of integration, C, is established from the condition (x_o, y_o) at the end $x = x_o$ of the interval of integration.

Alternatively, if definite integration between the limits x_o and x is employed

$$u(x) \cdot y(x) - u(x_o) \cdot y(x_o) = \int_{x_o}^x u(s) \cdot q(s) \cdot ds \qquad \text{(A.12)}$$

The integrating function $u(x)$ may be established from the requirement that multiplying the original equation by it should turn its left-hand side into an exact differential. That is

$$d[u(x) \cdot y(x)] = u(x) \cdot dy(x) + u(x) \cdot p(x) \cdot y(x) \cdot dx \qquad \text{(c)}$$

Evaluating the left-hand side of this expression, canceling like terms, and regrouping, produces that $\frac{du(x)}{u(x)} = p(x) \cdot dx$; with integral $ln\{u(x)\} = \int p(x)dx$; which leads to the integrating factor:

$$u(x) = e^{\int p(x)dx} \qquad \text{(A.13)}$$

Example A.3 Obtain the general solution of the differential equation $\frac{d}{dt}q(t) + \frac{1}{\tau}q(t) = p(t)$, if the solution passes through the point (t_o, q_o). That is, if $q(t_o) = q_o$.

---

[3] Elementary Differential Equations, E.D. Rainville, P.E. Bedient, 4th Edition, Collier Macmillan, London (1969), pp. 34–40.

Solution

Expressions of this type occur naturally in viscoelasticity and are used to determine the "state variable," q, in terms of the source function, p. By the present method, an integrating factor for this equation is obtained from (A.13) as $u(t) = e^{t/\tau}$; and noting that t is the independent variable, its general solution, in accordance to (A.11) or (A.12), is

$$q(t) = e^{-\frac{t}{\alpha}} \cdot \int_{t_o}^{t} e^{\frac{s}{\alpha}} \cdot p(s)ds + e^{-\frac{(t-t_o)}{\alpha}} \cdot q_o$$

Example A.4 The stress, σ, in viscoelastic system may be expressed in terms of the applied strain, $\varepsilon(t)$, in the form: $\frac{d}{dt}\sigma + a \cdot \sigma = \frac{d}{dt}\varepsilon$; $a \in \mathcal{R}$. Determine $\sigma(t)$ if the initial stress is $\sigma(0) = \sigma_o$ and the applied strain is $\varepsilon(t) = R \cdot t$.

Solution

The given stress–strain relation is a first-order ordinary linear differential equation in $\sigma(t)$. An integrating factor for it may be found from (A.13), as
$$u(t) = e^{\int a dt} \equiv e^{at}$$

This, and (A.11) or (A.12), produce that $\sigma(t) = e^{-at}[\sigma(0) + \int(e^{as} \cdot Rs)ds]$. Integrating this expression by parts, simplifying, and using the initial condition on σ, produces the result sought: $\sigma(t) = e^{-at}\sigma(0) + \frac{R}{a}[t - \frac{1}{a}(1 - e^{-at})]$.

A.9 Differential Operators

Differentiation may be considered as an abstract operation, which when applied to a given function returns another function. With this idea in mind, the first derivative, $\frac{df}{dt}$, of a function, f, with respect to some variable, t, may be regarded as the result of applying the derivative operator of first order, $\frac{d}{dt}$, to the function f. The derivative operator of first order is represented by any one of several symbols, as $\frac{d}{dt}$, ∂_t, D, D_t. Also, note that this operator is linear, because the operation of differentiation is homogeneous and additive; that is

By homogeneity is meant that for any function λ which does not depend on t:

$$\frac{d}{dt}(\lambda f) = \lambda \frac{d}{dt}(f) \tag{a}$$

• By additivity is meant that, for any two suitable functions, f_1 and f_2:

$$\frac{d}{dt}(f_1 + f_2) = \frac{d}{dt}(f_1) + \frac{d}{dt}(f_2) \tag{b}$$

Invoking the property of differentiation that the derivative of the first derivative of a function is the second derivative of the function; and so on, we can construct derivative operators of order higher than one. Thus, $\frac{d}{dt}\left(\frac{d}{dt}f\right) = \frac{d^2}{dt^2}(f), \ldots,$ $\frac{d}{dt}\left(\frac{d^{n-1}}{dt^{n-1}}f\right) = \frac{d^n}{dt^n}(f).$

We denote the derivative operator of order n by $\frac{d^n}{dt^n}$, ∂_t^n, D^n, D_t^n; and, for practical purposes, generalize it to include the derivative operator of order zero, to signify no derivative is taken: $\left(\frac{d^0}{dt^0}f\right) = 1 \cdot f \equiv f.$

The foregoing serves to define the general linear differential operator of order n with constant coefficients, or with coefficients which do not depend on the differentiation variable—as a sum of derivative operators of orders up to and including order n:

$$P \equiv p_0 \frac{d^0}{dt^0} + p_1 \frac{d^1}{dt^1} + p_2 \frac{d^2}{dt^2} + \cdots + p_n \frac{d^n}{dt^n} \equiv \sum_{i=0}^{n} p_i \frac{d^i}{dt^i} \tag{A.14}$$

where the p_i's are any suitably continuous functions which do not depend on t; and, as required by the order of the operator, $p_n \neq 0$.

By definition, linear differential operators can be added together (by additivity), or factored by constants or functions that do not involve the variable of differentiation (by homogeneity). Specifically, given any two linear operators, P, and Q, a suitably smooth function, f, of the differentiation variable, and any constant, λ, the following relation holds:

$$(P + Q)(\lambda f) = P(\lambda f) + Q(\lambda f) = \lambda P(f) + \lambda Q(f) \tag{c}$$

In addition, linear differential operators are both commutative and associative. These properties result from the fact that "the mth derivative of the nth derivative is equal to the nth derivative of the mth derivative." Because of their linearity, these operators are distributive as well. Thus, for linear operators P, Q and R, and function f_1 and f_2:

- Commutative property

$$P(Q(f_1)) = (PQ(f_1)) = (QP(f_1)) = Q(P(f_1)) \tag{d}$$

- Associative property

$$(PQR)(f_1) = (PQ)(Rf_1) = (P)(QRf_1) \tag{e}$$

- Distributive property

$$(PQ)(f_1 + f_2) = (PQ)(f_1) + (PQ)(f_2) \tag{f}$$

A.10 Linear Differential Equations of Higher Order

A differential equation is one involving an unknown function and some of its derivatives. If the derivatives appearing in the equation are ordinary derivatives, the equation is called ordinary; and if the derivatives are partial derivatives, the equation is a partial differential equation. The order of the highest-ordered derivative in the equation represents the order of the differential equation. Also, a differential equation is said to be linear if it is linear in the unknown function and its derivatives.

Using differential operator notation, an ordinary linear differential equation of order n may be expressed in the following symbolic form:

$$(f)y(x) = R(x) \tag{A.15}$$

In which, as in the previous section, the operator (f) is defined by

$$f \equiv a_0 D^0 + a_1 D^1 + a_2 D^2 + \cdots + a_n D^n \tag{A.16}$$

Clearly, for the equation to be linear, none of its coefficients can depend on the unknown, function y.

Only linear equations with constant coefficients are considered here. Also, if the function $R(x)$ is identically zero in the interval of interest, the differential equation is called homogeneous; otherwise, the equation is non-homogeneous.

A.10.1 Homogeneous Differential Equations

By definition, a homogeneous ordinary linear differential equation has the simple form:

$$(f)y(x) = 0 \tag{A.17}$$

Solutions to this type of differential equation are found by inspection. Indeed, noting that $D^k(e^{mx}) = m^k e^{mx}$, we try the function $y(x) = Ae^{mx}$ as a solution to (A. 17) and arrive at

$$(f)(Ae^{mx}) = \left(a_0 + a_1 m + a_2 m^2 + \cdots + a_n m\right) \cdot Ae^{mx} \equiv f(m) \cdot Ae^{mx} = 0 \quad \text{(a)}$$

After canceling out the factor Ae^{mx}, this produces the auxiliary equation:

$$f(m) \equiv \left(a_0 + a_1 m + a_2 m^2 + \cdots + a_n m^n\right) = 0 \quad \text{(b)}$$

The auxiliary equation (b) is a polynomial equation of degree n in the exponent m of the trial solution and has, in general, n roots, m_i. Once these roots are obtained, the solution of the homogeneous differential equation should be given, by its linearity, as the sum of the n independent solutions, $A_i e^{m_i x}$. Two cases are distinguished: non-repeated and repeated roots, as follows.

Non-repeated roots
In this case, the complete solution to (A.17) is given by the sum:

$$y(x) = \sum_{i=1}^{n} A_i e^{m_i x} \tag{A.18}$$

The coefficients A_i are established from the boundary (or initial) values of the function $y(x)$ and its derivatives up to order $n - 1$.

Repeated roots
When the auxiliary equation $f(m) = 0$ has repeated roots, the operator (f) has repeated factors. Assume, for definiteness that the auxiliary equation has p equal roots $m_i = b$. Then, the operator (f) must have a factor $(D - b)^p$, and Eq. (A.17) can be cast as

$$(f)y(x) = (g)(D - b)^p y(x) = 0 \tag{c}$$

where the operator (g) contains all the factors other than $(D - b)^p$. Clearly, any solution of

$$(D - b)^p y(x) = 0 \tag{d}$$

is also a solution of Eq. (c), and therefore of the original Eq. (A.17). In summary, we need to find p independent solutions of Eq. (d). This is easily done, noting that the functions $y_k = x^k e^{bx}$, $k = 0, 1,..., p - 1$, are linearly independent because the x_k's are linearly independent, and consequently, so are the proposed functions, y_k. Using this, we construct the general solution of (A.17), for the case of p repeated roots $m_k = b$, of its auxiliary equation, as

$$y(x) = \sum_{i=1}^{n-p} A_i e^{m_i x} + \sum_{k=0}^{p-1} B_k x^k e^{bx} \tag{A.19}$$

The same ideas and procedure are used when the auxiliary equations has multiple sets of repeated roots.

Example A.5 Find the general solution of the homogeneous equation $\frac{d^2 y}{dx^2} - 2\frac{dy}{dx} + 4 = 0$.

Solution
    The operator form of the given equation is $(D^2 - 2D + 1)y(x) = 0$. This leads to the auxiliary equation $m^2 - 2m + 1 = 0$, with roots $m_{1,2} = 1, 1$, for which the solution is $y(x) = Ae^x + Bxe^x$.

A.10.2 Non-Homogeneous Differential Equations

The general solution of the non-homogeneous ordinary linear differential equation in (A.15) may be expressed as the sum:

$$y(x) = y_h(x) + y_p(x) \tag{A.20}$$

of the complementary solution, y_h, to homogeneous Eq. (A.17), and some particular solution, y_p, to the original Eq. (A.15).

The solution of the homogeneous equation is obtained as described before, from the roots, m_i, of the auxiliary equation $f(m) = 0$. In what follows, a particular solution is sought using the method of undetermined coefficients. This method is applicable only to cases where the function $R(x)$ is itself a solution of some linear homogeneous differential equation, say: $(g)y_p(x) = R(x)$. Indeed, in such a case, the roots, m'_k, of the auxiliary homogeneous equation $g(m') = 0$ are obtained by inspection, from $R(x)$. In addition, the differential equation:

$$(g)(f)y(x) = 0 \tag{e}$$

has an auxiliary equation whose roots are the roots, m_i, of the auxiliary equation $f(m) = 0$, together with the roots m'_k of the auxiliary equation $g(m') = 0$. This means that the general solution of (e) contains the particular solution y_p, referred to in (A.20) and so, is of the form: $y(x) = y_h(x) + y_r(x)$. Then, $(f)y_r(x) = R(x)$, because $(f)y_h(x) = 0$. Removing y_h from the general solution of (e) leaves the function y_r, which for some specific numerical values of its coefficients, must satisfy the original non-homogeneous equation (A.15). The required coefficients of y_r are established from the requirement that $y_r = y_p$.

Example A.6 Find the general solution of the equation $\frac{d^2y}{dx^2} - 2\frac{dy}{dx} + 1 = 2e^{2x}$.

Solution

The roots of the complementary homogeneous equation were obtained in Example A.5, as $m_{1,2} = 1, 1$. The function e^{2x} on the right-hand side of the given equation corresponds to an auxiliary equation with root: $m'_1 = 2$. Therefore, the roots of the general solution $y = y_h + y_p$, have to be: $m_{1,2,3} = 1, 1, 2$; which requires that $y = A_1e^x + A_2xe^x + Be^{2x}$. The constant B is established from the requirement that $(f)y(x) = 2e^{2x}$, which produces $B = 2$. Hence, the general solution of the equation posed is $= A_1e^x + A_2xe^x + 2e^{2x}$.

A.11 The Divergence Theorem

This theorem is an example of an integral transformation. It concerns a function $f(x,y,z)$ which, together with its first partial derivatives, $\partial f/\partial x$, $\partial f/\partial y$, $\partial f/\partial z$, is single-valued and continuous inside a simply connected volume V—that is, one without holes— bounded by a close surface, S. Then, using the notation: $x_1 \leftrightarrow x$, $x_2 \leftrightarrow y$, $x_3 \leftrightarrow z$, there follows:

$$\int_S (n_1f_1 + n_2f_2 + n_3f_3)dS = \int_V \left(\frac{\partial}{\partial x_1}f_1 + f_2\frac{\partial}{\partial x_2} + f_3\frac{\partial}{\partial x_3}\right)dV \tag{A.21}$$

Fig. A.3 Domain of definition of function f, in Green-
Gauss theorem

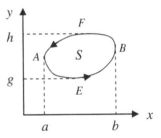

Here, n_i is the ith direction cosine or component of the outward unit normal to the surface S that encloses the volume, V, and always points away from the surface. Clearly, the n_i are thus functions of position on S. The surface S is required to be piecewise smooth, and such that it unambiguously defines and inside and an outside. The theorem is based on the following relationship involving the function f:

$$\int_S n_i f dS = \int_V \frac{\partial}{\partial x_i} f dV; \quad i = 1, 2, 3 \tag{A.22}$$

Proof To prove the Green-Gauss theorem, consider a function $f(x, y)$ defined on a plane area S, and having continuous first partial derivative with respect to x, and let C be a closed curve surrounding the plane area S, having the property that any straight line parallel to a coordinate axis cuts C in at most two points, as shown in Fig. A.3.

In addition, let $y_b(x)$ and $y_t(x)$ represent, respectively, the equations of the bottom and top curves, AEB and AFB, respectively. From (A.22), choosing $i = 2$, for definiteness, one can write that $\int_z dz \int_A \frac{\partial}{\partial y} f(x, y) dA = \int_z dz \int_C n_y(x, y) f(x, y) dS$; which, after canceling the common term $\int_z dz$, results in the expression to be proven:

$$\int_C n_y f dS = \int_A \frac{\partial}{\partial y} f(x, y) dA = \int_{x=a}^{b} \left[\int_{y=y_b(x)}^{y_t(x)} \frac{\partial}{\partial y} f(x, y) dy \right] dx$$

$$= \int_{x=a}^{b} f(x, y) \big|_{y=y_b(x)}^{y_t(x)} dx = \int_{x=a}^{b} [f(x, y_t) - f(x, y_b)] dx \tag{a}$$

$$= \int_{x=a}^{b} f(x, y_t) dx - \int_{x=a}^{b} f(x, y_b) dx = - \int_{x=a}^{b} f(x, y_b) dx - \int_{x=b}^{a} f(x, y_t) dx$$

$$= - \oint_C f(x, y) dx = - \oint_C f(x, y) \frac{dx}{ds} ds$$

As seen in Fig. A.4, the normal, $\hat{n}$, at any point on a smooth plane curve C, may be expressed as $\hat{n} = \hat{i} n_x - \hat{j} n_y = \hat{i} \sin\theta - \hat{j} \cos\theta$; whereby

Fig. A.4 Unit normal to a smooth plane curve

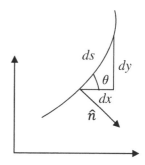

$$n_x = \frac{dy}{ds}; \ n_y = \frac{dx}{ds} \tag{b}$$

Using (b) on the right-hand side of (a) proves that

$$\int\limits_A n_y f dS = \int\limits_V \frac{\partial}{\partial y} f dV \tag{c}$$

The expressions for $i = 1$ and $i = 3$ in (A.22) can be proven similarly. Also, although for simplicity the proof was carried out assuming the closed curve C could be cut no more than at two points by any lines parallel to the coordinate axes, the theorem is valid for cases in which lines parallel to the axes meet the bounding curve, C, in more than two points. It is also easy to show that the Green-Gauss theorem applies just as well to multiply connected regions.[4] Because the quantity $\frac{\partial}{\partial x_1} f_1 + \frac{\partial}{\partial x_2} f_2 + f_3 \frac{\partial}{\partial x_3}$ under the volume integral (surface integral, in two dimensions) is the divergence of the function f, the theorem is also known as the divergence theorem of Gauss, or simply, divergence theorem.

A.12 The Unit Impulse Function

The unit impulse function, also known as the delta function and the Dirac delta function, was developed by the French mathematician, Jean-Baptiste-Joseph Fourier, but was first put to practical use by the engineer and theoretical physicist Paul Dirac, in the 1920s. There are several ways to define the unit impulse function, but it is always represented by the Greek letter δ and a real argument, such as x and t.

The delta function is usually defined as a special function whose value is infinite when its argument is zero and is identically zero everywhere else on the real axis; but such that its integral over the entire real axis equals 1. That is,

---

[4] M.R. Spiegel, Theory and Problems of Advanced Calculus, Schaum's Outline Series, McGraw-Hill (1963), pp. 202–205.

$$\delta(t) = \begin{cases} 0; & t \neq 0 \\ \infty; & t = 0 \end{cases} \tag{A.23}$$

$$\int\limits_{-\infty}^{+\infty} \delta(t)dt = \int\limits_{-\varepsilon}^{+\varepsilon} \delta(t)dt = 1, \quad \varepsilon > 0. \tag{A.24}$$

The delta function may also be defined as a symbolic or generalized function, in terms of its integral properties alone, as they affect a so-called testing function. A testing function is an arbitrary continuous function that vanishes identically outside some finite interval. For any such testing function, $\varphi(t)$, the δ-function is defined by the relation:

$$\int\limits_{-\infty}^{\infty} \delta(t)\varphi(t)dt = \varphi(t)|_{t=0} = \varphi(0) \tag{A.25}$$

Unlike the definition in (A.23) and (A.24), no value is assigned to the δ-function as a symbolic function. The integral itself has no meaning as an ordinary integral either. The integral and the δ-function are defined by the value $\varphi(0)$ assigned to the testing function (H.P. Hsu 1967). The following are some practical properties of the δ-function:

$$\int\limits_{-\infty}^{\infty} \delta(t - t_o)\varphi(t)dt = \varphi(t + t_o)|_{t=0} = \varphi(t_o) \tag{a}$$

$$\int\limits_{-\infty}^{\infty} \delta(c \cdot t)\varphi(t)dt = \frac{1}{|c|}\varphi(\frac{t}{|c|})|_{u=0} = \frac{1}{|c|}\varphi(0) \tag{b}$$

$$f(t)\delta(t) = f(0)\delta(t), \quad \text{where } f \text{ is continuous at } t = 0 \tag{c}$$

Proofs Properties (a) and (b) are proven through simple changes of variables. To prove (a) introduce the change of variable $u = t - t_o$ and use (A.25) to obtain the value of the testing function. The proof of (b) requires the change of variable $u = c \cdot t$, once for $c > 0$, and then again for $c < 0$. Use of (A.25), and the definition of absolute value, completes the proof. To prove (c), select a testing function $\varphi(t)$, and since $f(t)$ is continuous, write:

$$\int\limits_{-\infty}^{\infty} [f(t)\delta(t)]\varphi(t)dt = \int\limits_{-\infty}^{\infty} \delta(t)[f(t)\varphi(t)]dt$$

$$= f(t)\varphi(t)|_{t=0} = f(0)\varphi(0)$$

$$= f(0)\int\limits_{-\infty}^{\infty} \delta(t)\varphi(t)dt = \int\limits_{-\infty}^{\infty} [f(0)\delta(t)]\varphi(t)dt \tag{d}$$

Since the testing function $\varphi(t)$ is arbitrary, the integrals on the left- and right-hand sides are valid for all choices of $\varphi(t)$, which implies that the quantities under the integral signs be identical, proving (c).

A.13 The Unit Step Function

The unit step function, also known as the Heaviside step function, is typically denoted by H. It is defined to be identically equal to zero for negative values of its argument and identically equal to one when its argument is non-negative, but is undefined at the origin:

$$H(t) = \begin{cases} 0; & t < 0 \\ 1; & t > 0 \end{cases} \qquad (A.26)$$

The graphical representation of the unit step function is shown in Fig. A.5. From the figure, note that the derivative of the unit step function is everywhere zero, except at the origin.

As it turns out, the unit step function is a symbolic function which may be alternatively defined in terms of a testing function $\varphi(t)$ by means of the relation:

$$\int_{-\infty}^{\infty} H(t)\varphi(t)dt = \int_{0}^{\infty} \varphi(t)dt \qquad (A.27)$$

The unit step function and the δ-function are related through the generalized derivative, which is defined for any differentiable function $f(t)$ and testing function $\varphi(t)$, by means of the relation:

$$\int_{-\infty}^{\infty} f'(t)\varphi(t)dt = - \int_{-\infty}^{\infty} f(t)\varphi'(t)dt \qquad (A.28)$$

Indeed, setting $f(t) = H(t)$, this yields $\int_{-\infty}^{\infty} H'(t)\varphi(t)dt = - \int_{-\infty}^{\infty} H(t)\varphi'(t)dt$; which, according to (A.27), becomes: $\int_{-\infty}^{\infty} H'(t)\varphi(t)dt = - \int_{0}^{\infty} \varphi'(t)dt = -[\varphi(\infty) - \varphi(0)] \equiv \varphi(0)$

Fig. A.5 Geometric representation of the unit step function

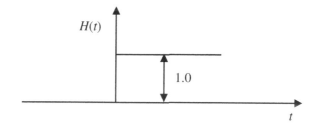

$H(t)$

1.0

t

Since: $\varphi(0) = \int_{-\infty}^{\infty} \delta(t)\varphi(t)dt$, in accordance with the definition of the δ-function given in (A. 25), it follows that

$$\int_{-\infty}^{\infty} H'(t)\varphi(t)dt = \int_{-\infty}^{\infty} \delta(t)\varphi(t)dt \tag{a}$$

Consequently, by the arbitrariness of the testing function, $\varphi(t)$:

$$H'(t) \equiv \frac{d}{dt}H(t) = \delta(t) \tag{A.29}$$

Note, however, that the derivative of the unit step function is zero both for negative and positive values of its argument.

A.14 Stieltjes Convolution

The Stieltjes convolution of two functions: $\varphi(t)$, which is continuous in $[0,\infty)$, and $\psi(t)$, which vanishes at $-\infty$, is defined as

$$\phi(t-\tau) * d\psi(\tau) \equiv \int_{-\infty}^{t} \phi(t-\tau)\frac{d}{d\tau}\psi(\tau)d\tau \equiv \phi * d\psi \tag{A.30a}$$

where the form on the far right is used when the argument is understood. Under the additional assumption that $\varphi(t) = 0$ for $t < 0$, one may split the interval of integration from $(-\infty, t)$ into $(-\infty, 0^-)\mathrm{U}(0^-, \mathrm{t})$ and write an alternate form for the convolution:

$$\phi(t-\tau) * d\psi(\tau) \equiv \int_{0^-}^{t} \phi(t-\tau)\frac{d}{d\tau}\psi(\tau)d\tau \tag{A.30b}$$

If, in addition $\psi(t) = 0$ for $t < 0$, then

$$\phi(t-\tau) * d\psi(\tau) \equiv \int_{0^-}^{t} \phi(t-\tau)\frac{d}{d\tau}\psi(\tau)d\tau = \phi(t)\psi(0^+) + \int_{0^+}^{t} \phi(t-\tau)\frac{d}{d\tau}\psi(\tau)d\tau \tag{A.31}$$

which is obtained by expressing the integration interval $(0^-, t)$ as $(0^-, 0^+)\mathrm{U}(0^+, t)$, and by integrating by parts the integral defined over $(0^+, t)$. In this case, as will be shown, the convolution integral is commutative, associative and distributive. Thus,

$$\varphi * d\psi = \psi * d\phi \qquad Commutivity$$

$$\varphi * d(\psi * d\theta) = (\phi * d\psi) * d\theta = \phi * d\psi * d\theta \qquad Associativity$$

$$\varphi * d(\psi + \theta) = \phi * d\psi + \phi * d\theta \qquad Distributivity$$

Proof of Commutivity Starting from the defining expression, integrating it by parts, and shifting the argument of the derivative operator, τ, to $(t - \tau)$; following this by a change of variable of integration from $(t - \tau)$ to s; and finally, changing the dummy variable s to τ again:

$$\phi * d\psi \equiv \phi(t)\psi(0^+) + [\phi(t - \tau)\psi(\tau)]_0^t - \int_{0^+}^t \psi(\tau)\frac{d}{d\tau}\phi(t - \tau)d\tau$$

$$= \psi(t)\phi(0^+) - \int_t^{0^+} \psi(\tau)\left[-\frac{d}{d(t - \tau)}\phi(t - \tau)\right][-d(t - \tau)] \tag{a}$$

$$= \psi(t)\phi(0^+) + \int_{0^+}^t \psi(t - \tau)\left[\frac{d}{d(\tau)}\phi(\tau)\right]d\tau$$

$$= \psi * d\phi$$

Proof of Associativity Setting $f = \psi * d\theta$ and $g = \varphi * d\psi$, the associative law may be written as

$$\varphi * d(\psi * d\theta) \equiv \phi * df = (\phi * d\psi) * d\theta \equiv g * d\theta \tag{b}$$

Now:

$$\phi * df = \int_{0^-}^\infty \phi(t - \tau)\frac{d}{d\tau}f(\tau)d\tau = \int_{0^-}^\infty \phi(t - \tau)\frac{d}{d\tau}\int_{0^-}^\infty \psi(\tau - s)\frac{d}{ds}\theta(s)dsd\tau \tag{c}$$

Differentiating the nested integral and effecting the change $\tau - s = u$ leads to

$$\varphi * df \equiv \int_{0^-}^\infty \phi(t - s - u)\int_{0^-}^\infty \frac{d}{du}\psi(u)\frac{d}{ds}\theta(s)dsdu$$

$$= \int_{0^-}^\infty \int_{0^-}^\infty \phi(t - s - u)\frac{d}{du}\psi(u)\frac{d}{ds}\theta(s)dsdu \tag{d}$$

In similar fashion:

$$g * d\theta \equiv \int_{0^-}^\infty g(t - s)\frac{d}{ds}\theta(s)ds = \int_{0^-}^\infty \int_{0^-}^\infty \phi(t - s - u)\frac{d}{du}\psi(u)\frac{d}{ds}\theta(s)dsdu \tag{e}$$

Hence, $\varphi * d(\psi * d\theta) = (\varphi * d\psi) * d\theta \equiv \varphi * d\psi * d\theta$, as was to be proven.

Proof of Distributivity The distributive property of the convolution integral emanates directly from the linearity of integration. Thus,

$$\varphi * d(\psi + \theta) = \int_{0^+}^{t} \phi(t - \tau) \frac{d}{d\tau}[\psi(\tau) + \theta(\tau)]d\tau$$

$$= \int_{0^+}^{t} \phi(t - \tau) \frac{d}{d\tau}\psi(\tau)d\tau + \int_{0^+}^{t} \phi(t - \tau) \frac{d}{d\tau}(\tau)d\tau \qquad \text{(f)}$$

$$= \varphi * d\psi + \varphi * d\theta$$

A.15 The Laplace Transform

The Laplace transform of a function $f(t)$, denoted by $L\{f(t)\}$ or $\bar{f}(s)$, is defined as

$$L\{f(t)\} \equiv \bar{f}(s) = \int_{0}^{\infty} e^{-st}f(t)dt; s = a + jb \qquad \text{(A.32)}$$

Provided

$$\int_{0}^{\infty} |f(t)|e^{-ct}dt < \infty; \quad \text{for some } c \in R, \ c > 0 \qquad \text{(A.33)}$$

In expression (A.32), s is a complex number, and for the Laplace transform of a piecewise continuous function $f(t)$ to exist, the function itself has to be of exponential order (that is, bounded by an exponential function of a real variable).

According to its definition, the Laplace transform converts a function from one space to another; in the present case, from the t-domain to the s-domain. The conversion from the transformed space back to the original space may be carried out by means of the inverse transformation, which is defined–from Fourier transform theory–as

$$L^{-1}\{f(t)\} \equiv f(t) = \frac{1}{2\pi i} \int_{\lambda-i\infty}^{\lambda+i\infty} e^{st}\bar{f}(s)ds \qquad \text{(A.34)}$$

Without going into any details, the evaluation of the line integral in (A.34) is carried out using analytic function theory. In practice, however, the definition given in (A.32) is used to construct a table of transforms and the table is then used in reverse to obtain the inverse transforms. This is possible because the Laplace transform of a function and the corresponding inverse transform are unique—except possibly for a so-called zero function. Either way, the important thing to bear in mind is that one can go from one space to the other and back, directly or indirectly.

As will be evident subsequently, the Laplace transform converts differential and integral operators into algebraic ones. The practical implication of this being that by applying the transform one can convert differential or integral equations into algebraic ones, which are easier to solve. Applying the inverse transform to the transformed solution will then lead to the solution in the original, physical space.

A.15.1 Properties of the Laplace Transform

A list of some properties of the Laplace transform, which are pertinent to the study of viscoelasticity, is provided next. For completeness of presentation, the proofs are given in each case.

Linearity

$$L\{a_1 f_1(t) + a_2 f_2(t)\} = a_1 L\{f_1(t)\} + a_2 L\{f_2(t)\} \equiv a_1 \bar{f}_1(s) + a_2 \bar{f}_2(s) \qquad (A.35)$$

The proof follows from (A.32) and the linearity of the integration operator:

$$L\{a_1 f_1(t) + a_2 f_2(t)\} = \int_0^t e^{-st}[a_1 f_1(t) + a_2 f_2(t)]dt$$

$$\equiv a_1 \int_0^t e^{-st} f_1(t)dt + a_2 \int_0^t e^{-st} f_2(t)dt \qquad (a)$$

Transform of Derivatives

$$L\{f^n(t)\} = s^n \bar{f}(s) - s^{n-1}f(0) + s^{n-2}f'(0) - \ldots - f^{(n-1)}(0) \qquad (A.36)$$

The proof is carried out by induction, starting with the transform of the first-order derivative, using integration by parts:

$$L\{f'(t)\} = \int_0^\infty e^{-st}f'(t)dt = -f(0) + s\int_0^t e^{-st}f(t)dt = s\bar{f}(s) - f(0) \qquad (b)$$

Having proven the formula valid for $k = 1$, we presume it valid for $k = n-1$, that is

$$L\{f^{n-1}(t)\} = s^{n-1}\bar{f}(s) - s^{n-2}f(0) - s^{n-3}f'(0) - s^{n-3}f''(s) - \ldots - f^{(n-2)}(0) \quad (c)$$

And proceed to prove that the formula is valid for $k = n$. To this end, take the Laplace transform of $f^{(n)}$, introducing the notation $g = f^{(n-1)}$, and using the transform of its first derivative: $L\{f^n(t)\} = L\{\frac{d}{dt}f^{n-1}(t)\} \equiv L\{g'(t)\} = s\bar{g}(s) - g(0)$. The proof is completed inserting the transform of $f^{(n-1)}$, and $g(0) = f^{(n-1)}(0)$.

Initial Value Theorem

$$\lim_{t \to 0} f(t) = \lim_{s \to \infty} s\bar{f}(s) \tag{A.37}$$

This theorem is proven by taking the limit as $s \to \infty$ of the Laplace transform of the first derivative of f, using the integral form of it and its value, $s\bar{f}(s) - f(0)$.

$$\lim_{s \to \infty} \int_0^\infty e^{-st} f'(t) dt = \lim_{s \to \infty} [s\bar{f}(s) - f(0)] \tag{d}$$

From this expression follows that $0 = \lim_{s \to \infty} [s\bar{f}(s) - f(0)]$; which completes the proof, because the continuity of f assures that $f(0) = \lim_{t \to 0} f(t)$.

Final Value theorem

$$\lim_{t \to \infty} f(t) = \lim_{s \to 0} s\bar{f}(s) \tag{A.38}$$

This statement is proven by taking the limit as $s \to 0$ of the transform of the first derivative, in manner similar to the initial value theorem.

$$\lim_{s \to 0} L\{f'(t)\} = \lim_{s \to 0} [s\bar{f}(s) - f(0)] = \lim_{s \to 0} \int_0^\infty e^{-st} f'(t) dt = \int_0^\infty f'(t) dt = f(\infty) - f(0)$$

$$\tag{e}$$

Implying the theorem, since, from the continuity of f: $f(\infty) = \lim_{t \to \infty} f(t)$.

Transform of Integrals

$$L\left\{ \int_0^t f(t) dt \right\} = \frac{1}{s} \bar{f}(s) \tag{A.39}$$

To prove this, we let $g(t) = \int_0^t f(t) dt$. In this manner, $g'(t) = f(t)$, $g(0) = 0$, and $L\{f(t)\} = L\{g'(t)\} = s\bar{g}(s) - g(0) \equiv sL\{\int_0^t f(t) dt\}.$, as asserted by (A.39). Likewise, mathematical induction would prove that for integrals iterated n times:

$$L\left\{ \int_0^t \int_0^t \cdots \int_0^t f(u) du \right\} = \frac{1}{s^n} \bar{f}(s) \tag{A.40}$$

Substitution

$$\bar{f}(s - a) = L\{e^{at} f(t)\} \tag{A.41}$$

This property is proven from the definition of the transform, by replacing the transform variable, s, with the shifted variable, $s - a$; thus

$$\bar{f}(s-a) = \int_0^\infty e^{-(s-a)t}f(t)dt \equiv \int_0^t e^{-st}\{e^{at}f(t)\}dt = L\{e^{at}f(t)\} \qquad (f)$$

Translation

$$L\{H(t-a)f(t-a)\} = e^{-as}\bar{f}(s); \quad a > 0 \qquad (A.42)$$

Indeed, $L\{H(t-a)f(t-a) = \int_0^\infty e^{-st}H(t-a)f(t-a)dt = \int_a^\infty e^{-st}f(t-a)dt$. The proof is concluded introducing the change of variable: $\tau = t-a$, and carrying out the integration:

$$L\{H(t-a)f(t-a) = \int_0^\infty e^{-st}[e^{-sa}f(\tau)]d\tau = e^{-sa}\int_0^\infty e^{-s\tau}f(\tau)d\tau \equiv e^{-sa}\bar{f}(s) \qquad (g)$$

Convolution

$$L\left\{ \int_0^t f(t-\tau)g(\tau)d\tau \right\} = L\{f(t)\}\cdot L\{g(t)\} \equiv \bar{f}(s)\bar{g}(s) \qquad (A.43)$$

This property, that the Laplace transform of the convolution of two functions is equal to the product of the transforms of the functions is also referred to as Borel's theorem. The proof uses the translation property of the Laplace transform to write

$$\bar{f}(s)\bar{g}(s) = f(s)\cdot \int_0^\infty e^{-s\tau}g(\tau)d\tau \equiv \int_0^\infty \{e^{-s\tau}f(s)\}g(\tau)d\tau$$

$$= \int_{\tau=0}^\infty \left\{ \int_{t=\tau}^\infty e^{-st}f(t-\tau)dt \right\} g(\tau)d\tau \qquad (h)$$

Interchanging the order of integration completes the proof:

$$\bar{f}(s)\bar{g}(s) = \int_{t=0}^\infty e^{-st}\left\{ \int_{\tau=0}^t f(t-\tau)g(\tau)d\tau \right\}dt \equiv L\left\{ \int_0^t f(t-\tau)g(\tau)d\tau \right\} \qquad (i)$$

A.15.2 Brief List of Laplace Transforms

- Dirac delta function:

$$L\{\delta(t)\} = 1 \qquad (j)$$

$$L\{\delta(t-a)\} = e^{-as} \qquad (k)$$

- Unit step function:
$$L\{H(t)\} = \frac{1}{s} \tag{l}$$

$$L\{H(t - a)\} = \frac{e^{-as}}{s} \tag{m}$$

- Exponential function:
$$L\{e^{at}\} = \frac{1}{s - a} \tag{n}$$

- Power function:
$$L\{t^n\} = \frac{n!}{s^{n+1}} \tag{o}$$

- Sine function:
$$L\{\sin(at)\} = \frac{a}{s^2 + a^2} \tag{p}$$

- Cosine function:
$$L\{\cos(at)\} = \frac{s}{s^2 + a^2} \tag{q}$$

A.15.3 Partial Fraction Expansion

As indicated before, a simple—and often effective—means to find the inverse transform of a function is to use a table of transformations in reverse. When the function that is to be transformed back to the physical plane appears as a quotient of two polynomials for which the inverse transform is not already available, it is usually advantageous to cast the given quotient of polynomials in term of its partial fractions.

Every rational function $\rho(x)$ can be expressed as the quotient of two polynomials in x. The division algorithm allows expressing $\rho(x)$ in the form:

$$\rho(x) = \varphi(x) + \frac{g(x)}{h(x)} \tag{A.44}$$

where $\varphi(x)$, $g(x)$ and $h(x)$ are polynomials in x, and the degree of the numerator, $g(x)$, is lower than that of the denominator, $h(x)$. If $h(x)$ can be expressed as the product of its irreducible factors, then it is possible to write $\rho(x)$ as the sum of $\varphi(x)$ and a given number of so-called partial fractions.

With the above in mind, let the known transform of a function $f(t)$ be given as

$$\bar{f}(s) = \frac{\bar{P}(s)}{\bar{Q}(s)} = \frac{p_0 + p_1 s^1 + p_2 s^2 + \cdots + p_n s^n}{q_0 + q_1 s^1 + q_2 s^2 + \cdots + q_m s^m} \tag{A.45}$$

To expand it into its partial fractions requires the m zeroes, r_i, of the denominator, so that

$$\bar{f}(s) = \frac{A_1}{s - r_1} + \frac{A_2}{s - r_2} + \cdots \frac{A_m}{s - r_m} \tag{A.46}$$

Which applies when all the roots of $\bar{Q}(s) = 0$ are distinct. A different expansion applies in the case of repeated roots. For instance, if r_1 occurs k times, the appropriate expansion becomes:

$$\bar{f}(s) = \frac{A_{11}}{(s - r_1)^1} + \frac{A_{12}}{(s - r_1)^2} + \cdots \frac{A_{1k}}{(s - r_1)^k} + \frac{A_2}{s - r_2} + \frac{A_3}{s - r_3} \cdots + \frac{A_{m-k}}{s - r_{m-k}} \tag{A.47}$$

In any case, the coefficients A_i correspond to the limiting process:

$$A_i = \lim_{s \to r_i} \left[\frac{P(s)}{Q(s)} (s - r_i) \right] \tag{A.48}$$

In practice, the partial fraction expansion of the ratio of two given polynomials is developed by first writing down the expansion based on the roots of the denominator, as in (A.46) or (A.47), multiplying by the polynomial in the denominator, equating coefficients of like powers of the transformed variable, and solving the resulting linear system in the coefficients. Also, when it is advantageous to leave the denominator of a partial fraction as a second-degree polynomial, its numerator must be a linear polynomial, in accordance with the requirements of partial fraction expansion indicated before. Thus, in analogy with (A.47), the partial fraction expansion of the rational polynomial $f(x) = \frac{p(x)}{(ax^2+bx+c)^n}$ would be given by the following decomposition:

$$\frac{p(x)}{(ax^2 + bx + c)^n} = \frac{A_1 x + B_1}{ax^2 + bx + c} + \frac{A_2 x + B_2}{(ax^2 + bx + c)^2} + \cdots + \frac{A_n x + B_n}{(ax^2 + bx + c)^n} \tag{A.49}$$

Example A.7 As an illustration, we obtain the inverse transform of the polynomial fraction:

$$\bar{f}(s) = \frac{4s + 7}{(s + 1)^2}$$

Solution

In this case, the denominator has two equal zeroes: $r = -1, -1$. Therefore, we seek the partial fraction expansion in the form: $\frac{4s+7}{(s+1)^2} = \frac{A_1}{s+1} + \frac{A_2}{(s+1)^2}$. Multiplying by the minimum common multiple of the expansion on the right-hand side and canceling the denominators on both sides of the resulting form: $4s + 7 = A_1(s + 1) + A_2$. This yields a linear system in A_1 and A_2, whose solution is $A_1 = 4$, and $A_2 = 3$; and thus, $\frac{4s+7}{(s+1)^2} = \frac{4}{s+1} + \frac{3}{(s+1)^2}$. The inverse Laplace transform of this expression can be easily obtained using the substitution property $f(s-a) = L\{e^{at}f(t)\}$ and the table of transforms in reverse.

$$\frac{4}{s+1} \equiv 4\frac{1}{s-(-1)} = 4f[s - (-1)]; \quad f(s) \equiv \frac{1}{s}$$
$$\Rightarrow L^{-1}\left\{\frac{4}{s+1}\right\} \equiv 4\{e^{-t}L^{-1}\{\frac{1}{s}\}\} = 4e^{-t} \cdot H(t)$$

$$\frac{3}{(s+1)^2} \equiv 3\frac{1}{[s-(-1)]^2} = 3f[s - (-1)]; \quad f(s) \equiv \frac{1}{s^2}$$
$$\Rightarrow L^{-1}\left\{\frac{3}{(s+1)^2}\right\} \equiv 3\{e^{-t}L^{-1}\{\frac{1}{s^2}\}\} = 3e^{-t} \cdot t$$

Therefore: $L^{-1}\left\{\frac{4s+7}{(s+1)^2}\right\} = L^{-1}\left\{\frac{4}{s+1}\right\} + L^{-1}\left\{\frac{3}{(s+1)^2}\right\} = (4 + 3t)e^{-t}$.

Example A.8 Find the partial fraction expansion of $\frac{x^5+1}{(x^2+4)^2}$

Solution

In this case, two facts need to be noted. Firstly, the given fraction is improper, as the degree of the numerator is larger than the degree of its denominator. Secondly, the denominator of the given expression is a quadratic polynomial and is repeated twice. Hence, according to (A. 44), we need to expand the denominator as $x^4 + 8x^2 + 16$, and use it to divide the numerator $x^5 + 1$ by it. The division algorithm then leads to the result: $\frac{x^5+1}{(x^2+4)^2} = x + \frac{-x^3-16x+1}{(x^2+4)^2}$. We use (A.49) and seek the partial fraction expansion of the second term, as $\frac{-x^3-16x+1}{(x^2+4)^2} = x + \frac{Ax+B}{x^2+4} + \frac{Cx+D}{(x^2+4)^2}$. Writing the partial fraction expansion in terms of the common denominator $(x^2 + 4)^2$ and collecting terms leads to $-x^3 - 16x + 1 = Ax^3 + Bx^2 + (4A + C)x + (4B + D)$. Equating coefficients of like powers of x produces: $A = -8$, $B = 0$, $C = 16$, $D = 1$. These values and the result of the division yield the partial fraction expansion: $\frac{x^5+1}{(x^2+4)^2} = x - \frac{8x}{x^2+4} + \frac{16x+1}{(x^2+4)^2}$.

A.15.4 Laplace Transform of Differential Equations

Any initial value problem which involves an ordinary linear differential equation of order n with constant coefficients may be converted into one in transformed space. Indeed, let the general differential equation in question be expressed as

$$\sum_{k=0}^{n} a_k \frac{d^k}{dt^k} y(t) = g(t)$$

$$\frac{d^0}{dt^0} y(t) \equiv 1 \tag{A.50}$$

$$\left. \frac{d^k}{dt^k} y(t) \right|_{t=0} = y(0); \quad k = 1, \ldots, n-1$$

Applying the Laplace transform, this equation takes the form:

$$\sum_{k=0}^{n} a_k \left\{ s^k \bar{y}(s) - \sum_{r=1}^{k} s^{k-r} y^{(r-1)}(0) \right\} = \bar{g}(s) \tag{A.51}$$

Pulling $\bar{y}(s)$ outside the summation sign and rearranging:

$$\bar{y}(s) = \frac{\bar{g}(s) + \sum_{k=1}^{n} a_k \sum_{r=1}^{k} s^{k-r} y^{(r-k)}(0)}{\sum_{k=0}^{n} a_k s^k} \tag{A.52}$$

The solution function, $y(t)$, may be obtained by Laplace transform inversion.

A.15.5 Laplace Transform of Integral Equations

An integral equation is one in which the unknown function appears inside an integral sign and, possibly, outside of it too. The general form of an integral equation in $v(t)$ is

$$\alpha(t) \int_{\tau=a}^{\tau=b} K(t,\tau) v(\tau) d\tau + \beta(t) v(t) = \gamma(t) \tag{A.53}$$

where α, β, and γ are arbitrary functions of the independent variable, t; $\alpha \neq 0$; and the lower limit of integration, a, is a constant. If both limits of integration are constant, the integral equation is called a Helmholtz integral equation; and when the upper limit is the independent variable, t, the equation is a Volterra integral equation. Equations for which the function $\gamma(t) = 0$ are called homogeneous; while equations in which the unknown appears only inside the integral (i.e., $\beta(t) = 0$) are called integral equations of the first kind; and integral equations of the second kind, otherwise.

In what follows, we concentrate on non-homogeneous Volterra integral equations of the second kind. There are two reasons for this: first, this type of equation always admits a solution (at least in principle); secondly, they occur naturally in viscoelasticity.

Without loss of generality, a Volterra integral equation of the second kind may be cast as

$$\gamma(t) = v(t) - \int_a^t K(t, \tau)v(\tau)d\tau \tag{A.54}$$

Where we have divided (A.53) throughout by $\beta(t)$, absorbed $-\alpha(t)/\beta(t)$ into K, and retained the notation $\gamma(t)$ for $\gamma(t)/\beta(t)$. Of particular interest is the case when $a = 0$, and the kernel $K(t,\tau)$ of the integral is a so-called difference kernel, when $K(t,\tau)$ may be expressed as $K(t,\tau) = K(t - \tau)$, such as occur for non-aging viscoelastic materials. The integral in (A.54) is then a convolution integral, so that, applying the Laplace transform, collecting terms, and re-arranging:

$$\bar{v}(s) = \frac{\bar{\gamma}(s)}{[1 - \bar{K}(s)]} \tag{A.55}$$

The solution to the original integral equation follows from the inverse transform of the expression in (A.55), as

$$v(t) = L^{-1}\left\{\frac{\bar{\gamma}(s)}{[1 - \bar{K}(s)]}\right\} \tag{A.56}$$

Example A.9 The relaxation modulus, M, and the creep compliance, C, of linear viscoelastic materials are related by $M(t)C(0) + \int_{0^+}^t M(t - \tau)\frac{d}{d\tau}C(\tau)d\tau$. Assuming – as happens to be the case–that $C(0) = 1/M(0)$, use the Laplace transform to solve this equation for the function $C(t)$, if $M(t) = M_1 e^{-t/\tau}$.

Solution

Direct application of the Laplace transformation, noting that the integral is the convolution of M and the time derivative of C, and using the pertinent properties of the transform yields $\bar{M}(s)C(0) + \bar{M}(s)[sC(s) - C(0)] = \frac{1}{s}$. Collecting terms and rearranging: $\bar{C}(s) = \frac{1}{s^2\bar{M}(s)} \equiv (1/M_1)\frac{s+1/\tau_1}{s^2} = (1/M_1)\{\frac{1}{s} + (1/\tau_1)\frac{1}{s^2}\}$. Using the list of Laplace transforms in reverse: $C(t) = \frac{1}{M_1}[H(t) + t/\tau_1]$. As a matter of curiosity, the material in question represents a fluid, among other things, because $C(\infty) \to \infty$. Also, $C(0) = 1/M_1$ is called the glassy compliance, and $M_1\tau_1$ is the material's viscosity.

A.15.6 Relationship Between the Laplace and Fourier Transforms

In general, the Laplace and Fourier transforms of an acceptable function[5] differ, but their definitions show considerable similarity:

$$L\{f(t)\} = \int_0^\infty f(t)e^{-st}dt; s = a + jb$$

$$\int_0^\infty |f(t)|e^{-ct}dt < \infty; c \in R; c > 0 \tag{A.57}$$

$$F\{f(t)\} = \int_{-\infty}^\infty f(t)e^{-j\omega t}dt$$

$$\int_{-\infty}^\infty |f(t)|dt < \infty \tag{A.58}$$

The question then arises if the two transforms can be made equivalent; and if so, under what conditions? The answer is that the two integral transforms will be equivalent for all functions which are absolutely integrable and vanish identically for all negative values of their argument, so that

$$\int_{-\infty}^\infty |f(t)|dt = \int_0^\infty |f(t)|dt < \infty \tag{r}$$

Since most functions in engineering, physics and viscoelasticity in particular, are causal, they vanish for all negative values of their arguments. The Fourier transform of any causal function, is given as

---

[5] In both instances, an "acceptable function" is one that is piecewise continuous and has a finite number of finite discontinuities on its interval of integration. As can be seen from their definitions, a function has to be of exponential order (bounded by an exponential function) on the positive real axis for its Laplace transform to exist, and absolutely integrable on the entire real axis for it to have a Fourier transform.

$$F\{f(t)\} = \int\limits_{-\infty}^{\infty} f(t)e^{-j\omega t}\,dt$$

$$= \int\limits_{-\infty}^{0} f(t)e^{-j\omega t}\,dt + \int\limits_{0}^{\infty} f(t)e^{-j\omega t}\,dt \qquad (A.59)$$

$$= \int\limits_{0}^{\infty} f(t)e^{-j\omega t}\,dt = \left.\int\limits_{0}^{\infty} f(t)e^{-st}\,dt\right|_{s=j\omega}$$

Hence,

$$F\{f(t)\} = L\{f(t)\}|_{s=j\omega}; \quad for: \quad f(t) \equiv 0, \quad \forall t < 0; \quad and: \quad \int\limits_{0}^{\infty} |f(t)|\,dt < \infty$$

$$(A.60)$$

This shows that the Laplace and Fourier transforms are equivalent for all absolutely integrable causal functions.

Example A.10 Using (60), find the Fourier transform of the exponential function defined by

$$f(t) = \begin{cases} e^{-at}; & t > 0;\ a > 0 \\ 0; & t < 0 \end{cases}$$

Solution

To apply (A.60), we must first make sure the given function satisfies the required conditions. By definition, the function f satisfies the causality condition that $f(t) \equiv 0$, $t < 0$. The function f is also absolutely integrable on the positive real axis, since $\int_0^\infty |e^{-at}|\,dt \equiv \int_0^\infty e^{-at}\,dt = \frac{1}{a} < \infty$. Hence, the Fourier transform of f can be obtained from its Laplace transform. From the list of Laplace transforms: $L\{e^{at}\} = \frac{1}{s-a}$. This and (A.60) yield $F\{e^{-at}\} = L\{e^{-at}\}|_{s=j\omega} = \frac{1}{s+a}|_{s=j\omega} = \frac{1}{j\omega+a}$

Example A.11 Demonstrate that the Fourier transform of the unit step function cannot be obtained from its Laplace transform.

Solution

The unit step function satisfies the causality condition, but the integral of its absolute value is infinite: $\int_0^\infty |u(t)|\,dt \equiv \int_0^\infty dt = \infty$. Since this violates the Dirichlet conditions, the Fourier transform of the unit step function cannot be derived from its Laplace transform.

A.16 Orthogonal Functions

A set of functions $\{f_k(t)\}$ is said to be orthogonal on an open interval $a < t < b$ if for any two function $f_i(t)$ and $f_j(t)$ in the set, the following relation holds

$$\int_a^b f_i(t)f_j(t)dt = \begin{cases} 0 & for\ i \neq j \\ m_i & for\ i = j \end{cases} \tag{A.61}$$

The set of sinusoidal functions $\{1, \cos(\omega t), \cos(2\omega t), \ldots, \cos(n\omega t), \ldots,$ $\sin(\omega t), \sin(2\omega t), \ldots, \sin(n\omega t), \ldots\}$ is an orthogonal set of functions on an open interval of total amplitude $p = 2\pi/\omega$. As may be shown by elementary calculus, the functions in this set are such that

$$\int_t^{t+p} \sin(k\omega t)dt = 0 \quad for\ all\ k \tag{A.62-a}$$

$$\int_t^{t+p} \cos(k\omega t)dt = 0 \quad for\ k \neq 0 \tag{A.62-b}$$

$$\int_t^{t+p} \sin(k\omega t)\cos(l\omega t)dt = 0 \quad for\ all\ k\ and\ l \tag{A.62-c}$$

$$\int_t^{t+p} \sin(k\omega t)\sin(l\omega t)dt = \begin{cases} 0 & for\ k \neq 0 \\ p/2 & for\ k = l \neq 0 \end{cases} \tag{A.62-d}$$

$$\int_t^{t+p} \cos(k\omega t)\cos(l\omega t)dt = \begin{cases} 0 & for\ k \neq 0 \\ p/2 & for\ k = l \neq 0 \end{cases} \tag{A.62-e}$$

The proofs of (A.62-a) and (A.62-b) are straightforward, as their integrals are proportional to a $\cos(k\omega t)$ and $\sin(k\omega t)$, respectively; and in addition, these functions are periodic, meaning that $\cos[(k\omega(t+p)] = \cos(k\omega t)$, and $\sin[(k\omega(t+p)] = \sin(k\omega t)$. Hence,

$$\int_t^{t+p} \sin(k\omega t)dt = -\frac{1}{k}\{\cos[(k\omega(t+p)] - \cos(k\omega t)\} = 0\ for\ all\ k \tag{a}$$

$$\int_t^{t+p} \cos(k\omega t)dt = \frac{1}{k}\{\sin[(k\omega(t+p)] - \sin(k\omega t)\} = 0\ for\ all\ k \neq 0 \tag{b}$$

To prove (A.62-c), we use the identity $\sin(A) \cdot \cos(B) = 1/2[\sin(A + B) + \sin(A - B)]$; so that

$$\int_t^{t+p} \sin(k)\cos(l\omega t)dt = \frac{1}{2}\int_t^{t+p} \sin[(k + l)\omega t] + \sin[(k - l)\omega t]dt$$

$$= \frac{1}{2}\frac{(-1)}{(k + l)\omega}\cos[(k + l)\omega t]_t^{t+p} + \frac{1}{2}\frac{(-1)}{(k - l)\omega}\cos[(k - l)\omega t]_t^{t+p}$$

$$= 0 \; if \; k \neq l$$

(c.1)

In the case $k = l$, we use the trigonometric identity $2\sin(A)\cos(A) = \sin(2A)$, to write

$$\int_t^{t+p} \sin(k\omega t)\cos(l\omega t)dt = \int_t^{t+p} \sin(k\omega t)\cos(k\omega t)dt$$

$$= \frac{1}{2}\int_t^{t+p} \sin(2k\omega t)dt$$ (c.2)

$$= \frac{-1}{4k\omega}[\cos(2k\omega t)]_t^{t+p}$$

$$= 0$$

The proof of (A.62-d) uses the identity: $\sin(A) \cdot \sin(B) = 1/2[\cos(A - B) - \cos(A + B)]$. Thus,

$$\int_t^{t+p} \sin(k\omega t)\sin(l\omega t)dt = \frac{1}{2}\int_t^{t+p} \cos[(k - l)\omega t] - \cos[(k + l)\omega t]dt$$

$$= \frac{1}{2}\frac{(1)}{(k - l)\omega}\sin[(k - l)\omega t]_t^{t+p} - \frac{1}{2}\frac{(1)}{(k + l)\omega}\sin[(k + l)\omega t]_t^{t+p}$$

$$= 0 \; if \; k \neq l$$

(d.1)

In the case $k = l$, we use the trigonometric identity $\sin^2(A) = 1/2 - 1/2\cos(2A)$, to write

$$\int_t^{t+p} \sin(k\omega t)\sin(l\omega t)dt = \int_t^{t+p} \sin^2(k\omega t)dt$$

$$= \frac{1}{2}\int_t^{t+p} [1 - \cos(2k\omega t)]dt \tag{d.2}$$

$$= \frac{1}{2}[t]_t^{t+p} - \frac{1}{4k\omega}[\sin(2k\omega t)]_t^{t+p}$$

$$= \frac{p}{2}.$$

The proof of (A.62-e) uses the identity: $\cos(A)\cos(B) = \frac{1}{2}[\cos(A+B) + \cos(A-B)]$. Thus,

$$\int_t^{t+p} \cos(k\omega t)\cos(l\omega t)dt = \frac{1}{2}\int_t^{t+p} \cos[(k+l)\omega t] + \cos[(k-l)\omega t]dt$$

$$= \frac{1}{2}\frac{(1)}{(k+l)\omega}\sin[(k+l)\omega t]_t^{t+p} \tag{e.1}$$

$$+ \frac{1}{2}\frac{(1)}{(k-l)\omega}\sin[(k-l)\omega t]_t^{t+p}$$

$$= 0 \text{ if } k \neq l$$

In the case $k = l$, we use the trigonometric identity $\cos^2(A) = \frac{1}{2} + \frac{1}{2}\cos(2A)$, to write

$$\int_t^{t+p} \cos(k\omega t)\cos(l\omega t)dt = \int_t^{t+p} \cos^2(k\omega t)dt$$

$$= \frac{1}{2}\int_t^{t+p} [1 + \cos(2k\omega t)]dt \tag{e.2}$$

$$= \frac{1}{2}[t]_t^{t+p} + \frac{1}{4k\omega}[\sin(2k\omega t)]_t^{t+p}$$

$$= \frac{p}{2}$$

A.17 Complex Numbers

A complex number, z, is an ordered pair of real numbers, x and y, which, for this reason, is usually written in the form $z = (x, y)$; just like a point –or vector– in two dimensions.

The collection of all ordered pairs of real numbers forms the set, $\mathbb{C}$, of complex numbers. The first component of the ordered pair (x, y) is termed the real part of the complex number and is denoted by $Re\{z\}$. For reasons which will become logical later on, the second number in the pair is referred to as the imaginary part or component, $Im\{z\}$, of the complex number. Using this notation, then

$$z = (x, y); \quad x = Re\{z\}; \quad y = Im\{z\} \tag{A.63}$$

The complex conjugate, $\bar{z}$, of a complex number z, is defined by the relation:

$$\bar{z} = (x, -y) \tag{A.64}$$

A.17.1 Elementary Operations with Complex Numbers

The arithmetic operations of addition –or subtraction– multiplication, and division of complex numbers are defined in terms of the corresponding operations for real numbers, as described next.

Addition
The addition of complex numbers is defined component-wise, by the simple rule:

$$z_1 + z_2 = (x_1, y_1) + (x_2, y_2) = (x_1 + x_2, y_1 + y_2) \tag{A.65}$$

From this definition follows that the complex number $(0,0) \equiv 0$ is the identity for complex addition; that is

$$\begin{aligned}
z + (0,0) = (x, y) + (0,0) &= (x + 0, y + 0) \\
&= (0 + x, 0 + y) \\
&= (0,0) + (x, y) \\
&= (0,0) + z
\end{aligned} \tag{A.66}$$

Another consequence of the definition of addition of complex numbers is that any complex number may be expressed uniquely as the sum two complex numbers; one with zero imaginary part, and the other with zero real part:

$$z = (x, y) = (x, 0) + (0, y) \tag{A.67}$$

Using this decomposition, z and $\bar{z}$ may be combined to express the real and the imaginary parts of a complex number, as

$$x = Re\{z\} = \frac{z + \bar{z}}{2}; \quad y = Im\{z\} = \frac{z - \bar{z}}{2} \tag{A.68}$$

The operation of addition –or subtraction– enjoys the same properties as the addition of real numbers. That is, the addition of complex numbers is associative, distributive, and commutative. This statement is proven using the definition of addition of complex numbers and the properties of addition of real numbers.

Example A.12 Prove that the addition of complex numbers is commutative.

Solution

To prove that the addition of complex numbers is commutative is equivalent to showing that the order of the addends in the sum of tow complex numbers can be interchanged without changing the result. To do this, examine the sum z_1+z_2 of two complex numbers:

$$z_1 + z_2 = (x_1, y_1) + (x_2, y_2) = (x_1 + x_2, y_1 + y_2)$$
$$= (x_2 + x_1, y_2 + y_1) = (x_2, y_2) + (x_1, y_1)$$
$$= z_2 + z_1$$

Example A.13 Prove that the addition of complex numbers is associative.

Solution

To prove associativity, consider the sum $(z_1+z_2) + z_3$:

$$(z_1 + z_2) + z_3 = (x_1 + x_2, y_1 + y_2) + (x_3, y_3)$$
$$= (x_1 + x_2 + x_3, y_1 + y_2 + y_3)$$
$$= (x_1, y_1) + (x_2 + x_3, y_2 + y_3) = z_1 + (z_2 + z_3)$$

For simplicity, the following representation of a complex number is used:

$$z = (x, y) = (x, 0) + (0, y) = x(1, 0) + y(0, 1) = x + j \cdot y \qquad \text{(A.69)}$$

Multiplication

The product of two complex numbers z_1 and z_2 is defined by the following rule:

$$z_1 \cdot z_2 = (x_1, y_1) \cdot (x_2, y_2) = (x_1 x_2 - y_1 y_2, x_1 y_2 + y_1 x_2) \qquad \text{(A.70)}$$

From this follows that the complex number $(1,0) \equiv 1$ is the identity element for complex multiplication. Indeed,

$$z \cdot (1, 0) = (x, y) \cdot (1, 0) = (x \cdot 1 - y \cdot 0, y \cdot 1 + x \cdot 0) = (x, y) = z \qquad \text{(A.71)}$$

and

$$(1, 0) \cdot z = (1, 0) \cdot (x, y) = (1 \cdot x - 0 \cdot y, 1 \cdot y + 0 \cdot x) = (x, y) = z \qquad \text{(A.72)}$$

In this notation, the letter "*j*" is used to represent the unit, $(0,1)$, of the imaginary numbers,[6] which, in accordance with the definition of the product of complex numbers, is such that

$$j^2 = (0, 1) \cdot (0, 1) = (-1, 0) = -1 \qquad \text{(A.73)}$$

In other words,

$$j = \sqrt{-1} \qquad \text{(A.74)}$$

---

[6] More commonly, the letter "*i*" is used to denote the complex unit, but we reserve this letter for use as a subscript.

Multiplication of two complex numbers enjoys the same properties as that of real numbers. For instance, that multiplication of complex numbers is commutative and associative results from the corresponding rules for real numbers.

$$z_1 \cdot z_2 = (x_1, y_1) \cdot (x_2, y_2) = (x_1 x_2 - y_1 y_2, x_1 y_2 + y_1 x_2)$$
$$= (x_2 x_1 - y_2 y_1, x_2 y_1 + y_2 x_1) \tag{a}$$
$$= z_2 \cdot z_1$$

$$(z_1 \cdot z_2) \cdot z_3 = (x_1 x_2 - y_1 y_2, x_1 y_2 + y_1 x_2) \cdot (x_3, y_3)$$
$$= (x_1 x_2 x_3 - y_1 y_2 x_3 - x_1 y_2 y_3 - y_1 x_2 y_3, x_1 x_2 y_3 - y_1 y_2 y_3 + x_1 y_2 x_3 + y_1 x_2 x_3)$$
$$= [x_1 (x_2 x_3 - y_2 y_3) - y_1 (y_2 x_3 + x_2 y_3), x_1 (y_2 x_3 + x_2 y_3) + y_1 (x_2 x_3 - y_2 y_3)]$$
$$= (x_1, y_1) \cdot [(x_2 x_3 - y_2 y_3), (y_2 x_3 + x_2 y_3)] = (x_1, y_1) \cdot [(x_2, y_2) \cdot (x_3, y_3)]$$
$$= z_1 \cdot (z_2 \cdot z_3)$$
$$\tag{b}$$

Additionally, from the definition of multiplication of complex numbers, one has

$$x = x \cdot 1 = x \cdot (1, 0) = (x, 0) \cdot (1, 0) = (x, 0) \tag{c}$$

$$y \cdot (0, 1) = [y \cdot (1, 0)] \cdot (0, 1) = (y, 0) \cdot [(1, 0) \cdot (0, 1)] = (y, 0) \cdot (0, 1) = (0, y) \tag{d}$$

Combining these results with the definition of addition given earlier, there results

$$z = (x, y) = (x, 0) + (0, y) = x(1, 0) + y(0, 1) \tag{e}$$

This expression shows that a complex number may be interpreted as the sum of two points or vectors in a two-dimensional Euclidean space. Indeed, the Argand plane is a two-dimensional plane in which the real part, x, of a complex number, z, is measured along the horizontal axis or real axis, and its imaginary part, y, is measured along the vertical or imaginary axis, as shown in Fig. A.6.

From its graphical representation, it can be seen that a complex number can be represented in so-called polar form, as

$$z = (x, y) = r \cdot \cos\theta + j \cdot r \cdot \sin\theta; r = \sqrt{x^2 + y^2} \tag{A.75}$$

$$\bar{z} = (\bar{x}, \bar{y}) = r \cdot \cos\theta - j \cdot r \cdot \sin\theta; \quad r = \sqrt{x^2 + y^2} \tag{A.76}$$

In these expressions, r, as the positive square root of nonnegative real numbers, x^2 and y^2, is a real number. The number r is called the modulus, or magnitude of the complex number z and is usually denoted as $|z|$. From the previous expressions, it is apparent that the modulus of any complex number is the same as the modulus of its conjugate; that is

$$r \equiv |z| = \sqrt{x^2 + y^2} = |\bar{z}|, \quad \forall z = (x, y) \tag{A.77}$$

Fig. A.6 Graphical representation of a complex num-
ber, z, and its conjugate, $\bar{z}$, in
the complex, Argand plane

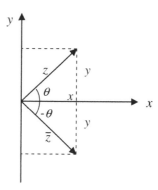

Division

The rule for the division z_1/z_2 of two complex numbers z_1 and z_2 may be readily
developed from the concept of the inverse, z^{-1}, of a non-zero complex number, z:

$$z \cdot z^{-1} = (1,0) \equiv 1 \tag{A.78}$$

At the same time, using the rule for multiplication of complex numbers and the
definition of complex conjugate, it follows that

$$
\begin{aligned}
z \cdot \bar{z} &= (x, y) \cdot (x, -y) \\
&= (x^2 + y^2, -xy + xy) \\
&= (x^2 + y^2, 0) \\
&= (r^2, 0) \equiv |z|^2
\end{aligned}
\tag{A.79}
$$

In other words,

$$\frac{z \cdot \bar{z}}{|z|^2} = 1 = (1,0) \tag{A.80}$$

Combining this with the definition of the complex multiplication inverse yields

$$z \cdot z^{-1} = (1,0) = \frac{z \cdot \bar{z}}{|z|^2} \tag{A.81}$$

Canceling out the common term z, and using that $|z|^2 = z \cdot \bar{z}$, leads to

$$z^{-1} = \frac{\bar{z}}{|z|^2} = \frac{1}{z} \tag{A.82}$$

Not only does this relation prove that the inverse of a non-zero complex number
is, as for real numbers, equal to its inverse, but also provides the rule: "the inverse
of a complex number is equal to its complex conjugate divided by it squared
modulus."

Example A.14 Find the inverse of the complex number $z = 3 - j$.

Solution

Apply the rule to obtain the inverse of a complex number: $z^{-1} = \frac{\bar{z}}{|z|^2}$ to get

$z^{-1} = \frac{1}{(3-j)} = \frac{3+j}{3^2+1^2} = 0.3 + 0.1j$

The same process yields the rule that "to obtain the quotient z_1/z_2 of two complex numbers, multiply the numerator and denominator of the fraction by the conjugate of the denominator." This is so, because

$$\frac{z_1}{z_2} = z_1 z_2^{-1} = z_1 \frac{\bar{z}_2}{|z_2|^2} = \frac{z_1 \bar{z}_2}{z_2 \bar{z}_2}$$

Example A.15 Find the quotient: $(2-3j)/(1+2j)$

Solution

Apply the rule for division to get $\frac{2-3j}{1+2j} = \frac{(2-3j)(1-2j)}{1^2+2^2} = -(0.8 + 1.4j)$

A.17.2 Euler's Formula

A useful representation of a complex number is by means of the so-called Euler formula. This formula expresses the exponential function of a complex argument in terms of the sine and cosine of the argument. Euler's formula may be easily derived by developing the function $e^{j\theta}$ in Taylor series, as follows:

$$\begin{aligned}
e^{j\theta} &= 1 + \frac{(j\theta)}{1!} + \frac{(j\theta)^2}{2!} + \frac{(j\theta)^3}{3!} + \frac{(j\theta)^4}{4!} + \cdots \\
&= 1 - \frac{\theta^2}{2!} + \frac{\theta^4}{4!} + \cdots + j \cdot \left(\frac{\theta}{1!} - \frac{\theta^3}{3!} + \right) \qquad \text{(A.83)} \\
&= \cos\theta + j \cdot \sin\theta
\end{aligned}$$

Combining this, with the polar representation $z = r[\cos(\theta) + j \cdot \sin(\theta)]$, gives

$$z = (x, y) = r\cos\theta + j \cdot r\sin\theta = re^{j\theta} \qquad \text{(A.84)}$$

In using the polar or the exponential forms of a complex number, it must be borne in mind that the cosine and sine functions are periodic functions with period 2π. Hence,

$$z = re^{j\theta} = re^{j(\theta + 2\pi \cdot k)}; \quad k = 0, 1, 2, \ldots \qquad \text{(A.85)}$$

A.17.3 Powers, Roots, and Logarithms of Complex Numbers

Powers, logarithms, and integral roots of complex numbers are easily calculated using their exponential representation and the corresponding rules for real numbers. Letting $z = r \cdot e^{j\theta}$ be an arbitrary complex number, the following expressions follow directly from the properties of real numbers:

Integral Powers

The *n*th power of a complex number is easily established from the polar form, and the rules of exponentiation of real numbers, as

$$z^n = \left(re^{j\theta}\right)^n = \left(re^{j(\theta+2\pi \cdot k)}\right)^n = r^n e^{jn(\theta+2\pi \cdot k)}; \quad k = 0, 1, 2, \ldots, n-1 \quad (A.86)$$

Radicals

The root of integral index *n* of a complex number is easily obtained from its polar form, and the rules of exponentiation of real numbers, as

$$z^{1/n} = \left(re^{j(\theta+2\pi k)}\right)^{1/n} = \left(re^{j(\theta+2\pi \cdot k)}\right)^{1/n} = r^{1/n} e^{\frac{j(\theta+2\pi \cdot k)}{n}}; \quad k = 0, 1, 2, \ldots, n-1 \quad (A.87)$$

Logarithms

The natural logarithm, $\log_e(z) \equiv \ln(z)$, of a complex number, *z*, can be established from the polar form of the complex number and the properties of logarithms of real numbers, as

$$\log_e z \equiv \ln\left(re^{j(\theta+2\pi k)}\right) = \ln(r) + j(\theta + 2\pi k); \quad k = 0, 1, 2, \ldots, n \quad (A.88)$$

The logarithm, $\log_b(z)$, of a complex number, z, in an arbitrary real base, *b*, may also be established from the properties of logarithms of real numbers, using that for any two bases, b_1 and b_2: $\log_{b_1} f = \log_{b_1}(b_2)\log_{b_2}(f)$, as

$$\log_b z = \log_b\left(re^{j\theta}\right) = \log_b(e) \ln\left(re^{j(\theta+2\pi k)}\right)$$
$$= \log_b(e)[\ln r + j(\theta + 2\pi k)], k = 0, 1, 2, \ldots, n \quad (A.89)$$

Trigonometric Functions

The circular functions $\sin(z)$ and $\cos(z)$ of a complex argument result from Euler's formula with complex argument, $e^{jz} = \cos(z) + j \cdot \sin(z)$. The sine function is obtained by subtracting e^{-jz} from e^{jz}, and the cosine function results from adding them. Thus,

$$\sin(z) = \frac{1}{2i}\left(e^{jz} - e^{-jz}\right); \quad \cos(z) = \frac{1}{2}\left(e^{jz} + e^{-jz}\right) \quad (A.90)$$

These relations may be recast in terms of the real and imaginary parts, *x* and *y*, of the complex number $z = x + jy$. Taking the sine function, for instance, leads to

$$\sin(z) = \sin(x + jy) = \frac{1}{2j}\left(e^{jx-y} - e^{-jx+y}\right)$$

$$= \frac{1}{2j}\left[e^{-y}(\cos x + j \cdot \sin x) - e^{y}(\cos x - j\sin x)\right] \qquad \text{(f)}$$

$$= \frac{1}{j}\left[-\cos x \frac{(e^{y} - e^{-y})}{2} + j \cdot \sin x \frac{(e^{y} + e^{-y})}{2}\right]$$

$$= \sin x \cdot \cosh y + j \cdot \cos x \cdot \sinh y$$

So that

$$\sin(x + jy) = \sin x \cdot \cosh y + j \cdot \cos x \cdot \sinh y \qquad \text{(A.91)}$$

In this expression, $\sinh(u) = \frac{1}{2}(e^{u} - e^{-u})$ and $\cosh(u) = \frac{1}{2}(e^{u} + e^{-u})$, are the usual hyperbolic functions of real analysis. In an entirely similar fashion, the cosine function leads to

$$\cos(x + jy) = \cos x \cdot \cosh y - j \cdot \sin x \cdot \sinh y \qquad \text{(A.92)}$$

Example A.16 Calculate *cos* (1+2j).

Solution:
Use: $\cos(x + jy) = \cos x \cdot \cosh y - j \cdot \sin x \cdot \sinh y$, with $x = 1$ and $y = 2$ *rad*, to obtain:

$$\cos(1 + 2j) = \cos 1 \cdot \cosh 2 - j \cdot \sin 1 \cdot \sinh 2$$

$$= (0.5403)(3.7622) - j(0.8415)(3.6269)$$

$$= 2.0327 - 3.0520j$$

The inverse trigonometric functions $\sin^{-1}z$ and $\cos^{-1}z$ are also readily defined. For instance, if $w = \sin^{-1}(z)$, then taking the *sine* of this expression leads to

$$z = \sin(w) = \frac{1}{2j}\left(e^{jw} - e^{-jw}\right) \qquad \text{(g)}$$

Multiplying throughout by $2j$, to remove the denominator, and multiplying the result by e^{jw}, and collecting terms, there results $e^{2jw} - 2jze^{jw} - 1 = 0$; which is a quadratic equation in e^{jw}, with roots $e^{jw} = jz \pm \sqrt{1 - z^2}$. The natural logarithm of this relation, using that $w = \sin^{-1}(z)$, yields

$$\sin^{-1}(z) = -j \cdot ln\left(jz \pm \sqrt{1 - z^2}\right) \qquad \text{(A.93)}$$

Note that the function $\sin^{-1}z$ is multi-valued because the natural logarithm is multi-valued. In an entirely similar way,

$$\cos^{-1}(z) = -j \cdot ln\left(z \pm \sqrt{z^2 - 1}\right) \qquad \text{(A.94)}$$

$$\tan^{-1}(z) = \frac{j}{2} \cdot \ln\left(\frac{j+z}{j-z}\right) = \frac{j}{2} \cdot \ln\left(\frac{1-jz}{1+jz}\right) \qquad (A.95)$$

The form in the far right of (A.95) results from the property that $j^2 = -1$, and the fact that

$$j + z \equiv j + (x + jy) = j[1 - j(x + jy)] \equiv j(1 - jz) \qquad (h)$$

and similarly,

$$j - z \equiv j - (x + jy) = j[1 + j(x + jy)] \equiv j(1 + jz) \qquad (i)$$

Appendix B
Elements of Solid Mechanics

B.1 Scalars, Vectors, and Tensors

A tensor is an ordered system of objects, which are typically numbers or functions. As such, tensors may be represented by matrices. In standard three-dimensional space, the number of components of a tensor of order n is simply 3^n. Accordingly, a tensor of order zero has only one component ($3^0 = 1$), arranged in a 1-by-1 matrix. In Euclidean 3-space, scalars or tensors of order zero are quantities, such as temperature, density, volume, which are fully described by one value. The matrix representation of a tensor of order one, also called a first-order tensor, has 3 components ($3^1 = 3$). These tensors are the well-known vectors of Euclidean geometry: 3-by-1 or 1-by-3 matrices. Continuing thus, the matrix representation of a second-order tensor has $3^2 = 9$ entries, corresponding to a 3-by-3 matrix. Tensors of fourth order have $3^4 = 81$ components.

It is common practice to use the terms scalar and vector, respectively, to refer to tensors of order zero and one; and to reserve the term "tensor" exclusively for tensors of order 2 and higher. Also, symbolically, a vector is represented placing a bar or an arrow over the stem letter, or by the use of bold type: $\bar{a}, \vec{a}, \boldsymbol{a}$.

The components of a tensor may be designated when the tensor is referred to a coordinate system. In Cartesian coordinates, for instance, the components of a vector, $\mathbf{A}$, may be expressed as A_x, A_y, A_z; or, as A_1, A_2, A_3, if the axes are numbered. The vector itself may be represented in several forms:

(a) As a point in space, $\boldsymbol{A} = (A_1, A_2, A_3)$.
(b) In explicit form, displaying the unit vectors, along the coordinate axes, using Gibbs' notation:[7] $\boldsymbol{A} = A_1\hat{e}_1 + A_2\hat{e}_2 + A_3\hat{e}_3 = \sum_{i=1}^{3} A_i\hat{e}_i$.
(c) In component indicial form, combining the summation convention—that a term is developed by summing over the range of its repeated indices: $A = A_i\hat{e}_i, i = 1, 2, 3$.

The magnitude of the vector $\mathbf{A}$ is the real number: $|A| = \sqrt{A_1^2 + A_2^2 + A_3^2}$.

---

[7] We typically represent unit vector with a hat, $\hat{e}_i$, to emphasize their magnitude is 1. However, they are also represented in bold face type, such as, $\mathbf{e_i}$.

D. Gutierrez-Lemini, *Engineering Viscoelasticity*, DOI: 10.1007/978-1-4614-8139-3,
© Springer Science+Business Media New York 2014

Vector addition is defined component-wise. Therefore: $\mathbf{A} + \mathbf{B} = \mathbf{C}$ is such that $C_1 = A_1 + B_1$, $C_2 = A_2 + B_2$, and $C_3 = A_3 + B_3$. Therefore, vector addition is commutative.

There are three types of multiplication of vectors: scalar or dot product, vector or cross product, and tensor product.

Scalar or Dot Product

The scalar product $\mathbf{A} \cdot \mathbf{B}$ of two vectors $\mathbf{A}$ and $\mathbf{B}$ is the scalar, α, defined by the relation:

$$\mathbf{A} \cdot \mathbf{B} = A_1 B_1 + A_2 B_2 + A_3 B_3 = A_i B_i = \alpha \tag{B.1}$$

where the summation convention is used in the expression on the far right. The scalar may be viewed pictorially as the product of the magnitudes of the two vectors, and the cosine of the –smallest– angle between them:

$$\mathbf{A} \cdot \mathbf{B} = A_1 B_1 + A_2 B_2 + A_3 B_3 = A_i B_i = \alpha = |\mathbf{A}||\mathbf{B}|\cos(\mathbf{A}, \mathbf{B}) \tag{B.2}$$

Using this definition, it is clear that the dot product $\mathbf{e_i} \cdot \mathbf{e_j}$ of any two unit base vectors of a rectangular Cartesian coordinate system is such that $\mathbf{e_i} \cdot \mathbf{e_j} = 1$, if $i = j$; and $\mathbf{e_i} \cdot \mathbf{e_j} = 0$, if $i \neq j$. This gives rise to a special symbol called the Kronecker delta, δ_{ij}, which is equivalent to the identity 3×3 matrix, and hence, to the second-order identity tensor:

$$\delta_{ij} = \begin{cases} 1, & if\ i = j \\ 0, & if\ i \neq j \end{cases} \tag{B.3}$$

Using this definition and Gibbs' notation, the dot product of two vectors, $\mathbf{A}$ and $\mathbf{B}$, may be computed as

$$\mathbf{A} \cdot \mathbf{B} = A_i \widehat{e}_i \cdot B_j \widehat{e}_j = A_i B_j \widehat{e}_i \cdot \widehat{e}_j = A_i B_j \delta_{ij} = A_i B_j \delta_{ji} = A_i B_i \tag{B.4}$$

This shows that, in particular, the scalar product of two vectors is commutative. The dot product is also associative and distributive with respect to vector addition.

Vector or Cross Product

The cross product $\mathbf{A} \times \mathbf{B}$ of two vectors $\mathbf{A}$ and $\mathbf{B}$ is a third vector, $\mathbf{C}$, that is perpendicular to the original vectors. The magnitude of the resultant vector may be viewed pictorially as the product of the magnitudes of the two vectors, and the sine of the—smallest—angle between them:

$$|\mathbf{A} \times \mathbf{B}| = |\mathbf{C}| = |\mathbf{A}||\mathbf{B}|\sin(\mathbf{A}, \mathbf{B}) \tag{B.5}$$

Using this definition, the vector or cross product, $\mathbf{e_i} \times \mathbf{e_j}$ of any two unit base vectors of a rectangular Cartesian coordinate system, is such that $\mathbf{e_i} \times \mathbf{e_j} = \mathbf{e_k}$, if the indices i, j, k are an even permutation of the coordinate directions, 1, 2, 3; and $\mathbf{e_i} \times \mathbf{e_j} = \mathbf{o}$ when i, j, k are an odd permutation of 1, 2, 3. This can be expressed more succinctly by means of the alternating symbol e_{ijk}, defined such that

$$e_{ijk} = \begin{cases} +1, & \text{when } ijk \text{ are an even premutation of } 123 \\ -1, & \text{when } ijk \text{ are an odd premutation of } 123 \\ 0, & \text{when any two subscripts are equal} \end{cases} \tag{B.6}$$

Using this definition and Gibbs' notation, the cross product of two vectors, **A** and **B**, may be computed as

$$\mathbf{A} \times \mathbf{B} = A_i \widehat{e}_i \times B_j \widehat{e}_j = A_i B_j \widehat{e}_i \times \widehat{e}_j = A_i B_j e_{ijk} \widehat{e}_k = C_k \widehat{e}_k = \mathbf{C} \tag{B.7}$$

Hence, in indicial tensor notation, the vector **C**, resulting from the cross product **A** × **B**, is given by $C_k = e_{ijk} A_i B_j$. The alternating symbol, having three free indices and being an ordered system of numbers, is a tensor of third order, called the alternating tensor. From the definition of the alternating tensor, follows directly that the cross product of two vectors is not commutative, as

$$\mathbf{A} \times \mathbf{B} = A_i B_j e_{ijk} \widehat{e}_k = -A_i B_j e_{jik} \widehat{e}_k \equiv -B_j A_i e_{jik} \widehat{e}_k = -\mathbf{B} \times \mathbf{A} \tag{B.8}$$

Tensor Product

The operation of lumping vectors in juxtaposition is called tensor product. The tensor product is also called the dyadic product; and the factors in the product, dyads. Second-order tensors result when two vectors are multiplied in this fashion. For instance, the tensor product **AB** of two vectors, **A**, and **B**, is the second-order tensor:

$$\overline{\overline{C}} = \mathbf{AB} = A_i \widehat{e}_i B_j \widehat{e}_j = A_i B_j \widehat{e}_i \widehat{e}_j \tag{B.9}$$

By this definition, the second-order identity tensor, represent by the letter I with a double over-bar, $\overline{\overline{I}}$, is the continued dyadic product:

$$\overline{\overline{I}} = \widehat{e}_i \widehat{e}_i \tag{B.10}$$

It is also clear that tensor multiplication is not commutative, as changing the order of the vectors changes the order of the components of the resulting tensor— indicated by the dyad $\widehat{e}_i \widehat{e}_j$. The tensor $\mathbf{BA} = A_i B_j \widehat{e}_j \widehat{e}_i$ is called the transpose of the tensor **AB**. By this definition, **AB** is itself the transpose of the tensor **BA**. The transpose of a given tensor is indicted by appending the superscript T to the stem symbol representing the tensor. Thus, the transpose of (**AB**) is $(\mathbf{AB})^T$. Higher-order tensors are defined by means of continued dyadic products. For instance, a third-order tensor may be written as $\overline{\overline{C}} = \widehat{e}_i \widehat{e}_j \widehat{e}_k C_{ijk}$.

Tensor or dyadic multiplication of any order is assumed to obey the following laws of operation:

- Associative law. The factors of a tensor product may be lumped in any manner without changing the result: $\overline{A}\,\overline{B}\,\overline{C} = (\overline{A}\,\overline{B})\,\overline{C} = \overline{A}\,(\overline{B}\,\overline{C})$.
- Distributive law. If $\overline{B}$ and $\overline{C}$ are tensors of the same rank, and $\overline{A}$ is any continued dyadic product, then $\overline{A}\,(\overline{B} + \overline{C}) = \overline{A}\,\overline{B} + \overline{A}\,\overline{C}$.

- Scalar multiplication law. A scalar, α, may be placed anywhere in a continued dyadic product without changing the result: $\alpha \overline{A}\,\overline{B}\,\overline{C} = \overline{A}\,\alpha\,\overline{B}\,\overline{C} = \overline{A}\,\overline{B}\,\alpha\,\overline{C} = \overline{A}\,\overline{B}\,\overline{C}\,\alpha$.

The dot and cross products of dyads are an extension of the definitions of these operations between two vectors. Thus,

$$\mathbf{AB} \cdot \mathbf{CD} = \mathbf{A}(\mathbf{B} \cdot \mathbf{C})\mathbf{D} = (\mathbf{B} \cdot \mathbf{C})\mathbf{AD} \tag{B.11}$$

$$\mathbf{AB} \times \mathbf{CD} = \mathbf{A}(\mathbf{B} \times \mathbf{C})\mathbf{D} \tag{B.12}$$

Two types of double dot and cross products are defined. In the first type of double product, the operations are resolved from right to left:

$$\mathbf{AB} \cdot\cdot \mathbf{CD} = \mathbf{A}(\mathbf{B} \cdot \mathbf{C}) \cdot \mathbf{D} = (\mathbf{B} \cdot \mathbf{C})(\mathbf{A} \cdot \mathbf{D}) = (\mathbf{A} \cdot \mathbf{D})(\mathbf{B} \cdot \mathbf{C}) \tag{B.13}$$

$$\mathbf{AB} \times\times \mathbf{CD} = (\mathbf{B} \times \mathbf{C})(\mathbf{A} \times \mathbf{D}) \tag{B.14}$$

In the second type of double product, the operations are resolved from left to right:

$$\mathbf{AB} : \mathbf{CD} = (\mathbf{A} \cdot \mathbf{C})(\mathbf{B} \cdot \mathbf{D}) \tag{B.15}$$

$$\mathbf{AB} {\overset{\times}{\underset{\times}{}}} \mathbf{CD} = (\mathbf{A} \times \mathbf{C})(\mathbf{B} \times \mathbf{D}) \tag{B.16}$$

Mixed dot and cross products are correspondingly defined. Thus, for instance:

$$(\mathbf{AB}) \times \cdot (\mathbf{CD}) = (\mathbf{A} \times \mathbf{C})(\mathbf{B} \cdot \mathbf{D}) \tag{B.18}$$

B.1.1 Coordinate Transformations

There are two types of coordinate transformations between reference systems having the same origin: rotation and reflection. A rotational transformation of a coordinate system preserves its right- or left-handedness. A reflection of a coordinate system about one of its coordinate planes, which is called an inversion, changes a right-handed system into a left-handed one, and vice-versa. The components of tensors have meaning relative to a reference system; hence, it is important to study the effect that coordinate transformations have on tensor components. We do this by considering how the mathematical description of tensors of various orders change in going from one reference system, (X_1, X_2, X_3), say, to another (X'_1, X'_2, X'_3), which is obtained from the first by an arbitrary rigid rotation, as indicated in Fig. B.1.

Rotation of Scalars
By definition, a scalar, or tensor of order zero, is a quantity that has the same value and hence mathematical description, in all coordinate systems. The value of a

Fig. B.1 Rigid rotation of
two coordinate systems

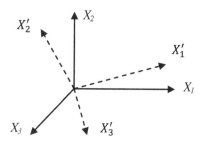

scalar, a, like mass or temperature, will only change if the units of measurement are changed. Thus, its values in the primed and unprimed systems are equal:

$$a' = a \tag{B.19}$$

Rotation of Vectors

As a mathematical entity, a vector exists irrespective of the coordinate systems used to represent it. The components of the same vector, however, change from one coordinate system to another. This is represented in Fig. B.2.

Using the summation convention with Gibbs' notation, an arbitrary vector $\mathbf{v}$, may be expressed in terms of the unit vectors of each of the two coordinate systems, and its corresponding components, as

$$\mathbf{v} = v'_j \widehat{e}'_j = v_i \widehat{e}_i \tag{B.20}$$

The relationship between the components in the primed and unprimed systems may be obtained post-multiplying this relation by $\widehat{e}'_k$ and using the definition of the unit tensor:

$$\mathbf{v} \cdot \widehat{e}'_k = v'_j \widehat{e}'_j \cdot \widehat{e}'_k = v'_j \delta'_{jk} \equiv v'_j \delta_{jk} = v'_k = v_i \widehat{e}_i \cdot \widehat{e}'_k \tag{a}$$

From this result, first note that—from the commutative property of the dot product:

$$\mathbf{v} \cdot \widehat{e}'_k = \widehat{e}'_k \cdot \mathbf{v} = v'_k \tag{B.21}$$

Also:

$$v'_k = \left(\widehat{e}_i \cdot \widehat{e}'_k\right) v_i = \left(\widehat{e}'_k \cdot \widehat{e}_i\right) v_i \tag{B.22}$$

where by definition of the dot product $\left(\widehat{e}_i \cdot \widehat{e}'_k\right) = \cos(X_i, X'_k) \equiv \cos(X'_k, X_i) = \left(\widehat{e}'_k \cdot \widehat{e}_i\right)$. Hence, the array $\left(\widehat{e}'_k \cdot \widehat{e}_i\right)$, which is a second-order tensor because it is described by two free indices, is the rotation matrix. We denoted it simply by $a_{k'i} \equiv \left(\widehat{e}'_k \cdot \widehat{e}_i\right)$. In this notation, the rotational transformation for vectors may be expressed as

$$v'_k = \left(\widehat{e}'_k \cdot \widehat{e}_i\right) v_i = a_{k'i} v_i \tag{B.23}$$

Fig. B.2 Graphical representation of a vector in two coordinate systems differing by a rigid rotation

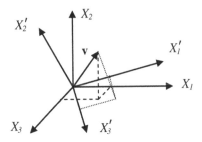

Had we chosen to post-multiply **v** by $\widehat{e}_k$, in the previous development, we would have obtained, respectively that

$$\mathbf{v} \cdot e_k = v_k \tag{B.24}$$

and

$$v_k = \left(\widehat{e}_k \cdot \widehat{e}_i'\right) v_i' = a_{ki'} v_i' \tag{B.25}$$

Using this and the corresponding expression for v_k', the following results

$$v_k = \left(\widehat{e}_k \cdot \widehat{e}_i'\right) v_i' = a_{ki'} a_{i'l} v_l \tag{B.26}$$

Because the left- and right-hand sides of this equation represent the same component, it follows that the rotation matrix is such that

$$a_{ki'} a_{i'l} = \delta_{kl} \tag{B.27}$$

To simplify, we drop the prime notation and adopt the convention that the first index in the rotation tensor, a_{kl}, always represents the resultant vector, and the second index, the vector being rotated. We then rewrite the previous relation as

$$a_{ki} a_{il} = \delta_{kl} \tag{B.27-a}$$

The left-hand side of this expression is the indicial form of the product of a matrix and its transpose; and the right-hand side is the indicial form of the identity matrix. Hence, in matrix notation,

$$\mathbf{A} \cdot \mathbf{A}^T = \mathbf{I} = \mathbf{A}^T \cdot \mathbf{A} \tag{B.27-b}$$

By definition of the inverse of a –non-singular– square matrix, it follows that

$$\mathbf{A}^{-1} = \mathbf{A}^T \tag{b}$$

A square matrix whose inverse is equal to its transpose is called orthogonal. Hence, the rotation matrix is orthogonal.

Rotation of Tensors

The effect of rotation of the coordinate system on a tensor is easily established from that of vectors, expressing the tensor in continued dyadic form using Gibbs' representation. For instance, let $\bar{\bar{T}} = T_{ij}\widehat{e}_i\widehat{e}_j$ represents a second-order tensor. Then, since the tensor is the same in all coordinate systems and the only thing that might change is its mathematical description in terms of components and unit vectors along the coordinate axes:

$$\bar{\bar{T}} = T'_{ij}\widehat{e}'_i\widehat{e}'_j = T_{ij}\widehat{e}_i\widehat{e}_j \tag{B.28}$$

Proceeding as before, we dot-multiply this expression by $\widehat{e}'_k$ and $\widehat{e}'_l$ and arrive at

$$T'_{ij} = a_{ik}a_{jl}T_{kl} \tag{B.29}$$

where to emphasize, $a_{ik} = (\widehat{e}'_i, \widehat{e}_k)$, and $a_{jl} = (\widehat{e}'_j, \widehat{e}_l)$.

In similar fashion, multiplying by $\widehat{e}_k$ and $\widehat{e}_l$:

$$T_{ij} = a_{ki}a_{lj}T'_{kl} \tag{B.30}$$

In this case, according to our convention, $a_{ik} = (\widehat{e}_i, \widehat{e}'_k)$, and $a_{jl} = (\widehat{e}_j, \widehat{e}'_l)$. Rotation of tensors of higher order is defined in an entirely similar fashion.

B.1.2 Isotropic Tensors

An isotropic tensor is one whose components are the same in every orthogonal coordinate system. By this definition,

(a) Every tensor of order zero, that is every scalar, is isotropic.
(b) There are no isotropic tensors of order one. That is no vector is isotropic.
(c) The identity tensor, δ_{ij}, is the only isotropic tensor of second order.
(d) The alternating tensor, e_{ijk}, is the only isotropic tensor of third order.
(e) The product of isotropic tensors is isotropic. Therefore, the tensors $\delta_{ij}\delta_{kl}$ and $e_{ijk}\delta_{kl}$ are isotropic tensors of fourth order.
(f) The most general isotropic tensor of order four may be expressed as a linear combination of fourth-order isotropic tensors, as

$$A_{ijkl} = \alpha\delta_{ij}\delta_{kl} + \beta\delta_{ik}\delta_{jl} + \gamma\delta_{il}\delta_{jk} \tag{B.31}$$

It may be written as the sum of symmetric and an anti-symmetric tensors, as

$$A_{ijkl} = \lambda\delta_{ij}\delta_{kl} + \mu\left(\delta_{ik}\delta_{jl} + \delta_{il}\delta_{jk}\right) + \nu\left(\delta_{ik}\delta_{jl} - \delta_{il}\delta_{jk}\right) \tag{B.32}$$

B.1.3 Tensors of Second Order

Second-order tensors enjoy the following properties:

- The double dot product, $\bar{\bar{T}} \cdot \cdot \bar{\bar{I}}$, of any second-order tensor $\bar{\bar{T}}$ with the identity tensor, $\bar{\bar{I}}$, is a scalar, $\text{tr}\{T\}$, called the trace of the tensor:

$$\bar{\bar{T}} \cdot \cdot \bar{\bar{I}} = T_{ij}\hat{e}_i\hat{e}_j \cdot \cdot \hat{e}_k\hat{e}_k = T_{ij}\hat{e}_i \cdot \delta_{jk}\hat{e}_k = T_{ij}\hat{e}_i \cdot \hat{e}_j = T_{ij} \cdot \delta_{ij} = T_{ii} \equiv tr\{\bar{\bar{T}}\} \quad (B.33)$$

- A second-order tensor $\bar{\bar{T}} = T_{ij}\hat{e}_i\hat{e}_j$ is said to be symmetric, if it is equal to its transpose; that is, if

$$\bar{\bar{T}} = T_{ij}\hat{e}_i\hat{e}_j = \bar{\bar{T}}^T = T_{ji}\hat{e}_j\hat{e}_i \quad (B.34)$$

That is, a tensor, $\bar{\bar{T}}$, is symmetric if its components satisfy the relations:

$$T_{ij} = T_{ji} \quad (B.35)$$

- A second-order tensor, $\bar{\bar{T}}$, may be expressed uniquely as the sum $\bar{\bar{T}} = \bar{\bar{T}}_S + \bar{\bar{T}}_A$ of a symmetric and an anti-symmetric tensor, $\bar{\bar{T}}_S$ and $\bar{\bar{T}}_A$, respectively. Where

$$T_{Sij} = \frac{1}{2}\left(T_{ij} + T_{ji}\right) \quad (B.36-a)$$

and

$$T_{Aij} = \frac{1}{2}\left(T_{ij} - T_{ji}\right) \quad (B.36-b)$$

As may be seen from these definitions: $\text{tr}\{\bar{\bar{T}}_A\} = T_{Aii} = 0$. Also, when $\bar{\bar{T}}$ is symmetric, this decomposition yields a zero antisymmetric tensor: $\bar{\bar{T}}_A = 0$.

- A second-order tensor, $\bar{\bar{T}}$, may be expressed uniquely as the sum $\bar{\bar{T}} = \bar{\bar{T}}_S + \bar{\bar{T}}_D$ of a symmetric and an anti-symmetric tensor, $\bar{\bar{T}}_S$ and $\bar{\bar{T}}_D$, respectively; where the symmetric tensor is chosen to be diagonal, and the antisymmetric tensor is defined as the tensor that is left over, so that

$$T_{Sij} = \frac{1}{3}T_{kk}\delta_{ij} \quad (B.37-a)$$

$$T_{Dij} = T_{ij} - \frac{1}{3}T_{kk}\delta_{ij} \quad (B.37-b)$$

In this decomposition, the symmetric tensor, $\bar{\bar{T}}_S$, is called the spherical tensor, and the antisymmetric tensor, $\bar{\bar{T}}_D$, is the deviatoric tensor. Also, even if the original tensor is symmetric, its deviatoric component is not zero: $\bar{\bar{T}}_D \neq 0$.

B.1.4 Principal Value Problem

The dot product of a tensor $\bar{\bar{T}} = T_{ij}\widehat{e}_i\widehat{e}_j$ with a vector $\mathbf{u} = u_j\widehat{e}_j$, produces another vector, $\mathbf{v} = v_i\widehat{e}_j$, for: $\bar{\bar{T}} \cdot \mathbf{v} = T_{ij}\widehat{e}_i\widehat{e}_j \cdot u_k\widehat{e}_k = T_{ij}u_k\widehat{e}_i\delta_{jk} = T_{ij}u_j\widehat{e}_i \equiv v_i\widehat{e}_i$

Thus,

$$T_{ij}u_j = v_i \tag{B.38}$$

If the resulting vector, $\mathbf{v}$, is parallel to the original vector, $\mathbf{u}$, then since $\mathbf{v} = \lambda\mathbf{u}$, one has

$$T_{ij}u_j = \lambda u_i \tag{B.39-a}$$

This expression may be written in the form:

$$\left(T_{ij} - \lambda\delta_{ij}\right)u_j = 0 \tag{B.39-b}$$

This represents a homogenous system of linear equations in the components u_j of $\mathbf{u}$. This system will possess a unique solution, only if its determinant is zero:

$$\left|T_{ij} - \lambda\delta_{ij}\right| = 0 \tag{B.40}$$

Since the matrix of a second-order tensor is a 3×3 square matrix, this equation represents a cubic equation in the parameter λ. It is customary to write this equation in the form:

$$\lambda^3 - I_T\lambda^2 + II_T\lambda - III_T = 0 \tag{B.41}$$

This equation is called the characteristic equation of the problem, and its roots are called the characteristic roots. The quantities I_T, II_T, and III_T are called first, second, and third (principal) invariants of the tensor. They are given by the following relations:

$$
\begin{aligned}
I_T &= T_{11} + T_{22} + T_{33} = T_{ii} = tr\{\bar{\bar{T}}\} \\
II_T &= -\left(T_{11}T_{22} + T_{22}T_{33} + T_{33}T_{11}\right) + T_{23}^2 + T_{31}^2 + T_{12}^2 \\
&= \frac{1}{2}\left(T_{ii}T_{jj} - T_{ij}T_{ij}\right) \\
III_T &= e_{ijk}T_{i1}T_{j2}T_{k3} = \det(T_{ij})
\end{aligned} \tag{B.42}
$$

The characteristic values are also called principal values, proper values, and quite frequently, "eigenvalues," borrowing the German word, eigen, for proper. By definition, to each eigenvalue, $\lambda_{(m)}$, there will correspond an eigenvector, $\mathbf{u}^{(m)}$. If the tensor is real and symmetric, its eigenvalues and associated eigenvectors will be real. The magnitude of the eigenvectors is indeterminate, because the resulting system for each eigenvalue $\lambda_{(m)}$

$$\left(T_{ij} - \lambda_{(m)}\delta_{ij}\right)u_j^{(m)} = 0, \quad no\ sum\ on\ m \tag{B.43}$$

is homogeneous; and for a non-trivial solution to exist, its rank is less than the number of equations it represents. For this reason, the eigenvectors are usually normalized as unit vectors. Taking a set of coordinate axes (that is, a "primed" coordinate system) along the eigenvectors, allows expressing each eigenvector, $\mathbf{u}^{(m)}$, as

$$\mathbf{u}^{(m)} = u_i^{(m)}\widehat{e}_i \equiv a_{mi}\widehat{e}_i \tag{B.44}$$

This defines the components of the eigenvector $\mathbf{u}^{(m)}$ as $u_i^{(m)} \equiv a_{mi}$. We use this to write the eigenvalue problem for eigenvector $\mathbf{u}^{(m)}$ in the form:

$$T_{ij}a_{mj} \equiv \lambda_{(m)}a_{mi}, \quad no\ sum\ on\ m \tag{B.45}$$

Multiplying both members of this expression by a_{pi}, using the orthogonality of the rotation tensor, and noting that $a_{pi}T_{ij}a_{mj} = T'_{pm}$ we arrive at the relation:

$$T'_{pm} \equiv \lambda^{(m)}\delta_{pm}, no\ sum\ on\ m \tag{B.46}$$

This expression proves that the transformation to the coordinate system represented by the eigenvector triad diagonalizes the original tensor, and that the elements along the main diagonal of the matrix representation of the tensor are the principal values of the tensor. That is, in the principal system, the matrix of any second-order symmetric tensor takes the diagonal form:

$$[T] = \begin{bmatrix} \lambda_{(1)} & 0 & 0 \\ 0 & \lambda_{(2)} & 0 \\ 0 & 0 & \lambda_{(3)} \end{bmatrix} \tag{B.47}$$

B.2 Bodies and Motion

A body B may be defined as a collection of material points. The simultaneous positions of the particles or material points of a body constitute a configuration of the body. For reference purposes, it is convenient to choose one configuration of the body to which all other configurations are compared. This special configuration is called the reference or initial configuration of the body. We use a rectangular coordinate system to describe the configurations of the body B and denote the particles by their position vectors. We use capital letter X to indicate a particle with position vector $\overrightarrow{X} = X_i e_i = (X_1, X_2, X_3)$ in the original configuration and lower case letter, x, to refer to the same particle with position vector $\overrightarrow{x} = x_i\widehat{e}_i = (x_1, x_2, x_3)$ in another configuration. Figure B.3 shows this convention.

Fig. B.3 Position of a particle in the reference and
another configuration

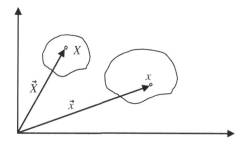

A configuration may be represented by a one-to-one (invertible) mapping, $\vec{\chi}$, as

$$\vec{x} = \vec{\chi}\left(\vec{X}\right); \quad x_i = \chi_i(X_1, X_2, X_3), i = 1, 2, 3 \tag{B.48}$$

With this, a motion of a body is a continuous sequence of configurations in time
and is denoted by

$$\vec{x} = \vec{\chi}\left(\vec{X}, t\right); \quad x_i = \chi_i(X_1, X_2, X_3, t), i = 1, 2, 3 \tag{B.49}$$

According to the previous definitions, the initial configuration is that occupied
by the body at time $t = 0$. Hence, in particular,

$$\vec{x} = \vec{\chi}\left(\vec{X}, 0\right); \quad x_i = \chi_i(X_1, X_2, X_3, 0), i = 1, 2, 3; \tag{B.50}$$

As a matter of terminology, the configuration of the body at the present time is
called current configuration. The displacement of a particle at the current time may
then be defined as the difference between its spatial coordinates at that time, x_i, and
its initial coordinates, X_i, as[8]

$$u_i(X_1, X_2, X_3, t) = x_i(X_1, X_2, X_3, t) - X_i \tag{B.51}$$

B.3 Analysis of Strain

Consider any two particles, $X^{(1)}$ and $X^{(2)}$, which in the reference configuration are
an infinitesimal distance apart, $dX_i = X_i^{(2)} - X_i^{(1)}$. After deformation, these two
particles will be an elemental distance $dx_i = x_i^{(2)} - x_i^{(1)}$ apart, and their
displacements $u_i^{(1)}$ and $u_i^{(2)}$ will also differ by a differential amount; that is

---

[8] In keeping with standard practice when using indicial notation, reference to the range of the
indices is omitted but shall be understood to run from 1 to the number of dimensions of the
Euclidean space of the discussions.

$du_i = u_i^{(2)} - u_i^{(1)}$. Using these facts and the displacement relationships $u_i = x_i - X_i$, introduced earlier, it follows that

$$dx_i = X_i^{(2)} - X_i^{(1)} + u_i^{(2)} - u_i^{(1)} = dX_i - du_i \qquad \text{(B.52)}$$

Recalling that the u_i are functions of the original position coordinates, X_k, it follows that

$$dx_i - dX_i = du_i = \frac{\partial u_i}{\partial X_j} dX_j \qquad \text{(B.53)}$$

Consequently,

$$\frac{dx_i - dX_i}{dX_j} = \frac{\partial u_i}{\partial X_j} \qquad \text{(B.54)}$$

Because of its two free indices, this expression represents a second-order tensor. Also, since the indices take values over the same range, the left-hand side of the relation measures the change in length in the coordinate direction indicated by the index of the numerator, per unit of original length along the coordinate indicated by its denominator. When the coordinates of numerator and denominator are the same, the ratio corresponds to the so-called direct strain in the selected coordinate direction. When the two indices are different, the ratio yields the shear strain, as measured by the change in displacement in one direction as one moves in another direction. This definition of shear strain gives rise to two different shear strains for each pair of coordinate directions, such as $\partial u_1/\partial X_2$ and $\partial u_2/\partial X_1$; and such shear strains would depend on rigid body motion. Only the sum $\frac{1}{2}(\partial u_1/\partial X_2 + \partial u_2/\partial X_1)$ is independent of rigid body motion. For this reason, the latter definition of shear strain is used, and the infinitesimal strain tensor, ε_{ij}, is defined as

$$\varepsilon_{ij} = \frac{1}{2}\left(\frac{\partial u_i}{\partial X_j} + \frac{\partial u_j}{\partial X_i}\right) \qquad \text{(B.55-a)}$$

Or, in explicit matrix form:

$$[\varepsilon] = \begin{bmatrix} \frac{\partial u_1}{\partial X_1} & \frac{1}{2}\left(\frac{\partial u_1}{\partial X_2} + \frac{\partial u_2}{\partial X_1}\right) & \frac{1}{2}\left(\frac{\partial u_1}{\partial X_3} + \frac{\partial u_3}{\partial X_1}\right) \\ & \frac{\partial u_2}{\partial X_2} & \frac{1}{2}\left(\frac{\partial u_2}{\partial X_3} + \frac{\partial u_3}{\partial X_2}\right) \\ Sym. & & \frac{\partial u_3}{\partial X_3} \end{bmatrix} \qquad \text{(B.55-b)}$$

As a second-order symmetric tensor, the strain tensor may be decomposed uniquely into spherical and deviatoric components, has three principal values and three principal directions, and becomes diagonal in the coordinate system defined by its principal directions.

B.4 Analysis of Stress

The forces acting on a body may be classified as internal and external. Internal forces are those originating inside the body. External forces originate outside the body. Forces may be further classified as body forces and surface forces. Body forces are external forces that act upon each volume element of a body. Surface forces may be external or internal. External surface forces act on the bounding surfaces of a body. Internal surface forces are due to interactions between material particles and thus act on the surfaces of internal volume elements. Surface forces are reckoned as force per unit surface area.

The body force per unit mass acting on an infinitesimal element dV of the body is denoted by the vector $\overrightarrow{b} = b_i \widehat{e}_i$. With this, the body force on the element dV is $\rho b_i \widehat{e}_i dV$, and the vector sum of all the body forces acting on a body of finite volume V is given by the volume integral:

$$\int_V \rho \widehat{b} dV \equiv \int_V \rho b_i \widehat{e}_i dV \equiv \widehat{e}_1 \int_V \rho b_1 dV + \widehat{e}_2 \int_V \rho b_2 dV + \widehat{e}_3 \int_V \rho b_3 dV \quad (B.56)$$

The surface force per unit area, also called surface traction, stress traction, or simply traction, is denoted by the vector $\overrightarrow{t} = t_i \widehat{e}_i$. The force acting on an elemental surface of area dS is $\overrightarrow{t} dS \equiv t_i \widehat{e}_i dS$; and the vector sum of the tractions across a finite surface S will be given by the surface integral:

$$\int_S \widehat{t} ds \equiv \int_S t_i \widehat{e}_i dS \equiv \widehat{e}_1 \int_S t_1 dS + \widehat{e}_2 \int_S t_2 dS + \widehat{e}_3 \int_S t_3 dS \quad (B.57)$$

Now assume the body in Fig. B.4 is subjected to some prescribed external surface tractions and body forces, which are functions of the position coordinates X_1, X_2, X_3.

At any point within the body, an elemental surface ΔS has a normal unit vector $\widehat{n}$. The stress traction, $\overrightarrow{t}^{(n)}$, at a point (X_1, X_2, X_3) is defined in terms of the surface force $\overrightarrow{F}$ that acts on ΔS, as the limit:

$$\overrightarrow{t}^{(n)} \equiv \lim_{\Delta S \to 0} \frac{\overrightarrow{\Delta F}}{\Delta S} = \frac{d\overrightarrow{F}}{dS} \quad (B.57)$$

where limit is assumed to exist. In this notation, the superscript n is appended to the surface traction, $\overrightarrow{t}$, to indicate the normal to the surface on which the traction acts. Of particular interest are surface traction vectors with normal parallel to the coordinate axes. Such traction vectors may be expressed as linear combinations of unit vectors along the coordinate axes, as

$$\overrightarrow{t}^{(i)} = t_j^{(i)} \widehat{e}_j = t_1^{(i)} \widehat{e}_1 + t_2^{(i)} \widehat{e}_2 + t_3^{(i)} \widehat{e}_3 \quad (B.58)$$

Fig. B.4 Surface tractions in a body

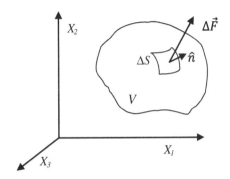

where by definition, $t_j^{(i)}$ represents the component of traction vector $\overrightarrow{t}^{(i)}$ along the positive direction of coordinate axis X_j. It is customary to write $\sigma_{ij} \equiv t_j^{(i)}$, and thus to express the traction vector as

$$\overrightarrow{t}^{(i)} = \sigma_{ij}\widehat{e}_j = \sigma_{i1}\widehat{e}_1 + \sigma_{i2}\widehat{e}_2 + \sigma_{i3}\widehat{e}_3 \qquad (B.59)$$

By convention then, the first subscript of σ_{ij} indicates the direction of the outward unit normal to the elemental area; and the second index represents the direction of the component of the traction vector. This sign convention is shown in Fig. B.5. Since the symbol σ_{ij} has two free indices, it represents a tensor of second order, which is called the stress tensor at the point in question.

Once the components of the stress tensor are available at a point, the traction vector acting on any other plane passing through the same point may be established. Indeed, using the continued dyadic form of the stress tensor and the properties of the Kronecker delta, expression (B.59) may be written as the inner or dot product of the unit vector e_i and the stress tensor:

$$\overrightarrow{t}^{(i)} = \widehat{e}_i \cdot \widehat{e}_k \sigma_{kj} \widehat{e}_j = \delta_{ik}\widehat{e}_k\sigma_{kj}\widehat{e}_j = \sigma_{ij}\widehat{e}_j \qquad (B.60)$$

By the same token, the surface traction vector acting at a point on a plane with a normal $\widehat{n} = n_i\widehat{e}_i$, would be given by

$$\overrightarrow{t}^{(n)} \equiv t_j^n \widehat{e}_j = \widehat{n} \cdot \underline{\sigma} = n_i\widehat{e}_i \cdot \widehat{e}_k \sigma_{kj}\widehat{e}_j = n_i\delta_{ik}\widehat{e}_k\sigma_{kj}\widehat{e}_j = n_i\sigma_{ij}\widehat{e}_j$$

From which, the components of the surface traction vector associated with a plane with normal $\widehat{n}$ are

$$t_j^{(n)} = n_i\sigma_{ij} \qquad (B.61)$$

For simplicity, we usually omit reference to the surface normal and simply express the traction vector as

$$t_j = n_i\sigma_{ij} \qquad (B.62)$$

Fig. B.5 Sign convention for the components of the stress tensor

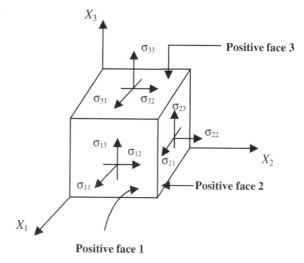

As a second-order symmetric tensor, the stress tensor may be decomposed uniquely into spherical and deviatoric tensors, has three principal values and three principal directions, and becomes diagonal in the coordinate system defined by the principal directions.

B.5 Constitutive Equations

The constitutive equations of linear elasticity are nine equations relating the nine components of the stress tensor to the nine components of the strain tensor, in the form:

$$\sigma_{ij} = M_{ijkl}\varepsilon_{kl} \tag{B.63}$$

These nine equations contain 81 material constants, but since the stress and strain tensors are both symmetric, $M_{ijkl} = M_{jikl} = M_{ijlk}$, the number of independent constants reduces to 36, because for each pair of indices ranging from 1 to 3, there are $3\cdot(3+1)/2 = 6$ distinct combinations, accounting for the $6\cdot6 = 36$ independent material constants. This allows for a convenient representation of (B.63) as a matrix equation relating a six-by-one column matrix of stresses to a six-by-one column matrix of strains, though a six-by-six matrix of elastic constants, as follows:

$$\begin{Bmatrix} \sigma_{11} \\ \sigma_{22} \\ \sigma_{33} \\ \sigma_{12} \\ \sigma_{13} \\ \sigma_{23} \end{Bmatrix} = \begin{bmatrix} M_{1111} & M_{1122} & M_{1133} & M_{1144} & M_{1155} & M_{1166} \\ M_{2211} & M_{2222} & M_{2233} & M_{2244} & M_{2255} & M_{2266} \\ M_{3311} & M_{3322} & M_{3333} & M_{3344} & M_{3355} & M_{3366} \\ M_{4411} & M_{4422} & M_{4433} & M_{4444} & M_{4455} & M_{4466} \\ M_{5511} & M_{5522} & M_{5533} & M_{5544} & M_{5555} & M_{5566} \\ M_{6611} & M_{6622} & M_{6633} & M_{6644} & M_{6655} & M_{6666} \end{bmatrix} \begin{Bmatrix} \varepsilon_{11} \\ \varepsilon_{22} \\ \varepsilon_{33} \\ \varepsilon_{12} \\ \varepsilon_{13} \\ \varepsilon_{23} \end{Bmatrix} \tag{B.64}$$

For elastic materials, the number of independent constants is further reduced from 36 to 21 due to the existence of a strain energy function—discussed subsequently. This extends the symmetry of the constitutive tensor to $M_{ijkl} = M_{jikl} = M_{ijlk} = M_{klij}$.

The specific coefficients entering the stress–strain relationships will depend on the directions of the reference axis. A material that allows certain types of transformation of reference axes without changing its elastic properties is said to possess elastic symmetry. If the material properties are independent of rotation about an axis, that axis is an axis of symmetry. If the properties remain the same under reflection in a plane, that plane is a plane of symmetry.

An elastic material with three orthogonal planes of symmetry –called orthotropic symmetry– requires only 9 independent constitutive coefficients. The corresponding constitutive equations take the form:

$$
\begin{Bmatrix} \sigma_{11} \\ \sigma_{22} \\ \sigma_{33} \\ \sigma_{12} \\ \sigma_{13} \\ \sigma_{23} \end{Bmatrix} = \begin{bmatrix} M_{1111} & M_{1122} & M_{1133} & 0 & 0 & 0 \\ & M_{2222} & M_{2233} & 0 & 0 & 0 \\ & & M_{3333} & 0 & 0 & 0 \\ & & & M_{4444} & 0 & 0 \\ & Sym. & & & M_{5555} & 0 \\ & & & & & M_{6666} \end{bmatrix} \begin{Bmatrix} \varepsilon_{11} \\ \varepsilon_{22} \\ \varepsilon_{33} \\ \varepsilon_{12} \\ \varepsilon_{13} \\ \varepsilon_{23} \end{Bmatrix} \quad (B.65)
$$

An elastic material for which every plane is a plane of symmetry is said to be isotropic. In this case, the fourth-order constitutive tensor, M_{ijkl}, is isotropic. As such, it may be represented as [see Sect. B.1.2]:

$$
M_{ijkl} = \lambda \delta_{ij} \delta_{kl} + \mu(\delta_{ik} \delta_{jl} + \delta_{il} \delta_{jk}) + \nu(\delta_{ik} \delta_{jl} - \delta_{il} \delta_{jk}) \quad (B.66)
$$

By the symmetry requirement, the last term drops out, so that

$$
M_{ijkl} = \lambda \delta_{ij} \delta_{kl} + \mu(\delta_{ik} \delta_{jl} + \delta_{il} \delta_{jk}) \quad (B.67)
$$

In other words, only two material property constants are needed to fully characterize the constitutive equations of isotropic materials. Combining this expression with (B.63), using the properties of the identity tensor, and simplifying, yields the constitutive equations for isotropic elastic materials, as

$$
\sigma_{ij} = \lambda \varepsilon_{kk} \delta_{ij} + 2\mu \varepsilon_{ij} \quad (B.68)
$$

In this expression, μ is the material's shear modulus; and both λ and μ are called Lame's constants. Also, as shown in a later section, when strains are small, $\varepsilon_{kk} = \varepsilon_{11} + \varepsilon_{22} + \varepsilon_{33} \equiv \varepsilon_V$ represents the volumetric strain.

Using the decomposition of the stress and strain tensors into their spherical and deviatoric parts, the constitutive equations may be cast in the following equivalent form:

$$
\sigma_S = 3\left(\lambda + \frac{2}{3}\mu\right)\varepsilon_S, \quad \sigma_S \equiv \frac{\sigma_{kk}}{3}, \quad \varepsilon_S \equiv \frac{\varepsilon_{kk}}{3} \quad (B.69\text{-}a)
$$

$$\sigma_{Dij} = 2\mu\varepsilon_{Dij}, \quad \sigma_{Dij} \equiv \sigma_{ij} - \sigma_S\delta_{ij}, \quad \varepsilon_{Dij} \equiv \varepsilon_{ij} - \varepsilon_S\delta_{ij} \qquad (B.69\text{-}b)$$

Note that in this form, the six constitutive equations of linear elasticity decouple into 7 one-dimensional constitutive equations, connected through their spherical and deviatoric components. Equation (B.69-a) between the spherical stress and spherical strain is usually presented in the following alternate forms:

$$\sigma_S = 3K\varepsilon_S \qquad (B.70\text{-}a)$$

$$-p = K\varepsilon_V, \quad -p \equiv \frac{\sigma_{kk}}{3}, \varepsilon_V \equiv \varepsilon_{kk} \qquad (B.70\text{-}b)$$

The material constant $K = \lambda + \frac{2}{3}\mu$, relating the hydrostatic pressure $-p \equiv \frac{\sigma_{kk}}{3}$, to the volumetric strain $\varepsilon_V \equiv \varepsilon_{kk}$ is properly termed volumetric or bulk modulus.

In practice, the tensile relaxation modulus, E, and the so-called Poisson's ratio, v, are easier to determine than λ, μ, or K. When E and v are used, the constitutive equations take on the following form:

$$\sigma_{ij} = \frac{E}{(1+v)}\left(\varepsilon_{ij} + \frac{v}{(1-v)}\varepsilon_{kk}\delta_{ij}\right) \qquad (B.71)$$

Only two independent constants are needed to fully characterize a linear elastic material, and there are several material constants available to do this—λ, μ, K, E, and v. The form of the constitutive equations will depend on which properties are chosen to express them. The elastic constants E, v, and K, are related to λ and μ, as follows:

$$E = 2(1+v)\mu, \quad v = \frac{\lambda}{2(\lambda + \mu)}, \quad K = \lambda + \frac{2}{3}\mu \qquad (B.72\text{-}a)$$

With these, the following interrelationships may be verified:

$$\mu = \frac{E}{2(1+v)} \qquad (B.72\text{-}b)$$

$$\lambda = K - \frac{2}{3}\mu = \frac{2\mu v}{1 - 2v} = \frac{vE}{(1+v)(1-2v)} = \frac{\mu(E - 2\mu)}{3\mu - E} \qquad (B.73\text{-}c)$$

$$K = \lambda + \frac{2}{3}\mu = \frac{2\mu(1+v)}{3(1 - 2v)} = \frac{E}{3(1 - 2v)} = \frac{E\mu}{3(3\mu - E)} \qquad (B.72\text{-}d)$$

$$E = \frac{\mu(3\lambda + 2\mu)}{\lambda + \mu} = \frac{9K\mu}{3K + \mu} = 2\mu(1+v) \qquad (B.72\text{-}e)$$

$$v = \frac{\lambda}{2(\lambda + \mu)} = \frac{3K - 2\mu}{6K + 2\mu} = \frac{E}{2\mu} - 1 \qquad (B.72\text{-}f)$$

There are isotropic elastic materials whose bulk modulus is orders of magnitude larger that their shear modulus. The practical implication of this is that their response to mechanical loads takes place by shearing only, with virtually no change in their volume. Such materials are called incompressible. As will be seen later on, volume preservation in linear materials requires that the volume strain, ε_V $= \varepsilon_{kk}$, be identically zero throughout the loading process. Hence, incompressibility enters as a constraint on the equations of motion. This means that one can add a spherical stress—uniform in all directions—of any magnitude to the stress field, without altering the strains. For incompressible materials, then, the stress tensor is determined from the strains, only up to a spherical stress. Since the spherical stress, $\sigma_S \equiv (\sigma_{kk})/3$, is the average of the normal stresses, its negative, "$-\sigma_S$," is called the hydrostatic pressure. Thus, $-p \equiv \sigma_S \equiv (\sigma_{kk})/3$. The constitutive equations for linear elastic isotropic incompressible materials may be written directly from Eq. (B.69-b), using $\sigma_S \equiv -p$, noting that $\varepsilon_{Dij} = \varepsilon_{ij}$, because $\varepsilon_{kk} \equiv 0$, and expressing σ_{ij} in terms of σ_{Dij}. Hence,

$$\sigma_{ij} = -p\delta_{ij} + 2\mu\varepsilon_{ij} \tag{B.73}$$

B.6 Conservation Principles

There are laws of physics which are obeyed by all substances in the bulk, be they elastic, viscous or viscoelastic, irrespective of whether their constitutive equations are linear or non-linear.[9] These laws proclaim the conservation of mass, linear and angular momenta, and energy. The expressions used to represent conservation principles are known as conservation or balance laws.

B.6.1 Conservation of Mass

The total mass of a system, which does not exchange mass with its surroundings, remains the same at all times. In particular, the total mass, $M(t)$, of a body at any time t, must be equal to its original mass, M_o. Using the material's density, $\rho(x)$ which measures mass per unit volume and letting the subscript 'o' denote the original value of a quantity, we write the equation of mass conservation, as[10]

$$\int_{V_o} \rho_o(X)dV_o = \int_V \rho(x)dV \tag{B.74}$$

---

[9] D.C. Leigh 1968.

[10] For simplicity of exposition, here, and in derivations that follow, we use X and x to denote a material particle as well as its position vector before and after deformation, respectively.

Changing the volume of integration from the current, deformed configuration at time t, to the original configuration:

$$\int_{V_o} \rho_o(X)dV_o = \int_V \rho(x)dV = \int_{V_o} J \cdot \rho(x)dV_o \tag{a}$$

Noting that the resulting expression must be valid for an arbitrary initial volume, we cancel the integral sign and write the balance of mass equation:

$$\rho_o(X) = \rho(x)J(x) \tag{B.75}$$

For clarity, we have omitted the indices that identify the coordinates, X_i, and x_j, and used instead the particle representation, X, and x. The additional implication that the motion of the particle X depends on time; that is, that $x = \chi(X, t)$ must also be kept in mind.

The Jacobian determinant, J, represents the ratio of the change in volume per unit original volume. Now, by the definition of small axial strain, as change in length per unit original length, a cube of sides of initial lengths l_{xo}, l_{yo}, l_{zo} subjected to strains ε_{xx}, ε_{yy}, ε_{zz} along its edges, will change its volume so that

$$\begin{aligned} V &= L_{xo}(1 + \varepsilon_{xx}) \cdot L_{yo}(1 + \varepsilon_{yy}) \cdot L_{zo}(1 + \varepsilon_{zz}) \\ &= L_{xo} \cdot L_{yo} \cdot L_{zo}[1 + \varepsilon_{xx} + \varepsilon_{yy} + \varepsilon_{zz}] + O(\varepsilon^2) \\ &= V_o[1 + \varepsilon_{xx} + \varepsilon_{yy} + \varepsilon_{zz}] + O(\varepsilon^2) \end{aligned} \tag{b}$$

where the symbol $O(\varepsilon^2)$ indicates terms of second order and higher. That is, to first order, the balance of mass may be alternatively expressed in terms of the direct strains, as

$$J = \frac{V}{V_o} = 1 + (\varepsilon_{xx} + \varepsilon_{yy} + \varepsilon_{zz}) \tag{B.76}$$

Two important consequences of this expression are that

(a) The change in volume per unit original volume, also known as volume strain, is equal to the sum of the direct strains in any three perpendicular directions:

$$\varepsilon_{vol} = \frac{V - V_o}{V_o} = \varepsilon_{xx} + \varepsilon_{yy} + \varepsilon_{zz} = \varepsilon_{11} + \varepsilon_{22} + \varepsilon_{33} \tag{B.77}$$

(b) Incompressible materials deform without changing volume: $J = V/V_o = 1$, and thus, for such materials, the sum of the direct strains in any three mutually perpendicular directions is zero:

$$\varepsilon_{vol} = \varepsilon_{xx} + \varepsilon_{yy} + \varepsilon_{zz} = \varepsilon_{11} + \varepsilon_{22} + \varepsilon_{33} = 0 \tag{B.78}$$

B.6.2 Conservation of Linear Momentum

Newton's second law of motion requires a balance between the external resultant load on a body and the rate of change of its linear momentum. The integral form of this law is due to Cauchy and gives rise to the equations of motion, valid for all materials in bulk. In terms of the Cartesian components of the stress tensor, σ_{ij}, and velocity field, v_i, these equations take the following form:

$$\frac{\partial}{\partial X_j} \sigma_{ij} + \rho b_i = \rho \frac{\partial}{\partial t} v_i \qquad (B.79)$$

When inertia terms are zero, as in static problems, or can be neglected, as under steady-state conditions, the acceleration term on the right-hand side is dropped:

$$\frac{\partial}{\partial X_j} \sigma_{ij} + \rho b_i = 0 \qquad (B.80)$$

Here, differentiation is understood with respect to the original coordinates, X_i, so that, for instance, the X-component of the equations of motion in unabridged form reads:

$$\frac{\partial}{\partial X} \sigma_{xx} + \frac{\partial}{\partial Y} \sigma_{yx} + \frac{\partial}{\partial Z} \sigma_{zx} + \rho b_x = \rho \frac{\partial}{\partial t} v_x \qquad (B.81)$$

The y- and z-component equations are written with corresponding permutations of the coordinates.

B.6.3 Conservation of Angular Momentum

Non-polar materials are defined as those for which the resultant internal moment on the surface of any volume element is zero. For such materials, the principle of conservation of angular momentum—the resultant external moment on a body is equal to the time rate of change of its angular momentum—yields the requirement that the stress tensor, σ_{ij}, be symmetric:

$$\sigma_{ij} = \sigma_{ji} \qquad (B.82)$$

Using unabridged notation, these expressions take the form:

$$\sigma_{xy} = \sigma_{yx}; \quad \sigma_{xz} = \sigma_{zx}; \quad \sigma_{yz} = \sigma_{zy} \qquad (B.83)$$

The balance of angular momentum reduces the number of independent components of the stress tensor from 9 to 6.

B.6.4 Conservation of Energy

The experimental fact that the total energy in a thermodynamically closed system remains constant represents the first law of thermodynamics. According to this experimental law, there is always a balance between all energy put into a system and all internal energy. The external energy consists of mechanical and thermal work, while the internal energy is separated into macroscopically observable kinetic energy and intrinsic energy. The rate form of the first law of thermodynamics, that $\partial E_{int}/\partial t = \partial E_{ext}/\partial t$, stipulates a balance between the intrinsic energy, on one side, and the stress and thermal power on the other:

$$\rho \frac{\partial e}{\partial t} = \sigma_{ij} \frac{\partial}{\partial t} \varepsilon_{ij} + \rho r - \frac{\partial q_i}{\partial X_i} \tag{B.84}$$

In this expression, the first term on the right-hand side is the sum total of the products of the stress components and the corresponding strain rates and is thus called stress power. The other two terms on the right-hand side represent the thermal power: r is due to the heat sources and the q_i are the components of the heat flux vector.

B.7 Boundary Value Problems

The isothermal boundary value problem of linear isotropic elasticity consists of the three equations of motion with their initial and boundary conditions. Taking the initial configuration of the body to be free of stress and at rest prior to the application of the loading—all field variables are identically zero for $t < 0$—the equations of motion and boundary conditions are

$$\frac{\partial}{\partial X_j} \sigma_{ji} + \rho b_i = \rho \frac{\partial^2}{\partial t^2} u_i, \quad X_k \in V$$

$$u_i(\vec{X}, t) = u_i^o(\vec{X}, t), \quad X_k \in S_u; \quad t \geq 0 \tag{B.85}$$

$$n_j(\vec{X}) \sigma_{ij}(\vec{X}, t) = t_i^o(\vec{X}, t), \quad X_k \in S_T; \quad t \geq 0$$

In these expressions, n_i represents the unit outward normal to the boundary, S, formed of S_u and S_T, where displacements and tractions, respectively, are prescribed. Here, n_i is not a function of time, which is consistent with the assumptions of small displacements and strains. Also, in a well-posed problem, S_u and S_T do not intersect $(S_u \cap S_T = \Phi)$; which means that one cannot prescribe different types of boundary conditions at the same point and in the same direction, at the same time.

The boundary value problem of linear elasticity, as posed in (a) consists of three equations in nine unknowns—three components of the displacement vector and six independent stress components. For a solution to exist, the system in (B.85) has to be complemented by other relations. The additional relationships are as follows:

The displacements, u_i, are related to the strains, ε_{ij}, through the six relationships that were used earlier to define the strain tensor:

$$\varepsilon_{ij} = \frac{1}{2}\left(\frac{\partial u_i}{\partial X_j} + \frac{\partial u_j}{\partial X_i}\right) \tag{B.86}$$

We now have three equations of equilibrium, six strain–displacement relations, and six stress–strain equations, for fifteen equations; and there are three displacements, six stresses and six strains, for fifteen unknowns. These are the field equations of isothermal linear elasticity. The uniqueness of solution of the boundary value problem posed is due to the fact that linear elastic materials have a positive definite strain energy function.

B.8 Compatibility Conditions

The strain–displacement relations relate the three components of the displacement field to the six components of strain. These expressions result in a unique set of strains for any prescribed set of displacements, but do not suffice in general to produce a unique set of displacements from an arbitrarily prescribed set of strains. The system of equations in the latter case, with six equations in three unknowns, is over-determined, thus preventing the six components of the strain tensor to be prescribed arbitrarily. The conditions that the strain tensor must satisfy to allow a unique displacement field upon integration of the strain–displacement relations are known as the integrability or compatibility conditions.

The equations of compatibility are obtained differentiating the strain–displacement relations twice and permuting indices. This process yields the following 81 equations, of which only six are independent:

$$\varepsilon_{ij,kl} + \varepsilon_{kl,ij} = \varepsilon_{ik,jl} + \varepsilon_{jl,ik} \tag{B.87}$$

Using the standard Cartesian coordinates x, y, z, for subscripts 1, 2, and 3, respectively, the equations of compatibility take the following unabridged form:

$$\frac{\partial^2}{\partial x^2}\varepsilon_{yy} + \frac{\partial^2}{\partial y^2}\varepsilon_{xx} = 2\frac{\partial^2}{\partial x\partial y}\varepsilon_{xy} \quad \frac{\partial^2}{\partial y\partial z}\varepsilon_{xx} = \frac{\partial}{\partial x}\left[-\frac{\partial}{\partial x}\varepsilon_{yz} + \frac{\partial}{\partial y}\varepsilon_{zx} + \frac{\partial}{\partial z}\varepsilon_{xy}\right] \tag{B.88}$$

Similar permutations of x, y, and z produce the other four independent relations. For two-dimensional problems in the x-y plane, the only non-trivially satisfied relation is the first one listed above.

On occasion, the compatibility conditions are required in terms of stresses. Since stresses and strains are connected through the constitutive equations, the integrability conditions in terms of stresses depend on material properties. They may be obtained by substituting the strain–stress constitutive equations in the compatibility equations. The resulting expressions are known as the Beltrami-Michell relations.

B.9 Energy Principles

Energy principles play an important role in the theory of elasticity. One reason for this is that the work performed on an elastic solid by external agents is stored in the solid in the form of internal energy, and this internal energy is fully recoverable[11] upon removal of the external agents. Because of this, energy principles are frequently used to derive methods of solution to elastic boundary value problems. Among the various energy principles that are available for elastic solids, we present the principle of virtual work, and the important theorems of minimum potential energy and minimum complementary potential energy.[12] Before discussing these principles, we introduce the concepts of *statically admissible* stress fields and *kinematically admissible* displacement fields.

Statically admissible field
A stress distribution or field is called statically admissible if it is continuously differentiable in the domain, V, occupied by the body, identically satisfies the equations of equilibrium inside V, and takes on the values assigned to the surface tractions on the portion, S_T, of the body, where tractions are prescribed.

Kinematically admissible field
A displacement field is called kinematically admissible if it is three times continuously differentiable[13] inside the domain, V, occupied by a body, identically satisfies compatibility in V, and takes on the values of the displacement field on the portion, S_u, of the body, where the displacements are prescribed.

B.9.1 Principle of Virtual Work

The principle of virtual wok, also referred to in the literature as the principle of virtual displacements, states that

Given any statically admissible stress field: $\{\sigma'_{ij}, t'_j = n_i\sigma'_{ij}, F_i \equiv \rho b_i\}$, and any kinematically admissible displacement field, u''_i then

$$\int_S t'_i u''_i dS + \int_V F_i u''_i dS = \int_V \sigma'_{ij}\varepsilon''_{ij} dV \qquad \text{(B.89)}$$

---

[11] This means that if the external agents that put work into an elastic solid were completely removed, the internal energy stored in the solid could be used to perform an amount of work equal to the work that was put into the solid in the first place.

[12] The principle of virtual work (or its generalization to virtual velocities), as well as the theorems of minimum potential energy, and minimum complementary potential energy have been applied to derive finite element methods.

[13] A kinematically admissible displacement field has to be thrice continuously differentiable because the compatibility conditions involve the second derivatives of the strains, and the strains are defined in terms of the first derivatives of the displacement field.

To prove the principle, we first transform the surface integral into a volume one using the Green-Gauss theorem [c.f. Appendix A]. Afterward, we invoke the equations of equilibrium: $\sigma'_{ji,j} + F_i = 0$; the strain–displacement relations: $\varepsilon''_{ij} = 1/2(u''_{i,j} + u''_{j,i})$; and the symmetry of the stress tensor, that $\sigma_{ij}, = \sigma_{ji}$, on account of which $\sigma'_{ji}u''_{i,j} = \sigma'_{ji}\varepsilon''_{ij}$. Since the proof of the principle does not involve constitutive equations, the principle is applicable to all materials in bulk.

B.9.2   *Principle of Minimum Potential Energy*

An elastic material may be defined as one for which there exists a single-valued positive definite potential function of the strains, $W(\varepsilon_{ij})$, such that

$$W(\varepsilon_{ij}) = \int_0^{\varepsilon_{ij}} \sigma_{kl}d\varepsilon_{kl} \tag{B.90}$$

The function W is also required to be path independent and convex in the strains; the latter requirement is in the sense that, for any two strain fields, ε_{ij} and ε^a_{ij} the following relationship must be satisfied:

$$W\left(\varepsilon^a_{ij}\right) - W\left(\varepsilon_{ij}\right) \geq \left(\varepsilon^a_{ij} - \varepsilon_{ij}\right)\frac{\partial W}{\partial \varepsilon_{kl}}\bigg|_{\varepsilon_{ij}} \tag{B.91}$$

For W to be path independent and a function of the final strains only, its total differential must be a perfect differential. This condition is satisfied if

$$\frac{\partial^2 W}{\partial \varepsilon_{ij}\partial \varepsilon_{kl}} = \frac{\partial^2 W}{\partial \varepsilon_{kl}\partial \varepsilon_{ij}} \tag{B.92}$$

or, equivalently,

$$\frac{\partial \sigma_{kl}}{\partial \varepsilon_{ij}} = \frac{\partial \sigma_{ij}}{\partial \varepsilon_{kl}} \tag{B.93}$$

This results follow directly from (B.90), since $dW\left(\varepsilon_{ij}\right) = \frac{\partial W}{\partial \varepsilon_{kl}}d\varepsilon_{kl} = \sigma_{kl}d\varepsilon_{kl}$ is a sum of differentials, and thus, a generalization of the two-dimensional form $df = Pdx + Qdy$, for which the proposition is easily proven[14]. The total differential of W also defines the explicit relationship between W and the stress field, as

---

[14] If $df = Pdx + Qdy$, is a perfect differential, then $df = \frac{\partial f}{\partial x}dx + \frac{\partial f}{\partial y}dy$, which implies that $P = \frac{\partial f}{\partial x}$ and $Q = \frac{\partial f}{\partial y}$. By the theorem of Schwartz for mixed partial derivatives, $\frac{\partial P}{\partial y} = \frac{\partial^2 f}{\partial y\partial x} = \frac{\partial^2 f}{\partial x\partial y} = \frac{\partial Q}{\partial x}$.

$$\sigma_{ij} = \frac{\partial W}{\partial \varepsilon_{ij}} \tag{B.94}$$

This relationship is what gives W its character as a potential function, since the stresses are calculated from it as the derivatives of the function with respect to the corresponding strains. Also, because W is measured per unit volume, it is referred to as the strain energy density.

The potential energy, Φ, of an elastic body subjected to conservative body forces, F_i, and surface tractions t_i^o, is defined for any kinematically admissible displacement field, u_i'', as

$$\Phi(u_i'') = \int_V W(\varepsilon_{ij}'')dV - \int_S t_i^o u_i'' dV - \int_V F_i dV \tag{B.95}$$

In this expression, $\varepsilon_{ij}'' = 1/2(u_{i,j}'' + u_{j,i}'')$. If $\sigma_{ij}, \varepsilon_{ij}$, and u_i represent the actual stresses, strains, and displacements, respectively, it is an easy matter to prove that, because of the convexity of the strain energy density, W:

$$\Phi(u_i'') \geq \Phi(u_i) \tag{B.96}$$

This equation represents the theorem of minimum potential energy, that among all the kinematically admissible displacement fields, the actual displacement field, which is also statically admissible by definition, minimizes the potential energy.

B.9.3 Principle of Minimum Complementary Potential Energy

In the previous section, we defined an elastic material as one for which a strain energy density, W, can be found, such that the stress–strain constitutive equations can be obtained from it by differentiation according to $\sigma_{ij} = \partial W/\partial \varepsilon_{ij}$. The existence of a unique inverse of these constitutive equations, together with the symmetry condition

$$\frac{\partial \varepsilon_{kl}}{\partial \sigma_{ij}} = \frac{\partial \varepsilon_{ij}}{\partial \sigma_{kl}} \tag{B.97}$$

guarantees there is a complementary strain energy density Y, defined by the relation:

$$Y \equiv \sigma_{ij}\varepsilon_{ij} - W \tag{B.98}$$

Since $dY \equiv d\sigma_{ij}\varepsilon_{ij} + \sigma_{ij}d\varepsilon_{ij} - dW = \sigma_{ij}d\varepsilon_{ij}$, the complementary strain energy density, Y, defines the strain–stress constitutive equations, through the relations:

$$\varepsilon_{ij} = \frac{\partial Y}{\partial \sigma_{ij}} \tag{B.99}$$

The complementary potential energy, Ψ, of an elastic body is defined for any statically admissible field σ'_{ij}, by the relation:

$$\Psi(\sigma'_i) = \int_V Y(\sigma'_{ij})dV - \int_{S_u} t'_i u_i dS \qquad (B.100)$$

Where the statically admissible surface tractions, t'_j, are computed from the statically admissible stress field, according to $t'_j = n_i \sigma'_{ij}$. If $\sigma_{ij}, \varepsilon_{ij}$, and u_i represent the actual stress, strain, and displacement fields, then it can be proven, using the convexity of the complementary strain energy density, Y, that:

$$\Psi\left(\sigma'_{ij}\right) \geq \Psi\left(\sigma_{ij}\right) \qquad (B.101)$$

This embodies the principle of minimum complementary potential energy that of all the statically admissible stress fields, the actual field, which is also kinematically admissible, minimizes the complementary potential energy.

Index

D. Gutierrez-Lemini, *Engineering Viscoelasticity*, DOI: 10.1007/978-1-4614-8139-3,   351
© Springer Science+Business Media New York 2014

Printed in the United States
By Bookmasters